ENCYCLOPAEDIA OF PALEONTOLOGY

ENCYCLOPAEDIA OF
PALEONTOLOGY

Vivek Sharma

ANMOL PUBLICATIONS PVT. LTD.
NEW DELHI - 110 002 (INDIA)

ANMOL PUBLICATIONS PVT. LTD.
H.O.: 4374/4B, Ansari Road, Darya Ganj,
New Delhi-110 002 (India)
Ph.: 23278000, 23261597

B.O.: No. 1015, Ist Main Road, BSK IIIrd Stage
IIIrd Phase, IIIrd Block
Bangalore - 560 085 (India)
Visit us at: www.anmolpublications.com

Encyclopaedia of Paleontology

First Published, 2006

ISBN 81-261-3027-X

PRINTED IN INDIA

Printed at Mehra Offset Press, Delhi.

Preface

Paleontology can be described as the science which treats the ancient life of the earth, or of fossils which are the remains of such life. This book is intended for all who have an interest in Paleontology, be it a student, a teacher or professional. Many of the terms used are technical, and few people can understand all of them. This Encyclopeadia of Paleontology is intended to help such understanding. It provides clear, authoritative definitions of terms associated with all aspects of Paleontology.

This Encyclopaedia of Paleontology broadly covers various aspects of Paleontology with illustrations. A very well written Encyclopaedia covering all aspects of theoretical and practical paleontology, including micro paleontology, invertebrate paleontology, vertebrate paleontology, spores, pollens, rock strata and much more.

We hope that professionals and students will find this reference volume useful. Although we could not include every concept in Paleontology, we have endeavoured to include those that our colleagues and we deemed most important. The aim of this volume is to provide a concise repository of number of major concepts in this field. In preparation of this book, the author has freely consulted large number of books and journals so no authenticity is claimed. Author is especially thankful to Anmol Publications Pvt. Ltd., New Delhi for shaping this book in its final form. Suggestions for further improvement of this book are not only welcome but also greatly appreciated.

Author

A

A type tetrad A tetrad in which the aborted cells (1-3 in number) are clearly associated with the fertile grain(s). Example: *Cyathodes juniperina* (Epacridaceae).

A, an A prefix indicating the absence of a feature, as for example in alete.

A.d. Abbreviation of the Latin *anno domini* meaning in the year of our lord. When used as a prefix or suffix to a date, it indicates the number of years elapsed since the supposed date of the birth of Christ.

A.S.L. Abbreviation for "above (mean) sea level"; the standard reference of elevation used in N.T.S. (national topographic series) maps; the vertical height of an object as measured from the mean elevation of high and low tides.

Aberrant. Non-typical; deviation from the class to which an artefact or phenomenon belongs.

Aboriginal Indigenous, Pertaining to the original occupants of a given region.

Abporal lacuna A lacuna in a lophate pollen grain situated at the end of an ectoaperture

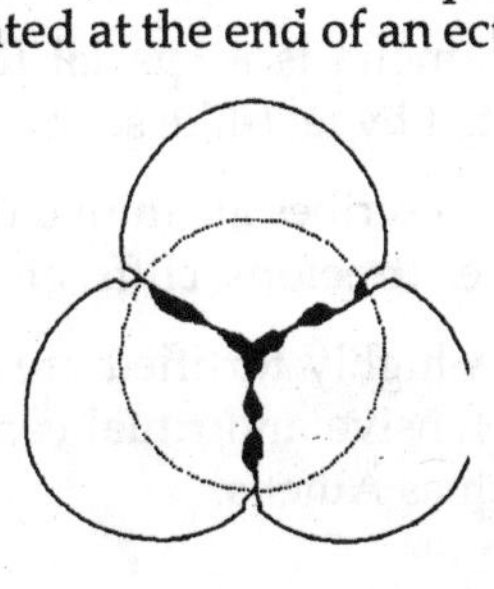

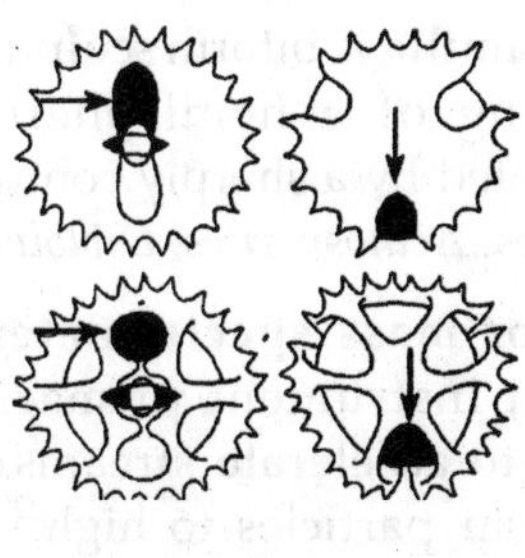

Fig. Abporal lacuna

that is divided by sexinous ridges into two or more lacunae.

Abrasive Stone Usually a sandstone slab used for grinding and polishing.

Absolute dating A dating method that determines an object's exact age, as opposed to its relative age; includes such techniques as dendrochronology and radiocarbon dating.

Abu Simbel Two temples located close to the border between Sudan and Egypt. They were constructed in the 13th century B.C.E. during the reign of Pharaoh Ramesses II.

Abyssal plain Large area of extremely flat ocean floor lying near a continent and generally over 4 km in depth.

Acalymmate Describing tetrads or polyads in which the sexine/ectexine of each monad is well differentiated, but does not form a single continuous envelope around the unit.

Antonym: calymmate.

Acanthodians A primitive group of Silurian to Permian jawed bony fishes, bearing bony spines in front of all but their caudal fins.

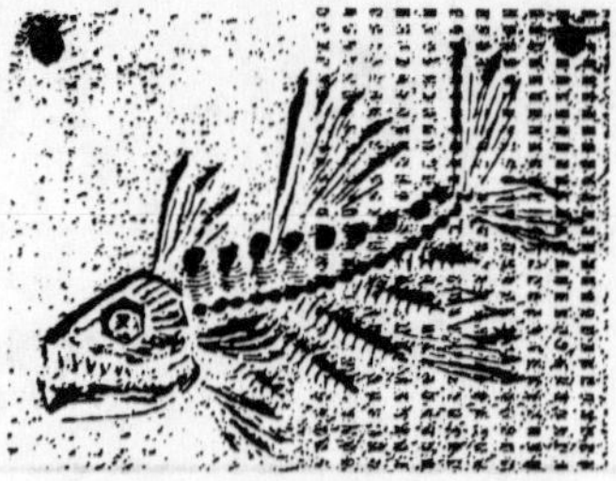

Fig. Acanthodians

Acanthomamilla A biform sculptural element consisting of a hemispheroidal base, surmounted by a sharply contracted spine. Examples: *Acinosporites, Diblolisporites.*

Accelerator mass spectrometer (AMS) A research instrument primarily used in physics to accelerate streams of charged subnuclear particles to high velocities in order to sort and analyze them. This technique is now also used to count carbon isotope atoms for radiocarbon dating. The advantage of this technique over the conventional radiocarbon method is that it requires a far smaller sample size and can potentially provide dates going back to around 100,000 B.P. At present, however, AMS dates generally are for events less than 60,000 years old.

Acclimatory adjustments Reversible physiological adjustments to stressful environments.

Accretion An object grown by the process of externally layering material one layer on top of another. Caliche and pearls are examples of accretions — accrete.

Acculturation The process by which a culture absorbs the traits or customs of another culture with which it is in direct contact.

Acephalous Society A society without a political head such as a president, chief, or king.

Acetolysis A widely used technique for preparing pollen and spore exines for study.

Achemenid Empire Persian empire named after its founder Achemens. The empire lasted from about 550 to 330 BCE when it was conquered by Alexander the Great.

Acheulian Archeologic culture attributed to *Homo erectus* of old world, 300,000 to 1.5 million years BP.

Achieved Status Social standing and prestige reflecting the ability of an individual to acquire an established position in society as a result of individual accomplishments/

Acolpate Without colpi.

Acquired trait A phenotypic characteristic, acquired during growth and development, that is not genetically based and therefore cannot be passed on to the next generation (for example, the large muscles of a weightlifter).

Acritarchs "An informal utilitarian, 'catch-all' category without status as a class, order or other supragenieric unit" comprised of "small microfossils of unknown and probably varied biological affinities consisting of a central cavity enclosed by a wall of single or multiple layers and a chiefly organic compositio". Precambrian - Recent, probably of algal affinity.

Acrolamella A leaf-like tapering segment on the proximal pole of a megaspore. Example: *Arcellites.*

An acrolamella is a special form of gula, represented by leaf-like segments.

Acronomic Describes an animal that dwells in high places (treetops, cliffs, etc.).

Acropolis A highly fortified area that served as the defensive and ritual center of Greek cities such as Athens.

Act The smallest unit of recurrent behaviour involving an artefact.

Actartif Any object manufactured, used or modified by humans.

Activity A set of related 'acts.'

Activity area (or activity sphere) That portion of an archaeological site which can be equated with a single activity such as flint knapping, butchering, or cooking.

Actuopalynology The study of extant palynomorphs which are either living, still retain their cell contents, or whose cell contents have been removed by maceratio Includes mellisopalynology, pollination ecology, aeroallergy, and criminology. Exclusive of paleopalynology.

Adaptation Any heritable characteristic of an organism that improves its ability to survive and reproduce in its environment. Also used to describe the process of genetic change within a population, as influenced by natural selection. Alternatively, some heritable feature of an individual's phenotype that improves its chances of survival and reproduction in the existing environment.

Adaptive landscape A graph of the average fitness of a population in relation to the frequencies of genotypes in it. Peaks on the landscape correspond to genotypic frequencies at which the average fitness is high, valleys to genotypic frequencies at which the average fitness is low. Also called a fitness surface.

Adaptive logic A behavior has adaptive logic if it tends to increase the number of offspring that an individual contributes to the next and following generations. If such a behavior is even partly genetically determined, it will tend to become widespread in the population. Then, even if circumstances change such that it no longer provides any survival or reproductive advantage, the behavior will still tend to be exhibited — unless it becomes positively disadvantageous in the new environment.

Adaptive radiation The diversification, over evolutionary time, of a species or group of species into several different species or subspecies that are typically adapted to different ecological niches (for example, Darwin's finches). The term can also be applied to larger groups of organisms, as in "the adaptive radiation of mammals."

Adaptive strategies A mode of coping with competition or environmental conditions on an evolutionary time scale. Species adapt when succeeding generations emphasize beneficial characteristics.

Adequate Synonym of spheroidal.

Adjustment The ability of humans to survive in stressful environments by nongenetic means.

Adolescent growth spurt A rapid increase in stature and other dimensions of the body that occurs during puberty.

Adult The period in an individual's life cycle after the eruption of the last permanent teeth.

Adze An axe-like implement in which the blade is hafted such that the cutting edge lies perpendicular to the handle after the fashion of a hoe. Used primarily for woodworking. trihedral adze. an adze with a triangular cross-section.

Aeolian (sometimes eolian) Sand, clay, silt, or mixed deposits that have been carried by the wind. Loess and sand dunes are typical aeolian deposits.

Aerial photography Vertical and oblique photographic imagery of the earth's surface taken from any point of advantage. The use of specialized films can render visible features which could not otherwise be detected. Topographic relief can be emphasized by photographing in the morning or early evening when shadows are most pronounced. Post human activity may be indicated by the patterning of different soil types or by the mosaic of vegetational types which they support. Cereals are particularly important in this regard, and study of such crop marks can reveal the presence of buried foundations and roads. Thus, aerial photography may be used not only as a technique for the discovery of new sites, but as a means of appreciating the context or setting of a previously-known site.

Aeroallergy The branch of medicine concerned with the seasonal occurrence, abundance, and allerogenic effect of spores and pollen.

Aeropalynology The study of palynomorphs found in the atmosphere.

Affiliative behavior Close-proximity behavior that includes touching, grooming, and hugging.

Affinal kin Persons related by marriage.

Agamo species A term for groups of primarily or secondarily uniparental (= agamic, apomiktic, asexual) organisms, that are morphologically so similar that they have been classified as a single species. These agamo-species represent entities that are fundamentally different from the biospecies of biparental organisms, since the relationships between the members of an agamo-species are exclusively hierarchical, contrary to the generally reticulate relationships within biospecies. Consequently, there is no difference between tokogenetic and phylogenetic relationships in all uniparental organisms, and furthermore there exists no particular cohesion by mate recognition mechanisms and gene flow. The only potential mechanism of cohesion might be the ecological concurrence between the individual organisms of an agamo-species. Nevertheless uniparental organisms form clones, that can be discovered by autapomorphies just like monophyletic groups.

Agate A banded or mottled chalcedony.

Agate basin A physiographic feature in wyoming, also known as moss agate arroyo which has given its name to the lanceolate projectile points recovered there, to the archaeological site from which they were recovered (occasionally including the nearby brewster site), to the complex of associated artifacts and the culture of the makers of the artifacts. In outline form, the points are neither notched nor stemmed and range in length from 5 to 15 cm. Flaking is horizontal and it is this trait that serves to distinguish them from the angostura morphological characteristics. Basal and/or lateral grinding may be present and maximum width tends to occur toward the distal end. Agate basin sites occur throughout the north american plains and adjacent areas. A small number of agate basin sites have been excavated in manitoba but many of the characteristic points occur in surface collections. The complex occurs within the late palaeo-indian (or plano) period and is believed to represent the material culture of nomadic bison hunting peoples.

Agave Sometimes called a century plant. Several species of the plant were used by Indians in the Southwest and Mexico. The plants vary greatly in size, but are characterized by a cluster of leaves spreading out at ground level from a short central stem. The narrow leaves are long and thick and terminate in a spine. At maturity, each plant sends up one long flowering stalk and then dies. Agaves grow at elevations of 3000 to 8000 feet. Species of agave are used in the manufacture of pulque and tequila, alcoholic beverages popular in Mexico. Raw agave is poisonous.

Age grade A group of people of the same sex and approximately the same age who share a set of duties and privileges.

Aggradation An accumulation of sediment resulting in the building up of a land surface. An example would be part of a river bank upon which sediments are regularly deposited during the spring flood.

Aging The uninterrupted process of normal development that leads to a progressive decline in physiological function and ultimately to death.

Agnathans A jawless, heavily armoured fish that appeared in the Devonian.

Agnostic A person who believes that the existence of a god or creator and the nature of the universe is unknowable.

Agonistic behaviour Behavior that involves fighting, threats, and fleeing.

Agrarian state The Fourth stage in the stage model , representing large regional systems or empires based primarily on non-mechanized agriculture and controlled by centralized and specialized bureaucracies.

Agriculture A subsistence mode which involves the use of machinery or domesticated animals in the cultivation of plants.

A-horizon The uppermost, often dark-coloured natural level in a soil profile characterized by roots, humus, and a lack of clay, iron, carbonates and soluble salts which have leached to lower levels.

AIMA Australasian Institute of Maritime Archaeology

Airlift Instrument like a giant vacuum cleaner used by underwater archaeologists to remove dirt and debris from underwater archaeological sites.

Al A suffix for of, relating to, or characterized by. Example: tectal.

Alabastron A traditional Egyptian oil jar made of alabaster. The Greeks made later versions of it out of clay.

Alberta A Plano projectile point style of the north-western plains. Specimens are as much as 20 cm in length, parallel-sided with blunt tips, and stemmed.

Albian European stage of the uppermost Lower Cretaceous, spanning the time between 107 and 95 million years ago.

Albinism A recessive abnormality that leads to little or no production of the skin pigment melanin.

Alete Describing a spore without a laesura.

Alga The cell walls of certain freshwater green algae are resistant to standard palynological procedures. The vegetative cells of the colonial forms *Pediastrum* and *Botryococcus* are frequently recognized in lacustrine sediment, as are the spores of *Spirogyra* and certain other Zygnemataceae. Round spores (leiospheres) are particularly abundant, but are seldom identified.

Algae An umbrella term for various simple organisms that contain chlorophyll (and can therefore carry out photosynthesis) and live in aquatic habitats and in moist situations on land. The term has no direct taxonomic significance. Algae range from macroscopic seaweeds such as giant kelp, which frequently exceeds 30 m in length, to microscopic filamentous and single-celled forms such as *Spirogyra* and *Chlorella*.

Algonkian (or algonquian). A grouping of related languages whose speakers were originally distributed from newfoundland to california and from the northern prairie provinces of canada to the american southeast. The term derives from the algonkin (or algonquin) people who resided in the ottawa and st. Lawrence river valleys. Algonkina languages and dialects in manitoba include cree, ojibwa or chippewa, and saulteaux.

Alidade A surveying instrument consisting of a sighting device and a scaled rule. Often used in conjunction with a plane table, the relative position of objects is determined by triangulation.

Alienation The fragmentation of individuals' relations to their work, the things they produce, and the resources with which they produce them.

Allele One of the alternative forms of a gene. For example, if a gene determines the color of eyes, one allele of that gene may produce brown eyes and another allele produce blue eyes. In a diploid cell there are usually two alleles of any one gene (one from each parent). Within a population there may be many different alleles of a gene; each has a unique nucleotide sequence.

Allen's rule A rule which states that among endotherms, populations of the same species living near the equator tend to have more protruding body parts and longer limbs than do populations farther away from the equator.

All-male party Among chimpanzees, a small group of adult or adolescent males.

Allogrooming Grooming another animal.

Allometric growth The pattern of growth whereby different parts of the body grow at different rates with respect to each other.

Allometry The relation between the size of an organism and the size of any of its parts. For example, an allometric relation exists between brain size and body size, such that (in this case) animals with bigger bodies tend to have bigger brains. Allometric relations can be studied during the growth of a single organism, between different organisms within a species, or between organisms in different species.

Allomorphs Forms contained in morphemes that differ in sound but not in meaning.

Allopatric speciation Speciation that occurs when two or more populations of a species are geographically isolated from one another sufficiently that they do not interbreed.

Allopatric species Species occupying mutually exclusive geographical areas.

Allopatry Living in separate places. Compare with sympatry.

Allophones Sounds that belong to the same phoneme.

Allotype Specimen designated from the type series that is the opposite sex of the holotype.

Alloying A technique involving the mixing of two or more metals to create an entirely new material, e.g. the fusion of copper and tin to make bronze.

Alluvial deposits Sediments laid down by streams in their channels or on flood plains.

Alluvial fan A fan- or wedge-shaped accumulation of silt, sand, gravel and boulders deposited by rapidly-flowing streams when they reach flatter terrain.

Alluvium A generally fine-grained mixture of sand, silt and mud deposited by flowing water.

Altamira Cave near the north coast of Spain discovered in 1868. The first site where Paleolithic Period cave paintings were found.

Altimeter A barometric device for determining elevations above sea-level.

Altithermal A climatic/geological period postulated by ernst antevs (1955) for the warmer, drier period which he contended characterized western north america between 7000 and 4000 years ago. The term Atlantic climatic episode is now more commonly used for this mid-post glacial interval.

Altricial Describes organisms that are born relatively undeveloped, requiring significant parental care for survival. Humans are an example of an altricial organism.

Altruistic act A behavior characterized by self-sacrifice that benefits others.

Alvarez, Luis Luis Alvarez (1911-1988) was a physicist who, with his son Walter Alvarez (a geologist), hypothesized that the a huge asteroid hit Earth the Earth 65 million years ago, causing a mass extinction. The Alvarez

Theory of Extinction is widely accepted. Luis Alvarez received a Nobel Prize in physics (1968) for his work on subatomic particles.

Alveolate Describing a type of sexine/ectexine structure, in which the infratectal layer is characterised by partitions forming compartments of irregular size and shape. Example: *Pinus* (Pinaceae).

Amarna General term used to refer to the reign of Akhnaten and surrounding years. Also modern name of the Egyptian city founded by Akhenaten.

Amarna letters A collection of clay tablets containing diplomatic correspondence of the Amarna Period.

Amb The outline of a pollen grain or spore seen in polar view.

The term does not necessarily coincide with the equatorial outline. The term is especially useful for the outline of spores seen in polar view, because of their heteropolar shape.

Amber A resin produced by certain trees since the Jurassic period. When secreted from the bark, resin forms a sticky orange fluid which insects and small creatures can become trapped in. Over time the resin hardens to form Amber, thus preserving itself and any insect or creature within. It was originally believed that mosquitoes trapped in this way, may contain the DNA necessary to resurrect a dinosaur. To date the evidence suggests that DNA could not survive for this length of time.

Ambilineal descent A descent ideology based on ties traced through either the paternal or the maternal line.

Amino acid Organic molecules that are building blocks of proteins. There are at least 20 different kinds of amino acids in living things. Proteins are composed of different combinations of amino acids assembled in chain-like molecules. Amino acids are primarily composed of carbon, oxygen, hydrogen, and nitrogen.

Amino acid dating Dating of organic materials based on racemization (chemical change) of amino-acids, which is both time and temperature dependent.

Amino acid racemization dating A method for dating organic matter that is based on the fact that amino acids progressively change to mirror image forms following the death of an organism—i.e., from L-amino acid to D-amino acid forms. Aspartic acid in organic samples is commonly used for this dating technique. Amino acid racemization could be considered to be a chronometric or a calibrated relative dating method.

Amino acid sequence A series of amino acids, the building blocks of proteins, usually coded for by DNA. Exceptions are those coded for by the RNA of certain viruses, such as HIV.

Ammonite Probably the most commonly associated fossil in the world. Ammonite's are an extinct group of marine 'cephalopod molluscs' that thrived during the Jurassic and Cretaceous period (170 - 65 million years ago). Ammonites consisted of a coiled shell within which the creature lived. These agile predators preyed on smaller marine creatures, including fish and shells. The closest living descendent of Ammonite's are Cuttle fish.

Fig. Ammonite

Ammonoid Extinct relatives of cephalopods (squid, octopi, and chambered nautiluses), these molluscs had coiled shells and are found in the fossil record of the Cretaceous period.

Ammonoids Cephlapods that evolved from nautiloids and are generally spiral-shaped shell with numerous chambers.

Amniotes The group of reptiles, birds, and mammals. These all develop through an embryo that is enclosed within a membrane called an amnion. The amnion surrounds the embryo with a watery substance, and is probably an adaptation for breeding on land.

Amoeba A microscopic, one-celled animal consisting of a naked mass of protoplasm.

Amphibians The class of vertebrates that contains the frogs, toads, newts, and salamanders. The amphibians evolved in the Devonian period (about 370 million years ago) as the first vertebrates to occupy the land. They have moist scale less skin which is used to supplement the lungs in gas exchange. The eggs are soft and vulnerable to drying, therefore reproduction commonly occurs in water. Amphibian larvae are aquatic, and have gills for respiration; they undergo metamorphosis to the adult form. Most amphibians are found in damp environments and they occur on all continents except Antarctica.

Amphora Large round ceramic container used for transportation and storage of goods. Used from antiquity until the 16th century or so. Used for wine, oil, olives, grain, etc, etc. Amphoras in a shipwreck can often tell the age and nationality of the wreck.

Amun Egyptian god associated with the state and the kingship during Egypt's New Kingdom.

Ana A prefix indicating the position of features, such as apertures, on the distal face. Examples: *Nymphaea* (Nymphaeaceae), *Sparganium* (Typhaceae).

Antonym: cata-.

Analogies Structures that are superficially similar and serve similar functions, but have no common evolutionary relationship.

Analogous structures Structures in different species that look alike or perform similar functions (e.g., the wings of butterflies and the wings of birds) that have evolved convergently but do not develop from similar groups of embryological tissues, and that have not evolved from similar structures known to be shared by common ancestors. Contrast with homologous structures. Note: The recent discovery of deep genetic homologies has brought new interest, new information, and discussion to the classical concepts of analogous and homologous structures.

Analogy A process of reasoning whereby two entities that share some similarities are assumed to share many others.

Analysis The process of studying and classifying artifacts, usually conducted in a laboratory after excavation has been completed.

Anasazi One of the three desert cultures that shaped life in the American Southwest from 300 B.C. to A.D. 1300. Developed a new way of building pueblos and the technique of farming on top of mesas. Used both hand-formed adobe bricks and stones to build their homes.

Anatolia The large peninsular region of Turkey, bordered by the Black Sea to the north and the Mediterranean Sea to the south, and constituting the westernmost point of Asia; also known as Asia Minor.

Anatomy 1. The structure of an organism or one of its parts.

2. The science that studies those structures.

Anazonasulculate Describing a pollen grain with a ring-like sulculus situated between the equator and the distal pole.

Antonym: catazonasulculate.

Ancestor. Forebear; one from whom another is descended.

Ancestral homology Homology that evolved before the common ancestor of a set of species, and which is present in other species outside that set of species. Compare with derived homology.

Anchor taxa Describes the taxa used to define a phylogenetic taxon or clade.

Ancillary sample Any non-artifactual materials collected by archaeologists to aid in dating, paleoenvironmental reconstruction, or other interpretations - e.g. carbon samples, soil samples, palynological samples etc.

Ancyrate Bearing sub-cylindrical or tapering processes which divide at their distal extrimities into anchor-shaped or multifurcate tips.

Anderson corner-notched A projectile point style defined by Macneish (1958) on the basis of his investigations in southeastern manitoba and attributed to the Anderson, Nutimik and Larter foci. As originally defined, these points are relatively long (30 to 68 mm) and narrow with straight bases and expanding stems. This designation is less commonly used than previously.

Anderson focus. The earlier (500 b.c. to a.d. 500) of the middle or initial woodland cultural-historical periods in Macneish's (1958) southeastern Manitoba chronology. It was later designated the Anderson phase by Mayer Oakes (1967). Hlady (1970) finally advocated the grouping of the Anderson with the other middle woodland focus due to a general absence of distinguishing artifactual traits, and further proposed that this new entity be designated the laurel phase. The suggestion met with universal acceptance and the earlier term is no longer used.

Andesite A fine-grained grey to green igneous rock composed primarily of minerals of the feldspar group — in particular andesine, amphibole and pyroxene.

Andrews, Roy C. Roy Chapman Andrews (1884-1960) was a US fossil hunter and director of the American Museum of Natural History. Andrews led four expeditions to Mongolia's Gobi desert between 1922 and 1925. Many important finds were made on these expeditions, including Protoceratops bones and eggs (the first dinosaur eggs found!), and the new dinosaurs Oviraptor, Pinacosaurus, Saurornithoides, and Velociraptor.

Angiosperms The collective name for flowering plants which evolved in the Late Jurassic period.

Angkor wat A complex of religious buildings in Cambodia (in southeastern Asia) that is considered one of the world's archaeological and architectural treasures. The complex combines a temple dedicated to Vishnu (a Hindu god) and a mausoleum (a large and stately tomb). Angkor Wat was built by Suryavarman II, who ruled the Khmer Empire from A.D. 1113 to 1145.

Angle of declination The degree difference between the direction of magnetic north and rotational north at a particular location.

Anglo Saxons A name used to describe the European warriors who invaded Britain around the 5th century A.D.; composed of two separate groups, the Angles and the Saxons.

Angostura A Plano projectile point style (previously termed "Long") named by R.P. Wheeler (in Wormington 1957) after the Angostura Basin in South Dakota. Angostura points, sometimes termed "Lusk" points, are long and narrow, lanceolate in outline form, rhomboidal in cross section, and have concave or straight bases.

Angulaperturate Describing an equatorially aperturate pollen grain with the apertures situated at the angles of the outline in polar view. Example: *Corylus* (Betulaceae).

Antonym: planaperturate.

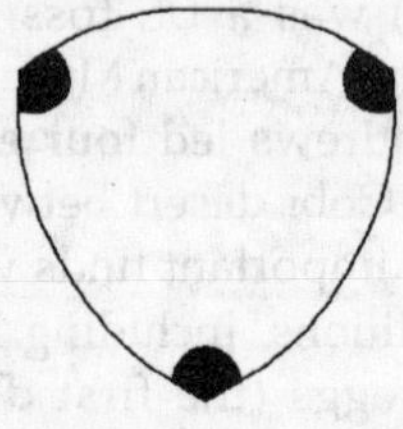

Fig. Angulaperturate

The term is useful for describing the position of apertures, but should be avoided as a description of equatorial outline.

Angustimurate With narrow muri.

Erdtman restricted the term to muri up to 1/5th of the diameter of the lumina.

Animal husbandry The breeding, care, and use of herd animals, such as sheep, goats, camels, cattle, and yaks.

Animatism Belief in an impersonal supernatural force.

Animist One who believes in animism, a belief that creatures, objects, and natural phenomena are inhabited by spirits.

Anisodiametric tetrad General term for tetrads which members are differing in size. Example: seed-megaspore tetrads.

Anisopolar Synonym of heteropolar.

Anneal. To temper or harden by exposure to heat. Some lithic materials may produce more regular planes of fracture subsequent to controlled annealing and some metals may be rendered less brittle.

Annealing In copper and bronze metallurgy, this refers to the process of heating and then cooling the material to remove stress from hammering.

Anning, Mary Mary Anning (1799-1847) was an early British fossil hunter who began finding fossils as a child, and supported herself and her family by finding and selling fossils. She lived on the southern coast of England, in Lyme Regis. Anning found the first fossilized plesiosaur and Ichthyosaurus. She found many important fossils, including Pterodactylus, sharks, and many other reptiles and fish.

Annulus An area of the exine surrounding a pore that is sharply differentiated from the remainder of the exine, either in ornamentation or thickness

Although the correct Latin spelling is anulus, the form used throughout the botanical literature is annulus.

Anteturma An artificial grouping of fossil spores and pollen in the turma-system of Potonié.

Turmae are grouped under two large headings the anteturmae Sporites and Pollenites.

Anthropocentricity The belief that humans are the most important elements in the universe.

Anthropoid A member of the suborder Anthropoidea; includes the New World monkeys, Old World monkeys, apes, and humans.

Anthropoidea Suborder of the order Primates that includes the New World monkeys, Old World monkeys, apes, and humans.

Anthropological linguistics The scientific study of human communication within its sociocultural context and the origin and evolution of language.

Anthropology The scientific and humanistic study of man's present and past biological, linguistic, social, and cultural variations. Major sub-fields include archaeology, physical anthropology, cultural anthropology, and linguistic anthropology.

Anthropometry The study of measurements of the human body.

Anthropomorphic "Man-like." Used to describe artifacts or art work decorated with human features or with a man-like appearance.

Antibacterial Having the ability to kill bacteria.

Antibiotics Substances that destroy or inhibit the growth of microorganisms, particularly disease-causing bacteria.

Antibiotic resistance A heritable trait in microorganisms that enables them to survive in the presence of an antibiotic.

Antinous Favorite companion of Emperor Hadrian.

Antonine wall Built during the early 140s AD. Northernmost Roman wall in Great Britain marked the edge of the territory of Hadrian's successor, Antonius Pius.

Anulus Orthographical variant of annulus.

Anvil A block of stone or metal upon which other materials are shaped or worked through striking.

Ape A common term that includes the lesser apes (the gibbons and siamang) and the great apes (the orangutan, common chimpanzee, bonobo, and gorilla).

Aperture Of a camera, the adjustable opening through which light passes to reach the film. The diameter of the aperture determines the intensity of light admitted. The pupil of a human eye is a self-adjusting aperture.

Aperture membrane The exine which forms the floor of an ectoaperture.

Depending on the type of ectoaperture these may be colpus membranes or pore membranes, usually representing the nexine. The membrane may be partially or completely covered by an operculum.

Apertures Any thinning or modification of the wall of spores or pollen that serves as an exit for its contents. In embryophyte spores these are the monolete or trilete laesura. In pollen these are the furrows (colpi) and pores. The arrangement and number of pores and furrows are a chief criteria for identifying pollen classes.

Apex A general term for the tip of an organ

In fossil spores applied to the tip or corner of a trilete spore.

Aphasia A language disorder resulting from brain damage.

Aphrodite Greek goddess of love and fertility. Known as Venus to the Romans.

Apicalfeld Synonym of apocolpial field.

Apiculate Describing the shape of a pollen grain that has slightly protruding polar caps. Example: *Adonis aestivalis* (Ranunculaceae).

Apiculate elements Projections from the general surface. Examples: bacula, pila, verrucae, spines.

Apocolpial field A region at the pole of a parasyncolpate pollen grain, delimited by the margins of the anastomosing colpi.

Apocolpium A region at the pole of a zonocolpate pollen grain delimited by lines connecting the apices of the colpi.

Synonym of polar area.

In the original definition the apocolpium was delimited by the polar limits of the mesocolpia, the more precise definition now used is derived from Iversen and Troels-Smith (1950, polar area).

Apocolpium index The ratio of the distance between the apices of two ectocolpi (d) of a zonocolpate pollen grain to its equatorial diameter (D).

Synonym of polar area index.

Apolar Describing pollen and spores without distinct polarity. Example: *Plantago* (Plantaginaceae).

Apomorph 1. Evolutionarily advanced ('derived') character state. The long neck of the giraffe is apomorphic; the short neck of its ancestor is plesiomorphic

2. In cladistics, a character state present throughout a clade but not present in any close outgroup of the clade —apomorphy — apomorphic.

Apomorphic (versus plesiomorphic) A derived character state is termed the apomorphic state. This property "apomorphic" is not generally valid, but always related to the compared character state and the hierarchical level in view: The scales of "reptiles" are apomorphic compared to the smooth and glandulated skin of amphibians, but plesiomorphic compared with the feathers of birds. Hairs are autapomorphic for mammals, but a symplesiomorphy for rodents. Shared derived similarities can be homologous (synapomorphies) or non-homologous (convergences). Klausnitzer & richter (1981) suggested the new (synonymous) terms "apotypic", "autapotypy" and "Synapotypy", since strictly speaking the suffix "-morphic" is only correct for morphological characters. The term "derived" should only be applied to character states, but never to organisms or taxa (taxa that originated from rather young splitting events and have developed numerous apomorphic states should rather be called "modern groups"), since all taxa always have a mixture of plesiomorphic and derived character states (compare living fossils). Considering the frequent phenomenon of reduction or reversals, derived states by no means have to be more sophisticated or more complex than the referring plesiomorphic state. Out of this reason the plesiomorphic state should better not be equalled with "primitive" state.

Apoporium An area at the pole of a zonoporate pollen grain that is delimited by a line connecting the borders of the pores.

The term contrasts with apocolpium and is most meaningful in pollen grains with large pores.

Applied anthropology The activity of professional anthropologists in programs that have as primary goals changes in human behavior believed to ameliorate contemporary social, economic, and technological problems.

Aptian European stage of the Lower Cretaceous, spanning the time between 114 and 107 million years ago.

Aquatic Living underwater.

Arable land Land fit for cultivation.

Aragonite $CaCO_3$; polymorph of Calcite. Precipitated by invertebrates at Earth's surface, readily reorders crystal lattice to Calcite arrangement.

Arbitrary levels An archaeological excavation technique in which the thickness of the layers removed is chosen for convenience. This method is generally used when a site does not possess natural stratigraphy and cannot, therefore, be excavated stratum by stratum.

Arboreal Describes an organism that spends most of its life in trees, off the ground.

Arc of exhumation Geomorphic zone inside and parallel to margin of ancient ice sheet in which preglacial strata have been largely stripped away to expose crystalline bedrock. Large lake/sea basins located within this arc: Great Lakes, Great Slave and Bear Lakes, Baltic Sea, Kattegat, *etc.*

Archaean The middle era of Precambrian time, spanning the period between 3.8 and 2.5 billion years ago. Life arose on Earth during the early Archaean, as indicated by the appearance of fossil bacteria in rocks thought to be about 3.5 billion years old. Its name means "ancient."

Archaeo-astronomy The systematic study of astronomical knowledge and lore of prehistoric peoples.

Archaeocyaths Conical, calcareous, Cambrian fossils with a long history of phylogenetic uncertainty and changing interpretations. There is now a strong consensus that archaeocyaths are sponges.

Archaeological context The physical setting, location, and cultural association of artifacts and features within an archaeological site.

Archaeological palynology The analysis of pollen, spores , and other palynomorphs from archeological sites. Applications include the reconstruction of prehistoric diet, funary practices, artifact function and source, archaeological feature use, cultivation and domestication of plants, and human impact on vegetatio

Archaeological record. The sum of all evidence concerning past events and peoples.

Archaeological recovery Removal of artifacts from archaeological context with full recording of their four dimensions of variability.

Archaeological sequence Artifacts, behaviors, or phases (periods) ordered in time.

Archaeological site Any location which shows evidence of the prior presence or influence of human beings.

Archaeological survey On-ground inspection of a study area for artifacts and sites.

Archaeologist Anyone with an interest in the aims and methods of archaeology. A professional archaeologist usually holds a degree in anthropology with a specialization in archaeology.

Archaeology The study of the prehistory or early history of societies and their cultures. Unlike paleoanthropology, the focus of archaeology is mainly on the material remains of culture rather than biological evolution.

Archaeology - cult The study of the material indications of patterned actions undertaken in response to religious beliefs.

Archaeozoology Sometimes referred to as zooarchaeology, this involves the identification and analysis of faunal species from archaeological sites, as an aid to the reconstruction of human diets and to an understanding of the contemporary environment at the time of deposition.

Archaic A term variously used to denote a stage of cultural development or a specific period in prehistory characterized by a particular way of life (i.e. Primary forest efficiency) and/or the nature of the tool-kit employed. In manitoba, the term is most commonly used in reference to a period (ca. 5000 b.c. to the time of Christ) in the prehistory of the forested portion of the province subsequent to the palaeo-indian and prior to the woodland. It is roughly contemporaneous with the middle prehistoric or meso-indian period or stage – the second of the two three-staged schemes of plains prehistory. By definition, archaic assemblages do not include ceramics – the latter constituting the major technological marker of the woodland period. By the end of the archaic period, the atlatl has largely replaced the spear as a hunting implement, but the bow has not yet been introduced; human groups remain mobile, but only seasonally so, operating within more restricted geographic areas; and plant foods have come to play a more important role in subsistance.

Archean The earliest dated pre-cambrian rock on earth, dating from 3.8 billion years ago.

Archetype The original form or body plan from which a group of organisms develops.

Archives 1. A collection of primary historical documents such as journals, diaries, maps and personal and business correspondence.

2. The institutional repository within which such collections are housed. Archival information and the findings of historic archaeology often complement one another, providing more information on a historic site or time period than could either method of inquiry alone.

Archosauria ("ruling lizards") Clade of animals, partly distinguished by an antorbital fenestra, that includes the Crocodylia, Pterosauria, Dinosauria, and several other closely related groups of animals –archosaur.

Arctic small tool tradition (astt). A grouping of archaeological complexes distributed across the north american arctic from alaska to greenland which date between roughly 3000 b.c. to a.d. 1000. The tradition is so named due to the extremely small, finely worked tools which these people manufactured. Notable examples include burins, micro blades and side-blades. Manitoba occurrences of this tradition include the pre-dorset and dorset "cultures" which some have reason to believe are ancestral to inuit culture. Sites containing these materials are restricted to the northeastern quarter of the province.

Arcus A locally thickened band of sexine that extends in a sweeping curve from one aperture to another. Example: *Alnus* (Betulaceae).

Areola A feature of ornamentation in which the sexine/ectexine is composed of circular or polygonal areas separated by grooves which form a negative reticulum. Examples: *Apama* (Aristolochiaceae), *Phyllanthus* spp. (Euphorbiaceae).

Ares Greek god of war. Known to the Romans as Mars.

Argillite A fine-grained, metamorphosed mud and claystone. The deep-red-colored argillite artifacts found at the Hardy Site may have come from the Mazatzal Mountains in central Arizona.

Argon-40/argon-39 dating ($^{40}Ar/^{39}Ar$) A radiometric dating method based on the changing ratio of argon-40 to argon-39 with the passage of time in volcanic rock or ash. This technique was derived from potassium-argon dating. The argon-40/argon-39 method is usually more accurate than potassium-argon dating and doesn't require as large a sample.

Arrow A dart-like projectile propelled by a bow. Feathers may be attached to stabilize the arrow in flight, and a stone, bone or metal tip may be fitted to improve its capacity for penetration.

Arrow weed A rank-smelling shrub that forms dense thickets in stream beds and moist saline soil. The plant occurs at elevations of 3000 feet or lower, from Texas to Southem California and from Utah to northern Mexico. In addition to its use as a wall-covering material, arrow weed stems were used for arrow shafts by Indians in the Southwest.

Arrowhead The pointed tip of an arrow. If the means of propulsion cannot with certainty be identified as a bow, the term projectile point is more properly used.

Art object Any artifact carrying, or consisting of, decorative or artistic elements.

Arthropoda A phylum of invertebrates which contain insects, crustaceans, trilobites, eurypterids, and arachnids.

Articulated Jointed, as in the case of bones; to be in their normal positions relative to one another when found in the course of excavation. The term may also be used of broken artefacts, the fragments (or shreds) of which lie in their original relative positions.

Artefact. Any object manufactured, used, moved or otherwise modified by human beings, including all waste materials and by-products of these processes. Occasionally, the term is used in the more restricted sense of a completed object as opposed to the associated detritus.

Artifact time marker An artifact type that was made by a particular culture during a limited time period. When discovered clearly in association with ancient humans in an archaeological site, they are an indication of at least the relative time of the occupation. When the independent dating of the artifact types is reliable, this can be considered a calibrated relative dating method.

Old World artifact types used as time markers .

Artifact type A category of artifacts whose attributes are similar: spoons, tables, and coffins, for example, are artifact types.

Artifacts Things that are intentionally made according to a cultural pattern or inadvertently modified as a result of culturally patterned behavior. Artifacts are usually relatively portable objects such as projectile points, ceramic pots, and baskets.

Artificial selection The process by which humans breed animals and cultivate crops to ensure that future generations have specific desirable characteristics. In artificial selection, breeders select the most desirable variants in a plant or animal population and selectively breed them with other desirable individuals. The forms of most domesticated and agricultural species have been produced by artificial selection; it is also an important experimental technique for studying evolution.

Arvilla complex A late woodland complex known almost exclusively from mound excavations along the red river in the Dakotas and Minnesota and in the central portion of the latter state. As defined by Johnson (1973) it is characterized by flexed primary and bundled secondary interment below linear or circular mounds. Grave goods include both utilitarian and ornamental objects. Among the latter are numbered beads, pendants and gorgets>.

Asclepius Greek god of medicine and healing.

Ascribed status Social standing or prestige which is the result of inheritance or hereditary factors.

Asexual reproduction A type of reproduction involving only one parent that ususally produces genetically identical offspring. Asexual reproduction occurs without fertilization or genetic recombination, and may occur by budding, by division of a single cell, or by the breakup of a whole organism into two or more new individuals.

Asia minor The peninsula of western Asia bordered by the Black Sea to the north, the Mediterranean Sea to the south, and the Aegean Sea to the west; the Turkish region of Anatolia.

Aspect In the midwestern taxonomic method devised by McKern (1939), an aspect referred to a number of foci which could be grouped on the basis of at least one shared trait, such as a pottery style. The term is less frequently used than previously.

Aspen parkland A vegetational zone which lies between the grasslands and the northern coniferous or boreal forest. It comprises a mosaic of grassland interspersed with groves of white birch, white spruce, balsam poplar and most commonly, trembling aspen.

Aspidote An orthographical variant of aspidate.

Aspis A prominently protruding thickening of the exine around a pore. Examples: *Betula* (Betulaceae), *Dorstenia* (Moraceae).

An aspis is a special form of annulus, represented by a thickening rather than a thinning.

Aspis channel Synonym of pore canal.

Assemblage The totality of the kinds and styles of artifacts and features which are associated with one another and with a single occupation; all evidence of human activity within a single component. A minority of writers use this term in reference to the material remains of a culture or associated with a phase.

Assimilation The gradual process by which a minority group takes on the characteristics, including customs and attitudes, of the prevailing culture in which it lives.

Association The joint occurrence of artifacts in a sealed natural or cultural unit such as stratum, tomb, grave or cache. It is often assumed that such artifacts were deposited at the same time by people of the same culture.

Assortative mating The tendency of like to mate with like. Mating can be assortative for a certain genotype (e.g., individuals with genotype AA tend to mate with other individuals of genotype AA) or phenotype (e.g., tall individuals mate with other tall individuals).

Assyria An ancient empire in Mesopotamia.

Astarte Goddess of love and fertility worshipped in various parts of the mideast. Her origin was Phoenician.

Asteroid A small rocky or metallic body orbiting the Sun. About 20,000 have been observed, ranging in size from several hundred kilometres across down to dust particles.

Astragalus Talus; a bone of the ankle.

Astrolabe Instrument used for celestial navigation.

Ate A suffix for possession of. Example: porate.

Atectate Describing pollen grains that have an exine with little or no internal structure. Example: *Degeneria* (Degeneriaceae).

The term was originally intended to describe walls of primitive angiosperms lacking columellae. Related terms include intectate, which is used for pollen grains with columellae but without a tectum, and etectate which is applied to pollen grains believed to have lost their tectum during phylogeny.

Athapap culture An archaeological culture defined by hlady (1967) on the basis of investigations at a number of sites on lake athapapuskow near flin flon. The complex included athapap lanceolate, evans lanceolate and baker's narrows corner-notched projectile points, biface blades, scrapers, drills and gravers. Hlady estimated the age of the culture at 2500 to 5000 years.

Athapaskan or athabascan. A grouping (or family) of native american languages within the nadene phylum. Athapaskan speakers were originally distributed from the arctic to the american southwest and as far west coastal california. The principal athapaskans in maitoba at contact were the chipewyan who occupied the extreme north of the province.

Atheism The doctrine or belief that there is no god.

Athena Greek goddess of wisdom and warfare, patron of Athens. The Romans called her Minerva.

Atlantis Legendary civilization described by ancient writers like Plato.

Atlatl The Aztec word most commonly used by archaeologists to refer to the spear thrower or throwing board. It consists of a board, approximately 70 cm in length with a longitudinal groove to receive the dart or spear and a hook at one end which makes contact with the butt end of the projectile. The atlatl is held by the end opposite the hook, and a small polished stone may be attached immediately behind the grip to lend balance. By serving essentially as an extension of the thrower's arm, a projectile can be propelled 60 per cent further than would otherwise be possible. An experienced user can maintain a practical degree of accuracy at distances on the order of 80 m.

Atomic mass The mass number of one atom of an isotope (e.g., carbon-14). It is the sum of the number of protons and neutrons in the nucleus of one atom.

Atomic number The number of protons in the nucleus of one atom of an element.

Atomistic (as applied to theory of inheritance) Inheritance in which the entities controlling heredity are relatively distinct, permanent, and capable of independent action. Mendelian inheritance is an atomistic theory because in it, inheritance is controlled by distinct genes.

Atrium A space within the aperture of a compound pore that has a much larger endopore than the ectopore, so that the pore canal widens towards the interior of the grain. Example: *Myrica* (Myricaceae).

Attic From the area around Athens. (Attica)

Attribute A property or quality of any archaeological object such as the length of a projectile point, the hardness of a potsherd or the colour of a bottle fragment. Theoretically an artifact possesses an infinite number of attributes, but an archaeologist will limit himself to those he believes to be diagnostic – those which will provide him with the information he is seeking.

Auricula Synonym of valva.

The term is mostly used for spores with extreme valva.

Australopithecine A group of bipedal hominid species belonging to the genus *Australopithecus* that lived between 4.2 and 1.4 mya.

Australopithecus afarensis An early australopithecine species that was bipedal; known fossils date between 3.6 and 2.9 mya (for example, Lucy).

Australopiths Extinct early humans who evolved 4 to 5 million years ago in Africa.

Autapomorph 1. Apomorph character state that is unique to a particular species or lineage in the group under consideration

2. in cladistics, an apomorphy that distinguishes a single clade –autapomorphy –autapomorphic.

Autocracy A form of government in which a single person possesses unlimited political power; despotism.

Autonomy The right of a nation to govern itself; independence.

Autosome Any chromosome other than a sex chromosome.

AUV Autonomous Underwater Vehicles are underwater robots that are not remote controlled and operate with artificial intelligence. Just like ROVs they are used instead of divers for difficult operations, e.g. on great depth.

Avalonia A separate plate in the Early Paleozoic consisting of much of Northern Europe, Newfoundland, Nova Scotia, and some coastal parts of New England.

Avebury Built around 2,500 B.C. Massive Late Neolithic stone circle in Wiltshire, UK.

Avian Of, relating to, or characteristic of birds (members of the class Aves).

Avocationals These are recreational scuba divers and amateur underwater archaeology groups who give invaluable help to underwater archaeologists. Examples are volunteer unpaid divers during investigations and diving clubs cooperating with archaeologists and maritime museums. The term may also be applied to volunteers in other archaeological disciplines.

Avonlea A term applied to a projectile point style and the phase with which it is associated in early late prehistoric plains prehistory. As defined by kehoe (1973), the avonlea point is small and well-made with v-or u-shaped side-notches above a generally concave base and small ears. In some cases, avonlea points may be easily confused with some of the other small side-notched points of this period. Other avonlea phase artifacts include lithic scrapers, bifaces, choppers and ceramic vessels. Avonlea is represented at a number of sites in southwestern manitoba where these people pursued a way of life focusing on the communal hunting of bison. The occupation of the province by avonlea people is estimated to have occurred between approximately a.d. 400 and 700.

Awl A pointed hand tool, frequently of bone, used for punching holes in leather.

Axe A heavy chopping tool of stone or metal which may be handheld (a handaxe) or hafted. In the latter instance, the head is attached such that the cutting edge parallels the handle.

Aztec The civilization that ruled the region now called Mexico between A.D. 1000 and 1500. The capital of the Aztec Empire was called Tenochtitlan.

B

B.C.E. An abbreviation used to denote dates that occurred "Before the Common Era" as a more neutral alternative to the "B.C." ("Before Christ") of the Christian calendar.

B.P. date A chronometric date that is measured in years before the present. By convention, scientists have assigned 1950 A.D. in the Gregorian calendar as the present.

Babiche Lacings, thread, thongs or netting made of sinew, gut or hide.

Babylonia A region of Southern Mesopotamia named after the city of Babylon.

Babylonians A group known as the Amorites moved into Mesopotamia around 1900 B.C. The Amorite king, Hamurabbi, conquered all of southern Mesopotamia, and the civilization became known as Babylonian. Babylon was its richest and most powerful city.

Bacchus Roman god of wine. Dionysos to the Greeks.

Backed Intentionally dulled along one edge. A blade may be backed in order to allow it to be held opposite the cutting edge.

Backfill Refill an excavational unit at the end of the investigations; the dirt used to accomplish this. The latter is also known as backdirt.

Bacteria Tiny, single-celled, prokaryotic organisms that can survive in a wide variety of environments. Some cause serious infectious diseases in humans, other animals, and plants.

Baculum A cylindrical, free standing exine element more than 1μm in length and less than this in diameter. Example: *Raistrickia saetosa.*

Bakker, Robert Robert Bakker is a US paleontologist and dinosaur artist who revolutionized people's concepts of dinosaurs in the late 1960's, drawing them as active animals standing upright and not dragging their tails. He named: Chassternbergia (1988), Denversaurus (1988), Drinker (1990, with others), Edmarka (1992, with others), and Nanotyrannus (with others, 1988).

Balanced reciprocity Gift giving that clearly carries the obligation of an eventual and roughly equal return.

Bale seal A small, labelled metal plate that was attached by wire to a bale. Because it had to be cut to reopen the bale, thefts during shipment were reduced.

Baleanopteridae Family of baleen whales; specifically, rorqual whales, which are characterized by pleated throats and advanced skulls.

Baleen Whalebone The term is more commonly used to refer to the bony substance within the mouth of the whale which is used to strain food. it is widely used by Eskimos for making tools and ornaments.

Baltic sea The world's largest brackish sea, located in northern Europe. The low salinity affects not only shipwrecks and other underwater artifacts, but also animal life, where the fish are of different species. The oceans have a salinity exceeding 3%, but the Baltic Sea has a salinity of 0.8% in the south, 0.3% in the north and 0.6% in average. Through currents there is a constant exchange of salt water from the Atlantic with brackish water from the Baltic. The heavier salt water stays in the deep, usually below 40 m depth (in the south) and 80 m depth (in the north). The lighter brackish water is always nearer surface. Between these layers there is also a constant exchange with the water movements – salt spreading up, and brackish water and oxygen diffusing down. For reasons unknown, perhaps climatic change, the Baltic Sea salinity is reducing.

Baltica A separate continental plate of the Early Paleozoic composed of the United Kingdom, Scandinavia, European Russia and Central Europe; named for the Baltic Sea.

Band In service's (1971) scheme, the least complex of the four levels of socio-economic integration. These kin-related societies are small, consisting of 30 to 1 00 people who tend to camp and travel together. There are no full-time "chiefs" and no real conception of ownership of either objects or territory. Division of labour is almost entirely determined by sex, and as most bands subsist by hunting and gathering, women gather plants and perhaps snare small animals near camp, while the men range further from home in search of game. The kinds of activities pursued, and sometimes the location of the camp may shift in accordance with the seasons as different foods become available in different locations and as game animals adjust their feeding patterns. In the majority of cases, members of band societies acquire spouses from outside of their own band (local exogamy). Married couples take up residence with the husband's band (virilocality) and their children are raised there (patrilocality) as a consequence.

Bannerstone A (usually) polished stone implement which may take a variety of forms. One of the most common is winged with a central hole. These may have served some ceremonial function or may simply be elaborate atlatl weights.

Barb A small point which faces the opposite direction to the main point on an object such as a fish hook or arrow tip.

Barremian European stage of the Lower Cretaceous, spanning the time between 118 and 114 million years ago.

Barsbold, Rinchen Rinchen Barsbold is a Mongolian paleontologist. He named: Adasaurus (1983), Ansermimus (1988), Conchoraptor (1986), the family Enigmosauridae (1983), Enigmosaurus (with A. Perle, 1983), Gallimimus (with H. Osmólska and E. Roniewicz, 1972), Garudimimus and the family Garudi-mimidae (1981), Harpymimus and the family Harpymimidae (with A. Perle, 1984), Ingenia (1981), the family Ingeniidae (1986), the family Oviraptoridae (1976), and the suborder Segnosauria (with A. Perle, 1980). Barsboldia (Maryanska et Osmolska, 1981) was named to honor Rinchen Barsbold.

Basal thinning The removal of flakes from the base of a projectile point or blade in a lengthwise fashion in order to facilitate hafting.

Basalt A fine-grained black, brown, gray or green rock consisting of feldspar, olivine, hornblende and augite. Often used for the manufacture of groundstone tools and ornaments.

Base The DNA molecule is a chain of nucleotide units; each unit consists of a backbone made of a sugar and a phosphate group, with a nitrogenous base attached. The base in a unit is one of adenine (A), guanine (G), cytosine (C), or thymine (T). In RNA, uracil (U) is used instead of thymine. A and G belong to the chemical class called purines; C, T, and U are pyrimidines.

Base line An arbitrary line established by stakes and string, or by surveying instrument, from which measurements are taken to produce a site-map, or to provide an initial axis for an excavation grid.

Basin and range province A geographic area extending from southern Oregon and Idaho to northern Mexico, and including most of western Arizona, the Great Basin of Utah and Nevada, and parts of eastern California. It is an area characterized by north-south trending mountain ranges interspersed by flat basins. The area was formed initially through block faulting during Tertiary times (15-20 million years ago), when, in a series of earthquakes, one section of land was lifted while the adjacent portion was lowered.

Basket A container manufactured by the weaving, coiling or twining of vegetal materials such as cane or straw.

Bas-relief Sculpture where figures project slightly from the background.

Bastion A projecting structure built onto a palisade for purpose of defence; any fortified place.

Batesian mimicry A kind of mimicry in which one non-poisonous species (the Batesian mimic) mimics another poisonous species.

Baulk. Unexcavated strip left standing between excavation units such that soil profiles remain in place for study and reference.

Bead Small disc-shaped, spherical or tubular artifact of bone, shell or glass which has been perforated such that it may be strung on a necklace.

Beaker people From the Late Neolithic to Early Bronze age (4000-2000 BC), named after their pottery. Styles of pottery known as funnel-beaker, protruding-foot beaker, and bell beaker.

Beamer A tool fashioned of wood or the longbone of a large animal. It consists of a sharpened edge which runs nearly along the full length of the tool. The ends serve as handles by means of which it is drawn towards the user. It is used in the treatment of hides.

Bear grass Also called sacahuista. Resembling clumps of large, coarse grass, this plant is found on mountain slopes around the Tucson Basin at elevations of 3000 to 6000 feet.

Beardman jug Common ceramic in the 17th and 18th centuries. May have contained wine or beer.

Bedding plane A surface separating layers of sedimentary rocks and deposits. A bedding plane marks termination of one deposit and beginning of another of different character, such as a surface separating a sandstone bed from an overlying mudstone bed. Rocks tend to breaks or separate along bedding planes.

Bedrock The solid layer of rock which underlies soil, gravel and other loose formations nearer the earth's surface.

Behavioral adjustment Cultural responses, primarily through technology, that make survival in stressful environments possible.

Behavioral sink A psychological state characterized by gross distortions of behavior.

Belemnite Belemnites were a diverse group of marine creatures that lived in the oceans and sea's across the globe. The creature consisted of a long bullet-shaped shell, within which the squid-like creature lived. Belemnite's were predators, feeding on smaller marine creatures, including fish and shells. In most cases the fossil remains only include the tail guard, located at the end of shell. This guard served as buoyancy and to protect the creature from other predators.

Benthic Term used to designate aquatic organisms that are bottom dwelling.

Bentonites A clay formed by the decomposition of volcanic ash, having the ability to absorb large quantities of water and to expand to several times its normal volume.

Bergmann's rule A rule which states that within the same species of endotherms, populations with less bulk are found near the equator while those with greater bulk are found farther from the equator.

Bergschrund Large crevasse usually found at head of cirque glacier.

Bering land bridge The vast tundra plain that was exposed between Asia and North America during the Last Glacial Maximum, about 21,000 years ago; it served as a migration route for people, animals, and plants. Also known as Beringia.

Beringia Landmass which existed in the Bering Strait between Alaska and Siberia during the last (Wisconsinan) Ice Age. At the height of the Wisconsin, sufficient water was "locked up" in the glaciers to cause a marked reduction in ocean levels. Thus, land was exposed in many coastal regions, and a "land bridge," over 1500 km wide was formed between Asia and North America. For a century, Beringia has been widely accepted as the most probable route of entry for early man into the New World. The land bridge likely flooded a number of times in accordance with climatic changes and fluctuations in sea level, but was finally submerged 10,000 years ago

Berriasian European stage of the lowermost Lower Cretaceous, spanning the time between 135 and 131 million years ago.

Besant A valley in southern Saskatchewan which has given its name to a projectile point style and the Late Prehistoric Period phase, horizon or culture within which it occurs. The side-notched points generally have convex edges, sharp shoulders and straight bases. The latter are often thinned and ground and maximum width tends to occur at the shoulder or base. Length ranges from approximately 15 to 80 mm. The remainder of the artifact complex consists of drills, perforators, gravers, scrapers, spokeshaves, mauls and abraders. Besant peoples pursued a way of life focusing the communal hunting of bison by means of (bison) jumps and (bison) pounds throughout most of the northern plains. Their diet was supplemented by fishing, fowling and the collection of shellfish. Many other aspects of the Besant Phase are controversial. Chief among these are whether or not Besant peoples made pottery and the nature of the relationship between Besant and the burial mounds of the Sonota Complex along the Missouri River in northern South Dakota. Although Besant is here classed as Late Prehistoric, the bow (one of the defining traits of this period) was not in use in the earlier portions of this phase.

Beta particles Energy charged subnuclear particles (consisting of an electron or a positron) given off by some isotopes when they decay or fission.

Bevelled surface One that meets two others at angles other than right angles.

B-horizon That natural level within a soil profile which directly underlies the surficial a-horizon and which contains the clay, iron oxides and carbonates which have leached down from it.

Biface A stone tool which has had flakes removed from both faces. No particular function is implied by this term as projectile points, knives and drills may all be bifacially worked.

Bifurcation A basis of kin classification that distinguishes the mother's side of the family from the father's side.

Big bang theory The theory that states that the universe began in a state of compression to infinite density, and that in one instant all matter and energy began expanding and have continued expanding ever since.

Bilateral Describing pollen and spores having a single, principal plane of symmetry.

According to Walker and Doyle (1975) bilaterally symmetrical objects could more correctly described as anisobisymmetric, but because of the familiarity of the term bilateral it was used by them.

Bilateral descent A descent ideology in which individuals define themselves as being at the center of a group of kin composed more or less equally of kin from both paternal and maternal lines.

Bilateral symmetry The condition, found in many organisms, where one half of the body or structure is the mirror image of the other.

Bilaterians A clade of animals whode members share: bilateral symmetry, are triploblastic (three tissue layers: ectoderm, mesoderm, endoderm), and with HOX genes in one or more clusters with the genes within a cluster arranged in the same order as the body parts they affect.

Biodiversity (or biological diversity) A measure of the variety of life, biodiversity is often described on three levels. Ecosystem diversity describes the variety of habitats present; species diversity is a measure of the number of species and the number of individuals of each species present; genetic diversity refers to the total amount of genetic variability present.

Bioengineered food Food that has been produced through genetic modification using techniques of genetic engineering.

Biogenetic law Name given by Haeckel to recapitulation.

Biogeography The study of patterns of geographical distribution of plants and animals across Earth, and the changes in those distributions over time.

Biological imperatives The basic human drives for food, rest, sexual satisfaction, and social contact.

Biological species concept The concept of species, according to which a species is a set of organisms that can interbreed among each other. Compare with cladistic species concept, ecological species concept, phenetic species concept, and recognition species concept.

Biological systematics The aim of biological systematics is the classification of the immense diversity of life into a general reference system, which enables communication about particular groups of organisms and the storage and retrieval of information about these groups. The entities of this biological classification are of utmost importance for all other biological sciences, since they are used as entities of generalisation for all known biological facts. Systematic biology consequently can be regarded as basis of all biological sciences, as well as their "crown", since it represents the most important integrating discipline which is using the results of all others (evolutionary theory, genetics, ecology, ethology, physiology, morphology, etc.) for the formation of the general reference system. On the basis of the Darwinian theory of evolution, only a consequent phylogenetic-systematic classification is suitable to satisfy the above mentioned claims, since only groups that reflect a true genealogical relationship (biospecies and monophyletic groups and clones) can be expected to allow proper generalisations of results that were achieved by the study of singular sample organisms. Biological systematics therefore should always be understood in terms of Phylogenetic Systematics, because otherwise it would rather represent an art or a kind of book-keeping than a natural science. Charles Darwin himself already remarked in his "origin of species" that he believes that the arrangement of the groups within each class, in due subordination and relation to the other groups, must be strictly genealogical in order to be natural. Since there can only exist one correct phylogenetic tree of life, a consequent phylogenetic (cladistic) classification will finally lead to a more stable classification, which is of course very desirable out of practical reasons.

Although the terms "classification" and "taxonomy" are more or less synonymous with "systematisation", the term classification is frequently restricted to all nomenclatorial matters (naming of new taxa, synonymies, Rules of Nomenclature, etc.), while the term taxonomy is often used in the sense of alpha-taxonomy, that means the description of new species level taxa and their subspecific subdivision. Phylogenetics is mostly understood as the reconstruction of the phylogenetic relationships between such species.

Biology The science concerned with the structure, function, distribution, adaptation and evolution of all living organisms including both plants and animals.

Bioluminescence The production of light by living organisms.

Biometrics The quantitative study of characters of organisms.

Biomineralization The formation of minerals by organisms. Biominerlization is a widespread phenomenon within nature whereby organisms form complex and genetically determined mineral structures for a myriad of reasons such as stiffening of soft tissues as in insect exoskeletons, mammal endoskeletons, and the formation of animal teeth for chewing food. In the process, minerals nucleate and grow on purely organic templates.

Biospecies A species in the sense of the "biological species concept", as a closed community of reproduction with a closed gene pool (ernst mayr: "A species is the most inclusive Mendelian population, sharing a common gene pool, which is reproductively isolated from other populations"). The biospecies is maintaining its cohesion with a common mating system (e.g. "sexual mate recognition system" *sensu* paterson) and is separated from other biospecies by "isolation mechanisms" (*sensu* dobzhansky). The biospecies concept can of course not be applied to uniparental organisms. A further problem is that the biospecies criterion is null-dimensional, that means it only allows to decide if two organisms that occur together at the same time and at the same place belong to the same species or not, but it does not allow to decide if allochronic populations (populations separated in time) or allopatric populations (populations separated in space) are conspecific or not. The biospecies concept in its original form does not give any clues to the spatio-temporal boundaries of a species.

In Phylogenetic Systematics the spatio-temporal boundaries of a biospecies ("time-biospecies" *sensu* brundin) are claimed, by the proposal that biospecies are no inventions of the human mind (logical classes) but real individual entities of nature (logical individuals), and therefore must have spatio-temporal boundaries ("species as individuals" *sensu* ghiselin and hull). However, it should be noted that all those so-called biospecies with allopatric populations rather represent sets of populations than intergrated wholes, since there are no cohesive processes between the allopatric populations (e.g. no gene flow). Anyway, the species boundaries in time can be regarded as the "birth" and "death" of a biospecies. An important consequence of the theory of Phylogenetic Systematics is that new species can only originate by speciation (splitting of stem-species into two or more new descendant species) but never by a successive phenotypic change of a single evolutionary line, which would mean a gradual transformation of one species into another, by a gradual shift of the gene-frequencies in an undivided gene-pool. The dismissal of a speciation by gradual transformation of species without multiplication of species by splitting, can be easily explained with the following example: A caterpillar that is transforming into a butterfly also remains the same individual during its whole life cycle, although its appearance drastically changes. New individuals of butterflies only originate by reproduction of this individual.

An individual species terminates either by extinction (extinction of all populations), or by disintegration by speciation (splitting into two or more new descendant species), in which the stem-species of course does no die in a literal sense but it looses its own individuality and hands it equally over to all its descendant species, like a cell that is dividing into two new cells. A stem-species is thus defined as portion of a phylogenetic tree between two successive splitting events. Species that are no stem-species became either extinct before they could produce descendant species, or they are still existing in the recent time horizon as potential stem-species which are not yet extinct nor disintegrated by speciation yet ("incomplete species"). According to this view so-called "surviving stem species" (= species that maintain their individual identity after budding of descendant species) are excluded by definition. Since species are recognized as individual spatio-temporal entities of nature, their ontological status as such is absolutely independent of any characters that might be necessary to recognize and distinguish them. Uniting a stem-species with only one of their descendant species, because of mere phenetic similarity, would be totally arbitrary and would furthermore automatically lead to the formation of paraphyletic groups that are dismissed by Phylogenetic Systematics for good reasons. One-egged twins are also not regarded as the same individual only because they look alike. Like a cell that is dividing into two new cells, a stem-species lives on in all of their descendants, irrespective of their appearance and similarity. Since an individual stem-species cannot maintain its individuality and at the same time being identical with several individual descendant species, its own existence necessarily ends with the splitting event. The well-known fact that biparental organisms survive the birth of their offspring without loosing their individual existence is no counter example, since the mechanisms of maintaining the individuality and the mechanisms of reproduction are fundamentally different from those of biospecies. The split of a stem-species into descendant species, caused by spatial separation and subsequent divergent evolution, is of course a continuous process (like the division of a cell), and consequently there is a fuzzy zone in which it will always be rather arbitrary to decide if two populations still belong to the same biospecies as subspecies or already constitute separate biospecies. This certain arbitrariness is by no means a weak point of the biospecies concept, or even a proof for the artificial status of species, but a direct corollary of all continuous natural processes.

A related problem is represented by allopatric populations that have retained the *potential* to interbreed, but do not actually interbreed anymore. Such populations might be classified as subspecies (or rather as semi-species) of the same (super-)species (which is of course different from a genuine biospecies) if they are still rather similar and represent a monophyletic group. If they have developed significant differences or do represent and paraphyletic assemblage they should be classified as separate (semi-)species, while the directly superordinated monophyletic assemblage should be classified as a super-species. Super-species are thus monophyletic assemblages of semi-species (eventually including some biospecies, too), while semi-species are distinct allopatric populations that have retained the potential ability to interbreed with other such populations. A subspecies is defined as a population within a biospecies that was once separated for a certain period of time and developed distinct characters in this period, but later came in contact with other populations of this biospecies again and is now interbreeding with these populations in a restricted area (hybridisation-zone). These additional terms can be very useful to indicate different stages of speciation. If a continuous morphocline of a character exists between temporally or geographically distant populations of a biospecies, this cline should not be arbitrarily partitioned into "pseudo-subspecies", although this was done quite often in the past (e.g. in case of the chrono-subspecies in paleontology).

Recently several further (cladistic) species concepts were endorsed by cladists of the Anglo-American region, such as the "phylogenetic species concept" (*sensu* cracraft), and the "monophyletic species concept" or "autapomorphic species concept". In these concepts species are defined either as "smallest units that are 100% diagnosable by a unique combination of characters" (phylogenetic species concept), or as "smallest units with autapomorphies" (monophyletic or autapomorphic species concepts). These concepts are all invalid from the viewpoint of Phylogenetic Systematics, since they are character based, and thus are confusing the tools of discovery of natural entities with the ontological justification of these entities. The "phylogenetic species concept" (*sensu* cracraft) can at least be regarded as an operational approach (discovery procedure) towards the recognition of hypothetical biospecies, since the interbreeding abilities within most species have to be postulated rather than studied, simply because of the practical obstacles regarding the existence of millions of species on earth. Contrary to the former concept, the so-called monophyletic and autapomorphic species concepts suffer from the severe problem that they exclude the existence of stem-species by definition, since these can neither be monophyletic, nor can they possess autapomorphies, since they inherit their "autapomorphies" to their descendant species, so that the former autapomorphies of the stem-species become the autapomorphies of the resulting monophylum and thus become symplesiomorphies of the stem-species. The mentioned cladistic species concepts are not only unacceptable with view to extinct species, but also for recent species, since these are all potential stem-species, too.

Biosphere The part of Earth and its atmosphere capable of sustaining life.

Biostratigraphy The relative dating of early human sites by association with index fossils uncovered in the same strata as human evidence. The assumption is that both the people and the species that is now an index fossil must have lived at about the same time.

Biped Any two-footed animal.

Bipedal A description given to the action of standing, walking or running on two legs.

Bipedalism Of hominids, walking upright on two hind legs; more generally, using two legs for locomotion.

Bipolar A technique used in stone tool manufacture in which the core is rested on an anvil while being struck with the hammer. The waves of force are therefore not only directed downward from the hammer, but also reflected back upward from the anvil. Hence the flake may appear to have been struck at both ends.

Bird arrow An arrow which has been purposefully blunted so that it will not damage the hides or animals or become imbedded in a tree and thus be lost.

Bird, Roland T. Roland T. Bird (1899-1978) was an American fossil hunter who rode aroung the USA on a Harley Davidson motorcycle looking for dinosaur fossils for the American Museum of Natural History. His most spectacular find was the Glen Rose Trackway, a beautiful set of 105-million-year-old fossil dinosaur footprints that he found in 1938 in Texas along the Paluxy River. He and Barnum Brown were also involved in excavating the Howe Quarry in Wyoming, which contained many dinosaur fossils.

Birdstone A polished stone object which resembles a bird in profile. Probably functioned as an atlatl handle or weight.

Bireticulate A two layered reticulum consisting of a suprareticulum supported by a microreticulate tectum. Examples: *Entelea arborescens* (Tiliaceae), *Phyllanthus oppositifolius* (Euphorbiaceae), *Salvia azurea* (Lamiaceae).

Bisaccate Describing pollen with two sacci. Example: *Pinus* (Pinaceae).

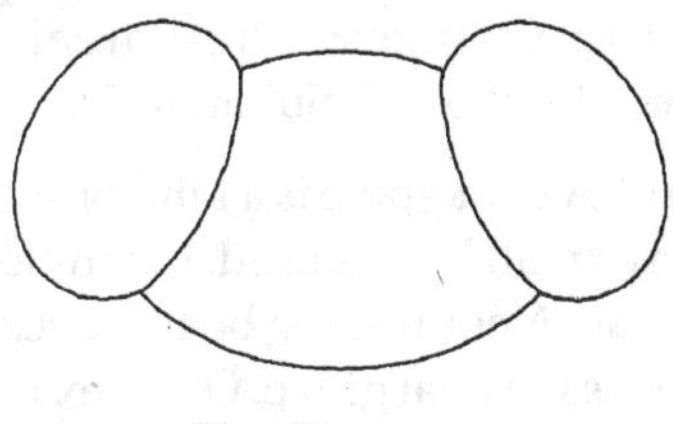

Fig. Bisaccate

Bison "jump" A site at which bison have been killed by being stampeded over a cliff. This ancient communal hunting technique was occasionally used in conjunction with a (bison) pound.

Bison jump A site at which bison have been killed by being stampeded over a cliff. This ancient communal hunting technique was occasionally used in conjunction with a (bison) pound.

Bison occidentalis A large, now extinct variety of bison that roamed the North American grasslands during the Holocene.

Bison pound A physiographic feature or a specially constructed enclosure into which bison were driven to be slaughtered.

Bit The cutting edge of an adze, axe, chisel, etc.

Bitterroot An archaeological phase or culture represented at a number of sites in the Columbia Plateau region in eastern Oregon and in southern and eastern Idaho which Swanson (1962) equates with the northern Shoshone. Projectile points of this complex are side-notched and essentially indistinguishable from those from plains environments to the east (termed Logan Creek or Simonsen), and from those of the Mummy Cave Complex of the eastern slopes of the Rocky Mountains from Alberta to Wyoming. Associated artifacts include conical and wedge-shaped cores, choppers, oval, trinagular and side-notched end scrapers, stemmed and corner-notched bifaces, perforators, manos, whetstones, bone awls and beads of stone and seeds. Fauna include deer, antelope, bison and sheep. Radiocarbon dates range from 5200 to 3650 B.C.

Bivalve A mollusk having two shells hinged together, as the oyster, clam, or mussel; or any animal with two halves to its shell such as an ostracode or brachiopod.

Black sea Inland sea connected to the Mediterranean through the Strait of Bosphorus.

Blackduck The name of a lake in minnesota which has lent its name to a distinctive late woodland ware as well as to the focus, phase, tradition, culture or horizon within which it occurs. The pots are round-based with constricted necks and flattened and thickened lips. Decoration occurs on the neck and rim, on the lip, and occasionally on the inner rim. The most common decorative elements are horizontal and oblique cord-wrapped stick impressions and exterior punctates. Method of manufacture was either by the paddle-and-anvil technique, or involved formation inside of a fabric container. As a consequence the undecorated portions of the vessels are either cord-impressed or fabric-impressed. Associated artifacts and features may include small triangular and side-notched projectile points, a variety of stone and bone hide-scraping tools, ovate knives, stone drills, smoking pipes, bone awls, needles, harpoons and spatulas, bear and beaver tooth ornaments and tools, small copper tools and ornaments and mound burials. Blackduck peoples were widely distributed from the shores of lake superior to the manitoba/saskatchewan border, and from central manitoba in the north to central minnesota in the south. The locations of these sites and the nature of the material remains within them indicate that these people exploited a variety of forest resources, possibly including wild rice as well as the resources of the grasslands – most notably bison.

Blackmore, susan A psychologist interested in memes and the theory of memetics, evolutionary theory, consciousness, the effects of meditation, and why people believe in the paranormal. A recent book, *The Meme Machine*, offers an introduction to the subject of memes.

Blade 1. The cutting edge of a tool.

2. A cutting tool.

3. That portion of a projectile point or knife which extends beyond the haft element.

4. A long, parallel-sided (prismatic, lamellar) flake core. These may be used as is, or used as the basis for the production of other tools. This highly sophisticated technique makes the most economical use of lithic resources.

Blank An incompletely manufactured stone tool which has the general outline of the intended final form. The rough fashioning of blanks at a quarry would obviate the necessity of transporting greater amounts of unmodified stone to camp or fashioning all stone tools at the source of the stone.

Blastoids A Mississippian fossil consisting of a ½- to 1-inch cup-like body that was attached at the base to a short stem. Each of these bodies were five-sided and had arms radiating from it. Usually only the fossilized bodies are seen today.

Blending inheritance The historically influential but factually erroneous theory that organisms contain a blend of their parents' hereditary factors and pass that blend on to their offspring. Compare with Mendelian inheritance.

Blowout Geological term used to refer to the large bowl-shaped depressions created by wind erosion in arid and semi-arid environments. As the top soil and occasionally some of the underlying strata are removed in this process, artifacts may be exposed.

Blytt Sernander sequence Classic sequence of pollen zones during late glacial and postglacial time of northern Europe, from oldest to youngest: Pre-boreal, Boreal, Atlantic, Subboreal, Subatlantic.

Boat grave A boat grave is a kind of ship burial, where a small boat is used. Examples of boat graves are Neolithic log boat graves, like the St Albans log boat grave. Other examples are planked boats used in Viking Age burials, perhaps they were simply poor man's/woman's versions of the larger ship burials.

Bodkin 1. An awl used for making holes in fabric.

2. A blunted, large-eyed needle.

Body shred Technically, a fragment of the body of a larger artefact. Most commonly, it refers to a fragment of a ceramic vessel which did not constitute part of the lip, rim, neck, shoulder or base.

Bog body Ancient human bodies preserved in bogs (waterlogged land filled with a substance called peat). Bog bodies have been found all over Europe, in bogs in Denmark, Germany, Ireland, Scotland, England, Sweden, and the Netherlands.

Bolide Very generally, an extraterrestrial body of indeterminant composition, in the 1-10-km size range, that impacts the earth at velocities faster than a bullet (20-70 km/sec = Mach 75), explodes upon impact, and creates a large crater.

Bonaparte, José F. José F. Bonaparte is an Argentinian paleontologist who has found and/or named many South American dinosaurs. He named: the family Abelisauridae and Abelisaurus (with F.E. Novas,1985), Alvarezsaurus (1991), Amargasaurus (with Salgado, 1991), Andesaurus (with Calvo, 1991), Argentinosaurus (1993, with R. Coria), Augustia (1998), Carnotaurus (1985), Guaibasaurus (with Ferigolo, 1998), Lapparentosaurus (1986), Ligabueino (1996), Mussaurus (with M. Vince, 1979), the clade Neoceratosauria (1990-1991), the family Noasauridae and Noasaurus (with J.E. Powell, 1980), Patogosaurus (1979), Piatnitzkysaurus (1979), Rayososaurus (1996), Riojasaurus (1969), Riojasuchus (1969), Saltasaurus (with J.E. Powell, 1980), Trialestes (1982), Velocisaurus (1991), Venaticosuchus (1971), and Volkheimeria (1979).

Bone The hard tissue, composed of both organic and inorganic materials, which makes up the skeletons of adult vertebrates. Because of their density, bones may survive in the archaeological record long after the decomposition of the soft tissue.

Bone bed A concentrated layer of articulated and disarticulated animal bones usually taken as an indication of a butchering and/or kill site. Typically found in association are weapons and butchering implements.

Bone grease The sweet marrow which is extracted by the smashing and boiling of bones. The grease floats and may be skimmed from the surface for immediate consumption, for storage or for use in pemmican.

Bony fishes Fish of the class Osteichthyes, characterized by a skeleton composed of bone in addition to cartilage, gill covers, and an air bladder.

Book of kells An illustrated manuscript of the four Christian Gospels (the New Testament books of Matthew, Mark, Luke, and John) created by monks in Scotland in about A.D. 800. The book is a masterpiece of Western art and includes amazing calligraphy (an artistic style of handwriting), colorful drawings of animals and people, and abstract designs. Some of the details are so fine that people can't see them with the naked eye.

Book of the dead The term Egyptologists use for the texts and illustrations that were buried with mummies to help them pass through the dangers of the underworld into the afterlife.

Borden designation. The standard archaeological site designation system in canada. The label consists of four letters (alternating upper and lower case) followed by a number, e.g. Eakv-l. The alphabetic prefix refers a block of 10 minutes by 10 minutes within a grid system which covers all of canada south of 62 n latitude. The numerical suffix indicates that this is the first site within this block to be designated.

Border cave, south Africa One of the earliest modern human sites on the planet, this rockshelter in the Lembombo Mountains was found by Louis Leakey to contain Homo sapiens skeletons dated around 70,000 years old.

Boreal archaic An archaeological tradition associated with the mixed coniferous-deciduous forests of the American Northeast. As defined by Byers (1959), it was characterized by stemmed and side-notched projectile points, thumbnail and keeled scrapers, expanding and side-notched-based drills or perforators, shouldered knives and a proliferation of ground and polished implements: spears, adzes, gouges, plummets, rods, tubes, bannerstones, semilunar knives and birdstones. It was believed that Boreal Archaic peoples employed a diversified economy involving fishing, hunting, shellfish collection and plant harvesting. This construct is no longer commonly used.

Boreal forest The technically correct term for the primarily coniferous forest which extends in a continuous arc from Alaska to Labrador and subsumes the Aspen Parkland – the transition between the coniferous forest and the grasslands to the south. The white and black spruces are the most common elements throughout, with tamarack, balsam fir, jackpine, alpine fir and lodgepole pine achieving more restricted distributions. Trembling aspen and balsam poplar are the most important deciduous species (Rowe 1972). The Boreal Forest is roughly equivalent to the taiga of ecologists.

Boreal Refers to high northern latitudes, for example the boreal forest is the conifer (spruce) forest extending across northern Canada and Scandinavia. Austral refers to high southern latitudes.

Boss A small mound-shaped node or protuberance. When used as a decorative element on pottery, they may be produced either by the impressing of a deep punctate on the opposite surface, or by the application and smoothing of small amounts of clay.

Botanist A person who pursues the scientific study of the structure, growth, and identification of plants.

Botany The science concerned with the study, classification, structure, ecology and economic importance of plants.

Bow A weapon consisting of a staff of elastic material such as wood, which is bent by a shorter piece of twine attached to each end. The tension thus imparted to the string is utilized to propel an arrow. In manitoba, the bow came to be used during the woodland or late prehistoric period. Composite bow. A bow the shaft of which is made of at least two different elements such as horn or sinew in addition to wood. Compound bow. A bow fashioned by fastening several pieces of wood together for increased power. Sinew-backed bow. A bow with sinew wrappings for added strength.

Bow drill A form of fire drill in which the stick is rotated with increased speed by virtue of the back-and-forth movement of a bow the string of which is looped around it.

Brachiopods (lampshells) These sea creatures have two shells (called valves). One valve is usually larger than the other. The top of one valve will curl over the top of the second. This gives them an "oil lamp" shape. Some shells are smooth while others have ridges and grooves that radiate out from the middle of the hinge. Brachiopods are common in rocks of Cambrian to Carboniferous age.

Brachycephalic. Round-headed; having a cephalic index of 80 or more.

Brakish water Mixture of seawater and freshwater. The low salt-rate usually excludes those organisms that eat wood on shipwrecks.

Breccia A composite rock composed of angular fragments of more ancient rocks bound together by a natural cement.

Brett-Surman, Michael Michael K. Brett-Surman (1950-) is an American paleontologist and author. Brett-Surman is the Museum Specialist for Dinosaurs at the Smithsonian Institutio He named the dinosaurs Secernosaurus (1979), Gilmorosaurus (1979), and Anatotitan. Brett-Surman grew up in Larchmont, New York. He was an undergraduate at the University of Boulder, Colorado, and went to graduate school at the University of California,

Berkeley, Johns Hopkins University, and George Washington University (while doing research at the Smithsonian Institute). Brett-Surman is the coeditor of The Complete Dinosaur (with James O. Farlow - Indiana University Press, 1997), coauthor of The World of Dinosaurs (with Thomas Holtz, illus. by James Gurney - Greenwich Workshop Books, 1998), coauthor of the Dinosaur Field Guide (with Thomas Holtz - Random House, 2001), and has many other publications. Brett-Surman was also the senior consultant for both issues of dinosaur stamps by the US Post Office (for artists J. Gurche and J. Gurney).

Brevi A prefix for short. Example: brevicolpate.

Breviaxe Pollen grains with a polar axis that is shorter than their equatorial diameter.

Synonym of oblate (s.l.).

Antonym: longiaxe.

Thomson and Pflug (1953) recognised Brevaxones a group of mid-Cretaceous and later angiosperm pollen grains that was contrasted with Longaxones. Breviaxy A phyletic series of pollen forms ranging from subspheroidal tricolpate, to tricolporate, to oblate tricolporate.

Brevissimi A prefix for very short.

Bridge A feature in which the margins of the colpi are raised in the equatorial region and connected with each other, forming a bridge over the ectocolpus and dividing it into two parts. Example: *Hemandradenia* (Connaraceae).

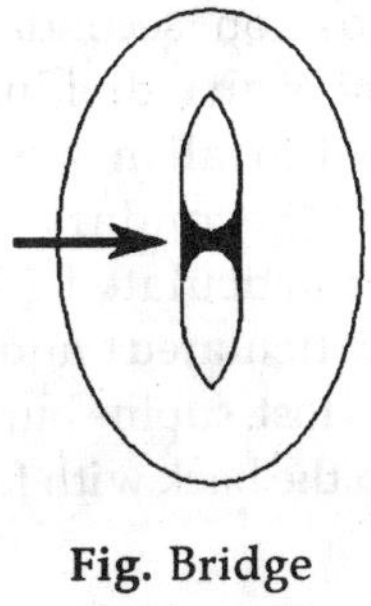

Fig. Bridge

Brochus A brochus consists of one lumen of a reticulum and half of the width of the adjacent muri.

Brodie, Edmund D., III A biologist who studies the causes and evolutionary implications of interactions among traits in predators and their prey. Much of his work concentrates on the coevolutionary arms race between newts that posess tetrodotoxin, one of the most potent known toxins, and the resistant garter snakes who prey on them.

Brodie, Edmund D., Jr. A biologist recognized internationally for his work on the evolution of mechanisms in amphibians that allow them to avoid predators. These mechanisms include toxins carried in skin secretions, coloration, and behavior.

Bronze Mixture of copper, tin, and other metals.

Bronze age The second age in Thomsen's three-age system, referring to the period when bronze tools were manufactured.

Brown, Barnum Barnum Brown (1873-1963) was a great US dinosaur hunter and assistant curator of the American Museum of Natural History. Brown discovered many dinosaurs, including the first T. rex specimens. He named: Anchiceratops (1914), Ankylosaurus (1908), Corythosaurus (1914), Hypacrosaurus (1913), Kritosaurus (1910), Leptoceratops (1914), Prosaurolophus (1916), Saurolophus (1912), and the family Ankylosauridae (1908). He co-named Pachycephalosaurus (1943) and Dromaeosaurus (with E. M. Schlaikjer, 1922).

Bruner, Jerome A psychologist and professor at Harvard and Oxford Universities, and a prolific author whose book, *The Process of Education*, encouraged curriculum innovation based on theories of cognitive development.

Brushed A method of modifying the surface of ceramic vessels by smoothing the still wet clay with a grass brush. This produces a heavily scored or striated appearance.

Bryophyta Bryophytes comprise the mosses (Class Musci), as well as liverworts (Class Hepaticae) and hornworts (Class Anthocerotae). They are believed to have been the first true plants, evolving from charophytes almost 500 million years ago. Unlike other plants, bryophytes do not have true organs, such as leaves, stems, or roots. In place of roots, most bryophytes have thin, hairy tubes called rhizoids that provide anchorage and nutrient uptake from the soil. The bryophyte life cycle is unique in having a dominant gametophyte generation. The actual green plant in mosses and worts is the gametophyte plant, while the sporophyte consists of simply an enclosed sporangium, typically atop a stalk.

Bryozoan A tiny marine invertebrate that forms a crust-like colony; colonies of bryozoans may look like scaly sheets on seaweed.

Bryozoans Aquatic, colonial animals with branching, mossy or fan-like growth. They resemble corals but have more complex nervous, muscular and digestive systems.

Buckland, William William Buckland (1784-1856) was a British fossil hunter, clergyman, and Oxford don (a Reader in Geology and Mineralogy) who discovered Megalosaurus in 1819 and named it in 1824. It was the first dinosaur ever described scientifically and the first theropod dinosaur discovered (theropods were the meat-eating dinosaurs). Buckland always collected his fossils in a large blue bag, which he carried around most of the time.

Buffalo chip A piece of dried bison dung used as fuel by Native Americans.

Bulb of percussion A bulb or boss-like feature on the ventral face of a flake immediately below the striking platform.

Bulbar scar A minute surface irregularity which is occasionally present on the bulb of percussion of a man-made flake.

Bull boat A simple tub- or bowl-shaped boat made by stretching a bison hide over a willow frame bound with thongs. Used by various North American Native peoples.

Burial 1. The covering-over of an object with earth.

2. The ceremonial entombment of a dead body beneath the ground or in a chamber.

3. The feature thus created consisting of the individual(s) and the context. bundle burial. the (re-)burial of bundled-up disarticulated, defleshed remains. extended burial. placement of the individual with arms at the sides and legs extended. flexed burial. placement of the individuals with arms and legs bent up against the body. intrusive burial. the excavation of a grave into a burial pit or mound constructed at an earlier period. Two individuals may thus appear to be in association although they are not contemporaneous. multiple burial. collective internment; the placement of two or more bodies within the same grave. platform burial. see scaffold burial. primary burial. placement of the dead in a grave with the flesh at least partially intact such that after further decomposition, the bones remain articulated. scaffold burial. placement of the dead on a scaffold above the ground where it may be defleshed by scavengers. The remains may be interred at a later date. seated burial. entombment of the deceased in a sitting position. secondary burial. the final interment of an individual subsequent to an earlier burial in which the flesh decomposed. Secondary burials are therefore not articulated (or frequently improperly articulated) and some bones may have been lost. supine burial. placement of the dead on the back with face and palms upward.

Burial mound Raised mass of earth or debris within or below which deceased individuals are placed.

Burin A generally small flake tool which bears a short, chisel-like cutting edge. They are believed to have been used for engraving or scoring bone, antler or ivory prior to splitting.

Burney, David A biologist whose research has focused on endangered species, paleoenvironmental studies, and causes of extinction in North America, Africa, Madagascar, Hawaii, and the West Indies.

C

C.E. An abbreviation used to denote dates that occur within the "Common Era," as a more neutral alternative to the "A.D." of the Christian calendar.

C-14 Abbreviation for "carbon 14"; a radioactive form (or isotope) of carbon used in radiocarbon dating. The numerical suffix indicates that the atom contains 14 particles within its nucleus as opposed to the 12 within the more common, stable (non-radioactive) isotope.

Cacao Seeds from which chocolate is extracted.

Cache An excavated pit, or mound of stones used to store and/or hide food or tools.

Cadastre (cadaster) A public record of the extent, value, and ownership of land within a district for purposes of taxation.

Caduceus A staff with two serpents coiled around it and a pair of wings at the top. Carried by the Greek god Hermes, known as Mercury to the romans.

Caecilians Wormlike, almost blind, tropical amphibians of the order Apoda.

Cairn A mound of stones serving as a monument or marker.

Calamites A Mississippian to Permian fossil plant. A scouring rush that grew to be 40-feet high, it had a ribbed, jointed trunk with leaf whorls at the joints.

Calcareous Of, containing, or like calcite (calcium carbonate).

Calcareous concretions A rounded mass of mineral matter occurring in sand stone, clay, etc., often in concentric layers around a nucleus.

Calcareous nannofossils Fossil remains of calcareous nannoplankton.

Calcareous nannoplankton Protists that normally produce coccoliths during some phase in their life cycle.

Calcined bone Burned bone reduced to white or blue mineral constituents.

Calcite A common compound (CaCO3) in rock formation and the main component of limestone. Calcite can be many colors and effervesces (bubbles) in hydrochloric acid

Calcium carbonate A mineral that forms naturally at the Earth's surface. Many creatures including Corals and Clams use this mineral to build and strengthen their skeletons.

Calendrical system System of measuring time that is based on natural recurring units of time, such as revolutions of the earth around the sun. Time is determined by the number of such units that have preceded or elapsed with reference to a specific point in time.

Calibrated relative dating Use of a relative dating method that measures somewhat irregular occurring natural phenomena that have been cross-dated with at least one chronometric technique so that the dates are somewhat comparable from sites in one region to another. Calibrated relative methods could be considered to be somewhere between ordinary relative methods and radiometric methods in terms of their ability to produce dates that closely approximate the actual date of a sample. Amino acid racemization and paleomagnetic dating are generally considered to be either chronometric or calibrated relative methods.

Caliche Nodules of calcium carbonate ($CaCO_3$) formed in soil. Caliche forms due to the rise and fall of mineral-rich groundwater during wet and dry seasons. When the water rises it deposits calcium carbonate into the soil which accretes into caliche nodules. The length of time it takes for certain size nodules to grow, and at what distance from the surface, is well know In paleontology caliche is therefore an excellent indicator of how long a layer of soil existed before it changed due to erosion or deposition.

Call system A repertoire of sounds, each of which is produced in response to a particular situation.

Callitrichidae Family of New World monkeys consisting of the marmosets and tamarins.

Callose wall Part of the special wall.

Calumet A peace pipe, usually elaborately decorated and often composed of both wood and stone elements.

Calymmate Describing tetrads or polyads in. which the sexine/ectexine of each monad is well differentiated and forms a continuous envelope around the unit. Example: *Acacia* (Mimosaceae).

Antonym: acalymmate.

Cambrian The period following the Pre-Cambrian, where most marine invertebrates evolved 545 million years ago. Most organisms before this period were not preserved in the fossil record.

Camera A cavity formed by the separation of two wall layers in spores that lacks an infrastructure.

Campanian European stage of the Upper Cretaceous, spanning the time between 84 and 72 million years ago.

Campbell strandline One of the major and certainly the most prominent of the now-extinct beaches created by glacial lake agassiz comprising shorelines and wave-cut escarpments. It was created between 7500 and 8000 b.c. when the lake occupied the manitoba lowlands.

Canaan A historical and Bibilical term used to describe the strip of land which includes most of present day Gaza Strip and Israel and the Western part of Jordan. The term was found on Egyptian writings from the 15th century BC.

Canaliculate Synonym of fossulate.

Cannibalism The consumption of human flesh by other humans for reasons of dire need or for ritual purposes. In the archaeological record, the forceful enlargening of the foramen magnum at the base of the skull (presumably for removal of the brains) and the smashing of long bones (for the extraction of bone grease) are often viewed as evidence of cannibalism. In at least some cases, however, it is possible that while the individual was thus prepared for consumption, they were only symbolically devoured.

Canoe A long, narrow open boat lacking sails and rudder. It is pointed at both ends and propelled by paddles.

Canopic jars Ancient Egyptian containers used to hold the internal organs that were removed from a dead person before mummification.

Capillus Synonym of fimbria.

Cappa The thick-walled proximal side of the corpus of a saccate pollen grain.

Cappula The thin-walled distal side of the corpus of a saccate pollen grain.

Synonym of leptoma.

Capsula An outer structure of a spore projecting at the equator and completely enclosing the spore body.

Caput The expanded apex (head) of a columella. Carapace The outer protective shell or shield protecting the body of creatures including crabs and insects.

Carbohydrates Organic compounds composed of carbon, oxygen, and hydrogen; includes the sugars and starches.

Carbon An element. Atoms of carbon are the building blocks of living cells.

Carbon 14 dating (C-14) See radiocarbon dating.

Carbon isotope ratio A measure of the proportion of the carbon-14 isotope to the carbon-12 isotope. Living material contains carbon-14 and carbon-12 in the same proportions as exists in the atmosphere. When an organism dies, however, it no longer takes up carbon from the atmosphere, and the carbon-14 it contains decays to nitrogen-14 at a constant rate. By measuring the carbon-14-to-carbon-12 ratio in a fossil or organic artifact, its age can be determined, a method called radiocarbon dating. Because most carbon-14 will have decayed after 50,000 years, the carbon isotope ratio is mainly useful for dating fossils and artifacts younger than this. It cannot be used to determine the age of Earth, for example.

Carbon sample A quantity of organic material, usually charcoal, collected for radiocarbon dating.

Carbonate compensation depth Depth in an ocean basin below which rapid solution of calcium carbonate takes place. Little calcareous ooze can accumulate below this depth, which varies from one ocean to another.

Carbonate rock A rock consisting primarily of a carbonate mineral such as calcite or dolomite, the chief minerals in limestone and dolostone, respectively.

Carboniferous A period of time extending from 354 - 290 million years ago. During this time large forests developed across the globe, laying down the organic material that would later form the worlds coal supply. At the same time, the first land dwelling vertebrates evolved.

Carbonization A process of forming fossils in which the weight of the sediments squeezes out the water and gas and leaves a residue of carbon (imprint).

Caribou lake complex A palaeo-indian artifact complex of the forested region of eastern manitoba consisting of lanceolate projectile points, trihedral adzes, and large, asymmetrical bifaces. With an estimated time-depth of 6000 to 4000 b.c., this complex is believed to represent the earliest inhabitants of this part of the province who came to rely on the resources of the forests as a result of altithermal xerothermy (steinbring and buchner 1980).

Caries Tooth decay The condition of the teeth of a skeleton is often an important clue to the diet and health of the individual.

Carnivore A flesh eating or predatory animal or plant.

Carnivores The name given to predators that lived on the flesh of other creatures.

Carnivorous Feeding largely or exclusively on meat or other animal tissue.

Carpal A bone of the human wrist, or one of the corresponding bones of the forelegs of other animals.

Carpenter, Kenneth Kenneth Carpenter (1949 -) is a paleontologist who works at the Denver Museum of Natural History (in Denver, Colorado, USA). Carpenter, Bryan Small, and Tim Seeber found the most complete Stegosaurus yet found in 1992, near Canon City, Colorado, USA. Carpenter named the dinosaurs Animantarx (Carpenter, Kirkland, Burge, and Bird, 1999), Cedarosaurus (Tidwell, Carpenter and Brooks, 1999), Gargoyleosaurus (Carpenter, Miles, and Cloward, 1998), Gojirasaurus (Carpenter, 1997), Maleevosaurus (Carpenter, 1992), Mymoorapelta (Kirkland and Carpenter, 1994), Niobrarasaurus (Carpenter, Dilkes, and Weishampel, 1995), Pectinodon (Carpenter, 1982). Carpenter has written many books on dinosaurs, including "Dinosaur Systematics," "Dinosaur Eggs and Babies," "The Dinosaurs of Marsh and Cope," and "The Morrison Formation - an Interdisciplinary Study."

Carpoids (formally homalozoa) Extinct subphylum (Phylum Enchinodermata) whose members have no trace of radial symmetry. The theca is depressed and asymmetrical. Homalozoa are known from Middle Cambrian to Middle Devonian rocks.

Carroll, Sean Developmental geneticist with the Howard Hughes Medical Institute and professor at the University of Wisconsin-Madison. From the large-scale changes that distinguish major animal groups to the finely detailed color patterns on butterfly wings, Dr. Carroll's research has centered on those genes that create the "molecular blueprint" for body pattern and play major roles in the origin of new features. Coauthor, with Jennifer Grenier and Scott Weatherbee, of *From DNA to Diversity: Molecular Genetics and the Evolution of Animal Design*.

Carrying capacity The point at or below which a population tends to stabilize.

Carson, Rachel A scientist and writer fascinated with the workings of nature. Her best-known publication, *Silent Spring*, was written over the years 1958 to 1962. The book looks at the effects of insecticides and pesticides on songbird populations throughout the United States. The publication helped set off a wave of environmental legislation and galvanized the emerging ecological movement.

Cartilaginous fishes Fish having a skeleton composed mostly of cartilage, as sharks and rays. Cartilage is gristle or a firm, elastic, flexible type of connective tissue.

Cartonnage Papyrus or linen soaked in plaster, shaped around a body. Used for Egyptian mummy masks and coffins.

Cartouche Elongated version of the hieroglyphic sigh W "shen" which means 'to encircle'. Two of the Pharaoh's five names were written inside the cartouche. The sign represents a loop of rope that is never ending, such as the arch of the sky and the world, to indicate that Pharaoh lead everything that the sun encircled.

Cast A replica of an organism created when minerals use the organism as a mold to create the replica. For example, a shell fills with minerals, the shell dissolves away and the cast (inside of the shell) is left behind.

Castellation A projecting or raised section on the rim of a pot.

Castle, W.E. An early experimental geneticist, his 1901 paper was the first on Mendelism in America. His *Genetics of Domestic Rabbits*, published in 1930 by Harvard University Press, covers such topics as the genes involved in determining the coat colors of rabbits and associated mutations.

Cata A prefix indicating the location of features, such as apertures, on the proximal face.

Antonym: ana-.

Catabolism The destructive metabolism of larger organic molecules into smaller constituents, usually with the release of energy (usually as ATP).

Catalogue The systematic list recording artifacts and other finds, recovered by archaeological research, including their description and Provenience.

Catalogue number A number assigned all items recovered by archaeological research to cross-index them to the catalogue.

Catarrhine nose A nose in which the nostrils open downward and are separated by a narrow nasal septum; found in Old World monkeys, apes, and humans.

Catarrhini Infraorder of the order Primates that includes Old World monkeys and the hominoids plus various extinct taxa.

Catastrophic age profile A mortality pattern based on bone or tooth wear analysis, and corresponding to a "natural" age distribution in which the older the age group, the fewer the individuals it has. This pattern is often found in contexts such as flash floods, epidemics, or volcanic eruptions.

Catastrophism The eighteenth-century theory that earthquakes, volcanic eruptions, and other natural disasters were responsible for the distribution of animal fossils and artifacts.

Catazonasulculate Describing a pollen grain with a ring-like sulculus situated between the equator and the proximal pole.

Antonym: anazonasulculate.

Cathaysian terranes A set of small landmasses that developed in tropical to subtropical latitudes on the eastern side of Pangea during the Permian and Triassic, includes modern North China (Sino-Korea), South China (Yangtze), Eastern Qiangtang, Tarim, and Indochina.

Cation-ratio dating This method aspires to the direct dating of rock carvings and engravings, and is also potentially applicable to Paleolithic artifacts with a strong patina caused by exposure to desert dust. It depends on the principle that cations of certain elements are more soluble than others; they leach out of rock varnish more rapidly than the less soluble elements, and their concentration decreases with time.

Catlinite A soft, red, easily worked stone of the Upper Missouri region which was commonly ground and polished into tobacco pipes. Also known as "pipestone".

Cattle complex An East African socioeconomic system in which cattle represent social status as well as wealth.

Cauldron A large kettle.

Cavate In **spore** terminology, a synonym for camerate.

Cavea A cavity between two layers of the exine extending to the colpus margin where the layers meet. Example: *Ambrosia* (Compositae).

Caverna Synonym of fastigium.

Originally a distinction was made between a separation of two layers of the ectexine, which was called a praecaverna, and a separation within the endexine, a postcaverna. The term is not used to describe recent pollen grains.

Cavium A subpolar chamber formed at the end of three anastomosing cavernae.

Cavus Synonym of cavea.

Cebid A member of the family Cebidae; the New World monkeys excluding the marmosets and tamarins.

Cebidae Family of New World monkeys that includes the squirrel, spider, howler, and capuchin monkeys, among others.

Cell The basic structural and functional unit of most living organisms. Cell size varies, but most cells are microscopic. Cells may exist as independent units of life, as in bacteria and protozoans, or they may form colonies or tissues, as in all plants and animals. Each cell consists of a mass of protein material that is differentiated into cytoplasm and nucleoplasm, which contains DNA. The cell is enclosed by a cell membrane, which in the cells of plants, fungi, algae, and bacteria is surrounded by a cell wall. There are two main types of cell, prokaryotic and eukaryotic.

Celt An ungrooved axe of metal or polished stone.

Celts A category of people who flourished from about 750 to 12 B.C. During this time, the Celts were the most powerful group in central and northern Europe. Although the Celts were composed of many different tribes, they shared similar languages, technology, customs, artistic styles, and beliefs. By A.D. 60, their power had been destroyed by the Romans. After that, only the Celtic tribes in the more remote areas of Europe, such as the British Isles, survived.

Cenomanian European stage of the lowermost Upper Cretaceous, spanning the time between 95 and 91 million years ago.

Cenotaph From the Greek word meaning; "empty tomb". A tomb built for ceremonial purposes that was never intended to be used for the interment of the deceased.

Cenote A natural waterhole. Cenote is a corruption by the Spanish of the Maya word dzonot, a large circular sink-hole created by the collapse of limestone caves. The water in cenotes is filtered through limestone and constituted one of the primary sources of drinking water for the Maya. Patterns of settlement among the early Maya often followed the location of cenotes.

Cenozoic The era of geologic time from 65 mya to the present, a time when the modern continents formed and modern animals and plants evolved.

Census A comprehensive survey of a population designed to reveal its basic demographic characteristics.

Central hall A frame house consisting of two rooms and an enclosed central hall.

Centralization Concentration of political and economic decisions in the hands of a few individuals or institutions.

Centromere A point on a chromosome that is involved in separating the copies of the chromosome produced during cell division. During this division, paired chromosomes look somewhat like an X, and the centromere is the constriction in the center.

Centrosymmetrical Synonym of radially symmetric.

Cephalic index A measure of the roundness of the skull calculated by dividing the maximum width of the brain case (usually just above the ears) by the maximum length (between the eyes to the back of the skull) and multiplying the resulting fraction by 100.

Cephalopod Cephalopods include squid, octopi, cuttlefish, and chambered nautiluses. They are mollusks with tentacles and move by forcing water through their bodies like a jet.

Cephalopoda An order of molluscs including Octopi, Cuttlefish, Squid, Belemnites, Nautiloids, Ammonoids and Ammonites.

Cephalopods A group of marine creatures distinguished by the presence of tentacles, including squid and octopus.

Ceramic Of or pertaining to pottery; a hard material made by drying and baking clay or some similar substance.

Cerberus Three headed dog that guarded the entrance to Hades.

Cercopithecidae Family that includes all the Old World monkeys, such as guenons, mangabeys, macaques, and baboons.

Cercopithecinae Subfamily that contains the Old World monkeys that are omnivorous and possess cheek pouches.

Ceremonial fund The portion of the peasant budget allocated to religious and social activities.

Cetotheridae A Family of primitive Mysticete whales.

Chac mool Maya stone reclining figure with a place for offerings on it's stomach.

Chaco canyon Site in New Mexico representative of the Anasazi culture that thrived there between A.D. 500 and 1300.

Chain A surveying chain, or long steel tape-measure, calibrated in meters or feet, used for site mapping and grid layout.

Chalcedony A microcrystalline form of quartz with crystals arranged in parallel strands. Chalcedony was commonly used for tool-making and còuld be either chipped or ground.

Chalk A variety of limestone made up in part of biochemically derived calcite, in form of skeletons or skeletal fragments of microscopic oceanic plants and animals mixed with fine-grained calcite deposits of biochemical or inorganic-chemical origin.

Character Any well-defined feature, whether anatomical, behavioral, biochemical, or genomic, of an organism.

Character displacement The increased difference between two closely related species where they live in the same geographic region (sympatry) as compared with where they live in different geographic regions (allopatry). Explained by the relative influences of intra- and inter-specific competition in sympatry and allopatry.

Character weighting The term character weighting is referring to a procedure that is allowing a choice between conflicting hypotheses of homology and monophyly, according to certain weighting criteria, by assigning higher weight to some characters than to other conflicting characters. The object of weighting is not the character as such (indeed true synapomorphies can not have different weights!), but the evidence for the correctness of the hypothesis of homology, that is involved in every character definition. Since statements of similarity are the primary cause for all assumptions of homology, the very different degree of complexity of these statements of similarity implies a different faith in the truth of the resulting hypotheses of homology, and consequently a different weight of the referring characters. The basic rational behind this statement can be easily explained by the following example: If there are two different pieces of paper, each with an identical single letter written with a typewriter on it, one would not necessarily assume that they were written by the same person , because the chance that two people independently type the same letter is relatively high (1:26). On the other hand, if there would be an identical poem written on these two pieces of paper, one would of course be quite shure that they were derived from the same source, simply because it is unlikely that two persons independently write exactly the same poem. Weighting does not mean that some synapomorphies are better than others, but only that we can and must have stronger faith in the correctness of those homology-hypotheses that are backed by better evidence, than in others that are backed by less good evidence (since no real probability is involved in statements about past events, the alternative expression "different truth probabilities of the homology hypotheses" should be avoided!). This kind of weighting does not need any knowledge of the evolutionary process, and it does not make any assumptions about it either, since it is exclusively based on properties of the characters and their pattern of occurence.

The most important criteria for an a priori weighting of characters (a weighting that is preceding the phylogenetic analysis) are the compatibility and structural complexity of the characters. Compatibility refers to the number of conflicting characters, what means that those characters that conflict with fewer other characters are regarded as stronger evidence, than characters that conflict with more other characters. Weighting on the basis of structural complexity means that simple structures that might easily evolve by convergence, or superficial similarities that might be based on an insufficient analysis, are regarded as weaker evidence, than characters that are so complex that they could hardly be non-homologous, and that are so well-investigated that the proposed similarity is not just superficial. Since there can be no complexity whatever in the mere absence of something (negative similarities), reductions are normally regarded as rather weak evidence (compare reversals). In molecular biology there is sometimes a further criterion available for a priori weighting, since the different probabilities of certain mutations are known in some cases (e.g. transitions are much more frequent than transversions, which was considered in the transition-transversion-parsimony-algorithm). The most important criterion for a posteriori weighting (weighting on the basis of the result of a phylogenetic analysis) is congruence, which is nothing else than parsimony. This means that characters that are less homoplastic in the resulting cladogram (characters with a CI closer to 1), are regarded as stronger evidence than characters that are more homoplastic (CI closer to 0) and thus imply many convergent origins or many convergent losses. All available weighting criteria should be used to estimate the relative weight of a character.

The representatives of so-called computer-cladism and Pattern-Cladism generally reject character weighting, or at least any sort of a priori weighting, because it shall be a much too subjective procedure. Nevertheless their dogma of (alleged) non-weighting is nothing but a beautiful dream, since the choice of characters and the delimitation of characters already involves so many subjective decisions, that these characters are already strongly weighted as soon as they are chosen and formulated. Furthermore the alleged non-weighting indeed represents an equal-weighting. This is even more problematic since an exactly equally good evidence for all involved homology hypotheses certainly represents one of the least likely cases one can think of. The issue of character weighting of course has most important consequences for the application of the principle of parsimony. The restriction of parsimony to a statistical analysis of the character pattern with a mere numerical minimization of homoplasies ("cladistic parsimony") is not only over-reductionistic, but even becomes rather absurd regarding the highly subjective impact of character choice and delimitation. Since nearly every character complex can be either lumped into a single character (e.g. "vertebrate eye"), or splitted into dozens of characters (retina, cornea, iris, ciliary muscle, etc.), a most parsimonious tree can be easily overthrown by a different formulation of the same characters, unless there is no conflicting evidence at all. Objective criteria for the delimitation of characters do not exist and almost certainly never will, because they are impossible out of theoretical reasons, since every subdivision of a continuum like a body always will have to be rather artificial and subjective, and thus more or less optional. Therefore, there can only be formalistic arguments but no sound scientific justification for the preference of a most parsimonious cladogram, only because it is a few steps shorter than alternative cladograms. The presented counter-arguments, simple as they are, definitely render any further discussions about parsimony algorithms (e.g. the pro and contra of three-taxon-parsimony), homoplasy indices and consensus procedures absolutely superfluous and ridiculous, even though such issues currently represent the majority of publications in journals like *Systematic Biology* and *Cladistics*. Consequently, the principle of parsimony (Ockham's Razor) must be understood in a much broader context, in such a way, that a cladogram that is some steps shorter than other cladograms, but has to interpret complex similarities as non-homologies, is regarded as less parsimonious than a cladogram that is a few steps longer, but treats these similarities as homologies.

The frequently heard argument by dedicated computer-cladists, that an analysis by hand should only be possible in cases of few characters without significant conflicts, while a computer-aided parsimony-analysis shall be far superior in cases of numerous characters with many conflicting evidence, is definitely ill-founded out of the reasons already explained above. Besides all this, a phylogenetic-systematic analysis is not done by hand anyway, but by brain, and therefore should be preferred over a computer-analysis without much brains. The apparent objectivity of computers is misleading, since the most important procedure is not the parsimony-analysis, but the character-analysis (careful study, choice and delimitation of the characters) in which no computers are involved anyway. One should always keep in mind one of the oldest wisdoms of computer enthusiasts: "garbage in - garbage out". Examples for the latter can frequently be found in the literature, which is full of nonsense phylogenies that are based on cladistic analyses of numerous "unweighted" characters. A central problem of computer-cladism is the fact that a parsimony-algorithm can even generate a fully resolved most parsimonious cladogram from an uninformative data set which is only containing very weak and extremely homoplastic characters. This "ability" has even been cherished by computer-cladists as "extraction of a cladistic signal from a noisy data set", while it is in reality nothing but an unwarranted transformation of noise into apparent information (even the best alchemists failed to make gold out of shit). If one cannot find convincing evidence for a phylogenetic tree in a "manual" analysis of a data set, the latter simply does not contain any useful phylogenetic information. Even if a correct tree could be calculated from this data set with a parsimony-analysis, the result would not be useful either, since the branchings are not supported by convincing evidence, what is often the case in published cladograms. Maybe the most fundamental difference between computer-cladistics and consequent Phylogenetic Systematics is, that the former is only striving to get trees from the available data, no matter how, while the latter is carefully searching for convincing evidence (strong characters) to reconstruct phylogenetic trees, that form the basis for other disciplines of evolutionary biology (historical biogeography, co-evolution, evolutionary scenarios, etc.).

Characters / character states (versus traits) Characters are shared similarities of properties (structures or behaviours) in different organisms, that are recognized and formulated, and that are supposed to be inheritable and homologous. The subdivision of the continuous body of an organism into discrete characters (incl. behavioural characters) will always remain to be rather subjective and optional, although by no means completely arbitrary! Differences between homologous characters in different organisms are called character states. Diagnostic characters (e.g. for keys) can include "primitive" (= plesiomorphic) and derived (= apomorphic) character states, while phylogenetic relationship can only be demonstrated by shared derived character states (synapomorphies). Contrary to the character state, that is only specifying a property, the character itself is not an observation but a hypothesis of homology, since it implies that all included character states are transformations of a shared ancestral state. The term "character" consequently requires a comparison of different organisms; - a single organism does not have characters but only certain properties. The biogeographic distribution, population by parasites or symbionts, the adaptation to certain hosts, habitats or resources, as well as acquired behaviours, do represent distinct properties of organisms, that may even have some correlation with their phylogenetic relationships, but they are no characters according to the above given definition, because they are not genetically inherited. Since they are still more important for evolutionary biology than the particular properties of single organisms, they are often called "traits" to express this difference without using the term "character". Even genetically inheritable features can be traits rather that characters if they are not fixed in a population, since in this case they do not diagnose distinct evolutionary lineages. Such traits should not be used for the reconstruction of phylogenetic relationships, but should be interpreted on the basis of phylogenetic trees that have been reconstructed by the use of genuine characters. This kind of procedure is often applied in biogeographic studies (vicariance-biogeography) and in studies of co-evolution (e.g. of hosts and parasites, or blossoms and pollinators).

Charcoal Carbon formed by heating organic matter in the absence of air; one of the preferred substances for radiocarbon dating.

Charon In Greek myth, the boatman who rowed the souls of the dead across the River Styx into the underworld.

Cheek pouch A pocket in the cheek that opens into the mouth; some Old World monkeys store food in the cheek pouch.

Chemistry The science concerned with the structure, properties, reactions and commercial application of substances.

Chernozem A rich, black organic soil well-suited to the growing of grasses, which is found in cool or temperate semiarid environments.

Chert A very fine grained rock formed in ancient ocean sediments. It often has a semi-glassy finish and is usually white, pinkish, brown, gray, or blue-gray in color. It can be shaped into arrowheads by chipping. It has often been called flint, but true flint is found in chalk deposits and is a distinctive blackish color.

Chiappe, Luis Luis M. Chiappe is a vertebrate paleontologist, Chairman of the Department of Vertebrae and Paleontology, and Associate Curator of the Natural History Museum of Los Angeles County. Chiappe has studied dinosaurs and the origins of birds. In 1997, Chiappe discovered a cache of thousands of 80-million-year-old fossilized titanosaur eggs at Auca Mahuevot in Patagonia, Argentina (titanosaurs are huge, long-necked, plant-eating dinosaurs). Many of the eggs contained fossilized dinosaur embryos. Chiappe wrote the books, "Tiniest Giants" (2001) and "SuperCroc" (2001).

Chiefdom The third stage in the 'stage model,' representing regional systems with mixed economies that are integrated through the institution of chief.

Chilam balam A series of books written by various Maya tribes in Spanish after the Spanish Conquest. The content probably came directly from Maya codices.

Chin, Karen Karen Chin is a paleontologist and ichnologist (studying trace fossils - coprolites in particular). In 1998, Dr. Chin studied the first fossilized T. rex dung (coprolites) that contained bits of Triceratops frill. She has also found traces of dung beetle tunnels in another dinosaur coprolite. Chin received her Masters Degree from Montana State University (working with Jack Horner), and her Ph.D. from University of California, Santa Barbara (in 1996).

Chinampas The areas of fertile reclaimed land, constructed by the Aztecs, and made of mud dredged from canals.

Chinking A mortar, usually composed chiefly of clay, used to plaster over gaps in walls or to bind bricks or stones.

Chipping station A restricted area of "floor" within an archaeological site which yields stone flakes to the virtual exclusion of other kinds of artifacts. Such features are frequently interpreted as places used for the chipping of stone.

Chitho A disc-shaped biface.

Chitin The polysaccharide coating of microforaminifera, scolecodonts, and the spores and hyphae of certain fungi. Highly resistant to forms of deterioration other than oxidatio

Chitinozoans Marine microfossils of uncertain affinity generally shaped like flasks, occurring individually or in chains. stratigraphic range from uppermost Cambrian to Devonia Usually assumed to be animal remains. (Traverse, 1995))

Chloroplast A structure (or organelle) found in some cells of plants; its function is photosynthesis.

Chol Maya language and ethnic group.

Cholera An acute infectious disease of the small intestine, caused by the bacterium *Vibrio cholerae* which is transmitted in drinking water contaminated by feces of a patient. After an incubation period of 1-5 days, cholera causes severe vomiting and diarrhea, which, if untreated, leads to dehydration that can be fatal.

Cholla Several species of spiny cactus having cylindrical stems and branches. The plants are found in many parts of semiarid and arid North America.

Chopper An axe-like tool, generally fashioned from a cobble or large pebble, and usually worked only on one face.

Chordate A member of the phylum Chordata, which includes the tunicates, lancelets, and vertebrates. They are animals with a hollow dorsal nerve cord; a rodlike notochord that forms the basis of the internal skeleton; and paired gill slits in the wall of the pharynx behind the head, although in some chordates these are apparent only in early embryonic stages. All vertebrates are chordates, but the phylum also contains simpler types, such as sea-squirts, in which only the free-swimming larva has a notochord.

C-horizon The bottom-most zone of a soil, consisting of unaltered natural sediments.

Chromosome A structure in the cell nucleus that carries DNA. At certain times in the cell cycle, chromosomes are visible as string-like entities. Chromosomes consist of the DNA with various proteins, particularly histones, bound to it.

Chrono species A morphological species concept that is unfortunately still very popular in paleontology. Although it is postulated that any time-slice of a chrono-species should reflect a hypothetical biospecies, this species concept has to be dismissed from the viewpoint of Phylogenetic Systematics as being rooted in typology, since the boundaries of these species in time are arbitrarily defined according to selected morphological character states. This species concept allows speciation by budding with surviving stem-species, and consequently is at odds with the basic principles of Phylogenetic Systematics. A chrono-species definitely does not represent a real entity of nature but only an invention of the human mind ("logical class").

Chronology 1. A listing of events in their proper order.

2. The sequence itself.

3. The methodology of placing events in order of occurrence.

Chronology building Devising a dated history for a region by combining numerous lines of evidence.

Chronometric date A date that places an event in its chronological position with reference to a universal time scale such as a calendar. Such dates usually are given in terms of the number of years before or after a calendar starting point. For instance, 1950 B.C. was 1950 years before the beginning date of the Gregorian calendar, which is commonly used today. Chronometric dating methods include the use of written records, dendrochronology, and radiometric methods.

Chronometric dating Placing an event or process with a range of dates on a calendrical time scale, usually by means of radiocarbon or potassium/argon techniques

Chronometry The art of measuring time accurately.

Cicatricose Describing spores marked with scars. Example: *Cicatricosisporites*.

Cicero Roman orator, died 43 BCE.

Cimmerian terranes An archipelago of small landmasses that developed in tropical and subtropical latitudes on the eastern side of Pangea during the Triassic, blocks that comprised it include modern Turkey, Iran, Afghanistan, Tibet, and Malaysia; also called Cimmeria.

Cingulum A thick outer structure of a spore that projects at the equator, but does not extend over the distal or proximal face. Example: *Densosporites anulatus*.

Circumaperturate Describing a pollen grain with equatorial apertures that are regularly arranged around a circular outline.

Circumpolar lacuna A lacuna in lophate pollen grain immediately adjacent to a polar lacuna that is not part of the apertural system.

Cirque Horseshoe-shaped basin at head of glacial valley, enclosed by high head and side walls, with low lip and smooth basin floor; often containing small lakes or glaciers today in alpine settings.

Cists Boxed burials (eg: some of the Neolithic graves at El Garcel, Almeria, Spain) are referred to as cists burials. The term simply comes from the German word 'Kiste' meaning a box or crate.

City-state City and surrounding countryside under it's influence. Main political entity of classical Greece.

Civilization A term used by anthropologists to describe any society that has cities.

Clack, Jenny A paleontologist at Cambridge University in the U.K., Dr. Clack studies the origin, phylogeny, and radiation of early tetrapods and their relatives among the lobe-finned fish. She is interested in the timing and sequence of skeletal and other changes which occurred during the transition, and the origin and relationships of the diverse tetrapods of the late Paleozoic.

Clade A hypothetical species of animal and all its descendants; a monophyletic taxo Clades should not be mistaken for actual "family trees" as clades are hypothetical groupings whereas "family trees" are real groupings. Clades replace the old system of Linnean rankings such as Class, Order, etc. An animal is a member of every clade that its ancestors are, thus making a bird a reptile (it is descended from a member of clade Reptilia) and a human a mammal (it is descended from a member of clade Mammalia). Clades carry both a definition, based on ancestry, and a diagnosis, based on synapomorphies. For example, the node-based clade Dinosauria is defined as the most recent common ancestor of "birds" and *Triceratops* (Holtz and Padian 1995) and all its descendants, and diagnosed by at least three sacral vertebrae, a partially to fully open acetabulum, and other synapomorphies. It should be noted that diagnoses of clades can change while definitions do not

Cladism Phylogenetic classification. The members of a group in a cladistic classification share a more recent common ancestor with one another than with the members of any other group. A group at any level in the classificatory hierarchy, such as a family, is formed by combining a subgroup at the next lowest level (the genus, in this case) with the subgroup or subgroups with which it shares its most recent common ancestor. Compare with evolutionary classification and phenetic classification.

Cladistic species concept The concept of species, according to which a species is a lineage of populations between two phylogenetic branch points (or speciation events). Compare with biological species concept, ecological species concept, phenetic species concept, and recognition species concept.

Cladistics (also known as *phylogenetic systematics* and *Hennigian systematics*) The whole field of phylogenetic reconstruction, approximation, description, and analysis using the principle of shared derived characters for determining relationships. Organisms are classified by analyzing the distribution of their characters and character states, determining which characters are synapomorphies, plesiomorphies, or homoplasies, and then grouping the organisms into clades using the synapomorphies. Although there were German and South American antecedents to cladistics dating back a century, cladistics as practiced today stems from the work of Willi Hennig in the 1950s and 1960s

Cladists Evolutionary biologists who seek to classify Earth's life forms according to their evolutionary relationships, not just overall similarity.

Cladogram (versus phylogenetic tree) A cladogram is a "phylogenetic argumentation scheme" in shape of a tree-like graphic (dendrogram), which is only specifying the relative degrees of phylogenetic relationship (sistergroup relationships) of the analysed taxa, as well as their monophyly, based on the evidence of the recognized synapomorphies and autapomorphies.

Contrary to such a cladogram, a genuine phylogenetic tree is also specifying direct ancestor-descendant relationships, just like a pedigree. Although such ancestor-descendant relationships of course exist, there is unfortunately more or less no possibility to identify an ancestor as such, since these cannot possess any exclusive characters (autapomorphies). A fossil that belongs to the stem-group of a monophyletic group, and is lacking any visible autapomorphies, consequently can either represent a direct ancestor, that is primarily lacking these autapomorphies, or it can be a member of an extinct side-branch of the stem-line, whose autapomorphies simply did not happen to become preserved in this fossil. Only in very rare cases of a continuous and complete fossil record over a considerable range of time at one locality (e.g. in some lacustrine sediments) there might be the possibility to postulate an ancestor-status, provided that no potential descendants are known from the same or even older age, and provided that an allopatric speciation (sister-species) could not be demonstrated. The groundplans of Phylogenetic Systematics are representing hypothetical reconstructions of the (undiscovered) stem-species.

Clan A unilineal descent group usually comprising more than ten generations consisting of members who claim a common ancestry even though they cannot trace step-by-step their exact connection to a common ancestor.

Class A ranked group within a stratified society-characterized by achieved status and considerable social mobility.

Classification The arrangement of organisms into hierarchical groups. Modern biological classifications are Linnaean and classify organisms into species, genus, family, order, class, phylum, kingdom, and certain intermediate categoric levels. Cladism, evolutionary classification, and phenetic classification are three methods of classification.

Clastic Refers to rock or sediments made up primarily of broken fragments of pre-existing rocks or minerals.

Clava A club-shaped element of the sexine/ ectexine that is higher than 1µm, with diameter smaller than height and thicker at the apex than the base.

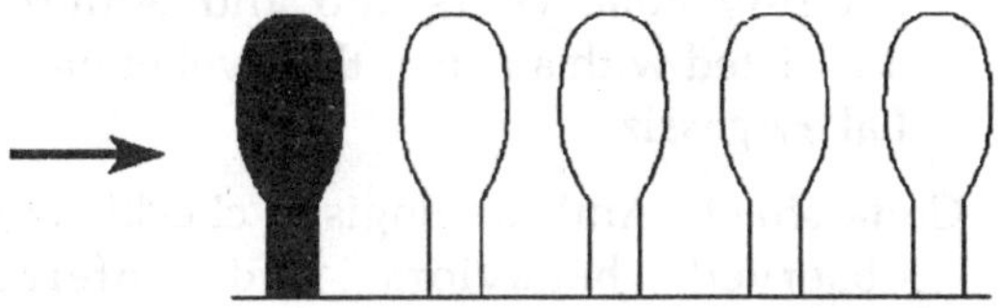

Fig. Clava

Clay Extremely fine (less than 0.01 mm in diameter) particles produced by the weathering of certain rocks. Its primary constituent is hydrated aluminum silicate, but numerous impurities, such as quartz, mica, calcium carbonate, alkalies, iron compounds, humus, and sand may also be present. Clay is plastic when moist, but hardens when dried and is used in the manufacture of ceramics.

Claystone Extremely fine grained mudrock.

Clearwater lake. A lake approximately 17 km north of the pas, manitoba which has given its name to a distinctive late woodland pottery type as well as to the complex and phase within which it occurs. The pots are round-based with constricted necks and generally outflaring rims. Exterior surfaces are fabric-impressed and exterior decoration is usually restricted to a single row of punctates which produce interior bosses. Lips are generally flattened and decorated in a great variety of ways. Associated tools include side-notched and triangular projectile points, scrapers, bifaces, gravers, celts, net sinkers, slate grinding stones, split bone awls, long bone flakers, bone spatulas, bird bone tubes, bone beads, shaft straighteners and red ochre (meyer 1978). Believed by many to be the handiwork of the prehistoric and protohistoric cree, the clearwater lake complex is widely distributed throughout the boreal forest of central saskatchewan, manitoba and northwestern ontario

Cleaver A large core tool with a straight, sharp edge at one end.

Cliff dwellings Shelters or villages built along the edges of cliffs.

CLIMAP Stands for "climate, long range investigation, mapping and prediction", a group project to monitor climatic change in ocean cores and to map the climatic patterns.

Climate The history of rain, snow, and temperature for an area. The average weather.

Cline A geographic gradient in the frequency of a gene, or in the average value of a character.

Clone A set of genetically identical organisms asexually reproduced from one ancestral organism.

Closed corporate community A community that strongly emphasizes community identity and discourages outsiders from settling there by restricting land use to village members and prohibiting the sale or lease of property to outsiders.

Clovis A town in New Mexico which has lent its name to a distinctive type of Paleo-Indian or Early Prehistoric Period projectile point as well as to the complex (also known as the Llano Complex) and culture within which it occurs. The highly distinctive projectile points are concave-based and highly variable in size, ranging from approximately

3 to 12 cm in length. One or both faces may be fluted with the channel flake extending one-half or less of the length of the point. Most Clovis sites are either surface finds of isolated projectile points or kill sites and hence the full nature of he complex is not known. Associated artifacts include a variety of scraping tools, blades, hammerstones, chopping tools and foreshafts and defleshers of bone (Frison 1978). Clovis points are distributed from the arctic to Mexico, and from California as far east as Nova Scotia. Radiocarbon dated sites range in age from 8500 to approximately 10,000 B.C. Where perishable materials are preserved and an association can be demonstrated, faunal remains are nearly invariably those of the mammoth. Clovis points are rare in Manitoba due to the fact that most of the province was glaciated or beneath the waters of glacial Lake Agassiz during the Clovis period. The small area in southwestern Manitoba which would have been available for occupation at that time probably did not support the kind of vegetation upon which mammoths depended for food (Pettipas 1975).

Cluster analysis A multivariate statistical technique which assesses the similarities between units or assemblages, based on the occurrence or non-occurrence of specific artifact types or other components within them.

Coadaptation Beneficial interaction between 1. A number of genes at different loci within an organism,

2. Different parts of an organism, or

3. Organisms belonging to different species.

Coal Rock formed from the fossilised remains of plants. The largest volume of coal originates from the Carboniferous period (354 million years ago).

Coaperturate Describing permanent tetrads in which the apertures of neighbouring monads join. Examples: *Erica* (Ericaceae), *Periploca* (Periplocaceae).

Cobble A medium-sized stone (larger than a pebble but smaller than a fieldstone) which has been rounded and occasionally polished by erosion.

Coccoliths Microscopic structures of varying shape and size that are made of calcite, are secreted by calcareous nannoplankton, and are found in marine deposits from the Triassic period to the Recent. Coccoliths range in size from one to thirty-five micrometers in size.

Cochrane re_advance A surging of the Wisconsinan ice sheet which occurred roughly 8000 years ago and which is associated with a rise in the level of glacial Lake Agassiz.

Code sheets Anthropologists' checklists of observed behaviors and inferred motivations for or attitudes toward them.

Codon A triplet of bases (or nucleotides) in the DNA coding for one amino acid. The relation between codons and amino acids is given by the genetic code. The triplet of bases that is complementary to a condon is called an anticodon; conventionally, the triplet in the mRNA is called the codon and the triplet in the tRNA is called the anticodon.

Cody A town in Wyoming which has lent its name to a distinctive style of Palaeo-Indian knife as well as a complex consisting of at least two forms of Plano projectile points (Eden and Scottsbluff) and possibly a third (Alberta). The knives are either single-shouldered or parallel-sided with a transverse blade. Associated artifacts include a variety of side- and end-scrapers, drills, knives, spokeshaves, gravers, perforators and denticulates. Cody Complex sites are more or less restricted to grassland environments and where preservation is good, they contain the remains of now-extinct forms of bison. In Manitoba, Cody artifacts occur above the Manitoba escarpment in the extreme southwestern corner of the province. Elsewhere, they have been radiocarbon dated between 5900 B.C. and 7900 B.C. (if Alberta is included) or 7100 B.C. if it is not.

Coelacanth Although long thought to have gone extinct about 65 million years ago, one of these deep-water, lungless fish was caught in the 1930s. Others have since been caught and filmed in their natural habitat.

Coelom Body cavity within which internal organs can develop, which is completely lined with tissue of mesodermal origin.

Coenose An assemblage of fossils

Coevolution Evolution in two or more species, such as predator and its prey or a parasite and its host, in which evolutionary changes in one species influence the evolution of the other species.

Coffin texts Texts written inside coffins of the Egyptian Middle Kingdom that are intended to direct the souls of the dead past the dangers and perils encountered on the journey through the afterlife. More than 1,000 spells are known.

Cognate word Words in different languages which are similar in terms of meaning and structure by virtue of descent from a common ancestral language.

Cognitive Relating to cognition, the mental processes involved in the gathering, organization, and use of knowledge, including such aspects as awareness, perception, reasoning, and judgement. The term refers to any mental "behaviors" where the underlying characteristics are abstract in nature and involve insight, expectancy, complex rule use, imagery, use of symbols, belief, intentionality, problem-solving, and so forth.

Cognitive anthropology The study of how peoples of different cultures acquire information about the world (cultural transmission), how they process that information and reach decisions, and how they act on that information in ways that other members of their cultures consider appropriate.

Cognitive imperative The human need to impose order on the world by mental processes.

Cognitive processes Ways of perceiving and ordering the world.

Coil fracture A potsherd, the shape of which reveals that it was a section of one of the coils used to manufacture the vessel.

Coiling A method of ceramic vessel manufacture which involves the stacking of rings of clay. The coils are later smoothed-over by hand or paddled to complete the finish and to bind the coils to one another.

Co-influence sphere An area within which human groups interact due to trade, conflict, migration, the nature of local resources and the manner in which various groups exploited them. As the basis for a research design, the Co-Influence Sphere Model emphasizes interaction as opposed to unilineal chronology, and relies upon cultural comparisons beyond the immediate research area as a basis upon which to draw conclusions.

Colbert, Edwin Edwin (Ned) Harris Colbert (September 28, 1905-November 15, 2001) was an American vertebrate paleontologist who named Staurikosaurus (1970) and Scutellosaurus (1981). Colbert discovered a Lystrosaurus (a dicynodont) in Antarctica; this cemented the continetal drift theory. He also found the huge dinosaur bonebeds at the Ghost Ranch in New Mexico in 1947. In 1955, Colbert suggested that Pachycephalosaurs' thick skulls may have been used as battering rams. Colbert published many papers and books on paleontology and dinosaurs, including *Evolution of the Vertebrates* (1955) and *Men and Dinosaurs: The*

Search in Field and Laboratory (1968) . Colbert was the curator of the American Museum of Natural History and later, the Museum of Northern Arizona. The dinosaur Nedcolbertia (1998) was named to honor Colbert.

Cold hammering Fashioning metal without the use of heat sufficient to melt it. In prehistoric Manitoba this was restricted to copper and recent evidence indicates that temperatures of up to 1000C were often applied to render the substance less brittle.

Collagen A protein which occurs in bone and may be used for radiocarbon dating.

Collateral flaking When flakes on a chipped stone artifact extend to the middle from both edges forming a medial ridge. The flakes are at right angles to the longitudinal axis, and regular and uniform in size.

Collecting The removal of materials in archaeological context from one settlement by the residents of another.

Collection 1. The total array of artifacts from a single site or area.

2. The total array of artifacts in the possession of an individual or institution.

Colluvial deposits Deposits formed on slopes near sources of sediment such as mountains.

Colluvium A mixture of rock fragments and debris occurring at the foot of a slope.

Colobinae Subfamily of Old World monkeys that includes the langurs and colobus monkeys; species that are specialized leaf eaters, possessing a complex stomach and lacking cheek pouches.

Colossus of rhodes A massive bronze statue of the sun god Helios located on the Greek island of Rhodes. It was built around 290 B.C. and was destroyed by an earthquake around 226 B.C. One of the Seven Wonders of the Ancient World, the Colossus was 110 feet high and was once thought to have straddled the entrance to the city harbor (a fact which scholars now know would have been impossible).

Colpodiporate Synonym of diploporate.

Colpoid Apertures more or less similar to colpi but less clearly defined in outline.

Colporoidate Describing pollen grains with colpi and indistinct ora.

Colpororate A compound aperture characterised by an ectoaperture, a shorter lolongate mesoaperture and a lalongate endoaperture. Example: *Sonchus* (Compositae).

This term descibes a compound aperture with a mesoaperture.

Colporus A compound aperture consisting of an ectocolpus with one or more endoapertures.

The term is most commonly used in its adjectival form, colporate.

Colpus An elongated, aperture with a length/breadth ratio greater than 2.

Some authors restrict the term to an meridional aperture (and contrast this with a distal or proximal sulcus), but since shape is the defining criterion the term is equally applicable to other locations (for example, pantocolpate pollen). The precise meaning is often indicated by the use of a prefix, as for example in, ectocolpus, endocolpus, syncolpate.

Colpus equatorialis Synonym of endocingulum.

Colpus membrane The aperture membrane of a colpus.

Colpus transversalis Synonym of endocolpus.

Columella A rod-like element of the sexine/ectexine, either supporting a tectum or a caput.

The difference between a baculum and a columella in current usage is, that a baculum is always a free standing element of sculpturing, whereas a columella is part of the structure.

Columellae layer Synonym of infratectum.

Combed Treated or modified by the application of a toothed instrument of wood, bone, metal, etc. A comb may be used to smooth and/or decorate pottery or to arrange and disentangle hair.

Commissure The slit or line of dehiscence in the laesura.

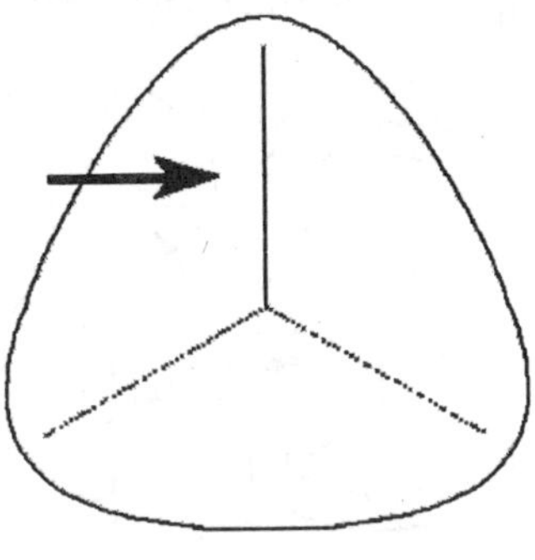

Fig. Commissure

Communal cult A society with groups of ordinary people who conduct religious ceremonies for the well-being of the total community.

Communication The transmission and reception of some stimulus or message. In relation to animal life, communication occurs when one animal transmits information to another animal.

Community The behavioral component comprised of groups of households whose members frequently interact.

Community identity An effort by speakers to identify themselves with a specific locality and to distinguish themselves from outsiders.

Comparative biology The study of patterns among more than one species.

Comparative method The study of adaptation by comparing many species.

Compass A device for determining directions. It uses a magnetic needle that points to magnetic north.

Competition Two or more organisms attempting to occupy or use the same niche or resource in an environment at the same time.

Complex A grouping of related and/or associated traits features and artifacts which comprise a complete process, activity or cultural unit. Thus the laurel complex consists of the sum total of all evidence in the archaeological record which pertains to laurel peoples; whereas the laurel burial complex would only include laurel burial mounds, grave goods, burial style, etc.

Component The archaeological evidence pertaining to a single group of people (more specifically a single focus) at an archaeological site. A site containing only one occupation is a single component site, while one which was reoccupied is termed a multicomponent site.

Composite aperture Synonym of compound aperture.

Compound aperture An aperture with two or more components that are situated in more than one wall layer.

Computer tomography A technique that uses X ray or ultrasound to provide images of layers of solid objects, such as pottery or the human body. The images are processed by a computer to create two- and three-dimensional pictures of the object.

Concave. Incurvate, as the interior surface of a sphere.

Conceptual The major assumptions or underlying premises of a field of research.

Concerted evolution The tendency of the different genes in a gene family to evolve in concert; that is, each gene locus in the family comes to have the same genetic variant.

Conchoidal Literally, "conch-like"; shaped like the exterior surface of a clam shell. The term is used to describe the fracturing properties of certain kinds of stone. In fine-grained materials such as flint, a fractured surface will exhibit roughly circular ridges radiating outwards from the point of impact.

Concordant pattern A pattern in a tectate pollen grain in which the arrangement of the columellae is the same as that of the elements upon the tectum. Example: *Lilium* (Liliaceae).

Antonym: discordant pattern.

Concretion A compact mass of mineral matter, usually spherical or disk-like in shape and embedded in a host rock of different composition. They form by precipitation of mineral matter about a nucleus such as a leaf, trilobite exoskelton or a piece of shell of bone.

Conglomerate Siliclastic sedimentary rocks with an average grain size over 2mm.

Congo craton A separate continental plate that rifted from the supercontinent Rodinia in the Late Precambrian; contained a large part of north-central Africa.

Coniacian European stage of the Upper Cretaceous, spanning the time between 90 and 88 million years ago.

Conjunctate With bacula or columellae wich are branched proximally into two or more parts.

Conodont A jawless fish that had tiny, tooth-like phosphate pieces that are abundant in the fossil record, these were the earliest known vertebrates.

Conoidal Literally, "cone-like". The term is most commonly used to describe the shape of ceramic vessels with pointed bases and straight profiles to the shoulder.

Conquistador A name given to the 16th-century Spanish explorers who came to the New World.

Conservation The scientific process of cleaning—and often repairing and/or restoring—an artifact in order to preserve it for further study and/or display.

Conservation archaeology A sub-field of archaeology which focuses on the preservation of archaeological resources. This position encourages the stabilization and preservation of archaeological sites as opposed to their immediate excavation.

Contact area Area on the proximal face of a spore interpreted as having been formed in contact with the other members of the tetrad. Example: *Retusotriletes pythovii.*

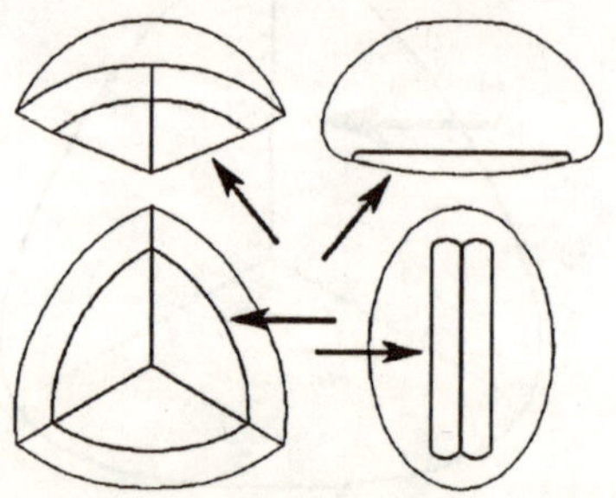

Fig. Contact area

Contemporary Living or happening at the same time.

Context. The immediate environment of an artifact including its association with other artifacts and features as well as its position within the site stratigraphy.

Continental crust Solid, outer layers of the earth, including the rocks of the continents. The part of the crust that directly underlies the continents and continental shelves. Averages about 35 km in thickness, but may be over 70 km thick under largest mountain ranges.

Continental drift The process by which the continents move as part of large plates floating on Earth's mantle.

Continental rise That portion of the continental margin that lies between the abyssal plain and the continental slope. The continental rise is underlain by crustal rocks of the ocean basin.

Continental shelf That portion of the continental margin that extends as a gently sloping surface from the shoreline seaward to a marked change in slope at the top of the continental slope . Seaward depth averages about 130 m.

Continental slope That portion of the continental margin that lies between the continental shelf and the continental rise. It is relatively steep, i.e., 3o - 6o degrees. The continental slope is underlain by crustal rocks of the continent.

Contrivance An object or characteristic used or modified to do something different from its usual use.

Conus Cone-shaped elements on the surface of spores in which the height is less than two times the basal diameter and the apex is pointed, blunt or rounded. Example: *Lophotriletes mosaicus*.

The term is only used in spore terminology and is more or less synonymous with spine.

Convergence The process by which a similar character evolves independently in two species. Also, a synonym for analogy; that is, an instance of a convergently evolved character, or a similar character in two species that was not present in their common ancestor. Examples include wings (convergent in birds, bats, and insects) and camera-type eyes (convergent in vertebrates and cephalopod mollusks).

Convergent Describes characters that are similar in structure and function but have arisen separately rather than from a common ancestor. Convergence is common in groups that become adapted to a similar habitat or way of life; the similar body shapes of the North American Gray Wolf and the Tasmanian Wolf, each evolving from different ancestors (the former is a placental mammal, the latter is a marsupial mammal), resulted from secondary adaptation to similar habitats —convergence.

Convergent evolution The evolution of species from different taxonomic groups toward a similar form; the development of similar characteristics by taxonomically different organisms.

Convex Bulging outwards; excurvate as in the case of the exterior surface of a sphere.

Conway morris, simon Paleobiologist and professor in the Department of Earth Sciences at Cambridge University in the U.K. His research centers around the early evolution of the metazoans, and he is a leading authority on Cambrian and Precambrian fossils. Conway Morris established a link between the Ediacaran fossils, a Burgess Shale fernlike frond *Thaumaptilon*, and the modern seapens, colonial animals related to the corals.

Copal An incense of Mesoamerica.

Cope, Edward drinker Edward Drinker Cope (1840-1897) was a US paleontologist who named over one thousand species of fossil animals (some of these were duplicates), including Dimetrodon. He named the following dinosaurs: Agathaumas (1872), Amphicoelias (1877), Camarasaurus (1877), Coelophysis (1889), Cionodon (1874), Diclonius (1876), Dysganus (1876), Dystrophaeus (1877), Hypsibema (1869), Monoclonius (1876), Paronychodon (1876), Pteropelyx (1889), Tichosteus (1877), and others. He also named the dinosaur families: Camarasauridae (1877), Compsognathidae (1875), Hadrosauridae (1869), Iguanodontidae (1869), and Scelidosauridae (1869). The dinosaur Drinker was named by R. Bakker, P. Galton, Siegwarth & Filla in 1990 as a tribute to Cope.

Cope's rule The evolutionary increase in body size over geological time in a lineage of populations.

Copper sheathing Used underwater (below the waterline) on wooden ships to repel marine organisms.

Coprolite Fossilized, desiccated< or otherwise preserved dung or human faeces. Study of coprolites can yield information on the diet, environment and habits of early peoples.

Coprolite The fossilized waste (dung; fecies) matter of animals.

Coprolite Fossilized, desiccated< or otherwise preserved dung or human faeces. Study of coprolites can yield information on the diet, environment and habits of early peoples.

Copropalynology The study of palynomorphs in coprolites or excrement.

Coptic The Afro-Asiatic language of the Copts, which survives only as a liturgical language of the Coptic Church; of or relating to the Copts, the Coptic Church, or the Coptic language.

Coral A small, soft-bodied animal that secretes a hard exterior skeleton. Corals can be found individually (solitary) or in huge colonies that form coral reefs. Corals thrive in warm, shallow seas. One species of coral looks like the horn of a cow. Another species looks like a human brain. Their common names are derived from these images.

Corballed arch A false arch constructed by putting ceiling tiles closer together on each successive layer until a capstone could be laid.

Cord-impressed, cord-marked, cord-roughened The impressing, marking or roughening of the surface of a ceramic vessel while the clay is still wet as a means of decorating or shaping the pot. The cord of either vegetal or animal fibre may be used as is or wrapped around a stick or paddle . In manitoba, these techniques are most common within the late woodland period.

Core 1. The stone from which flakes have been removed; the nucleus. A "prepared" core is one which has been specially modified in such a way as to control the shape of subsequent flakes. The core itself may be modified into a tool (core tool). core, conical. a cone-shaped core with the flat surface serving as the striking platform. core, polyhedral. a generally sphere-shaped core with many faces. core, wedge-shaped. a core in which flakes are removed from two faces, thus rendering it a wedge-shaped appearance.

2. A generally thin, cylindrical sample of soil or tree growth-rings.

Core sample Cylindrical core retrieved from a hollow steel pipe that is drilled into rock or sediment and brought to the surface for study.

Coria, Rodolfo Rodolfo Coria is an Argentinian paleontologist from the Carmen Funes Museum in Neuquen, Argentina who co-named Argentinosaurus (with J.F. Bonaparte, 1993), Gasparinisaura (with Salgado, 1996), Giganotosaurus (with Salgado, 1995), and Quilmesaurus (2001).

Corona An equatorial or subequatorial extension of a spore, resembling a cingulum, but divided into fringe-like elements (fimbria). Example: *Reinschospora speciosa*.

Corprolite Fossilized dung.

Corpus The body of a saccate pollen grain or camerate spore.

Corrugate Synonym of rugulate.

Cortex. The weathered, outer surface or rind of unmodified stone; patina.

Cosmic radiation Streams of highly penetrating charged particles (composed of free electrons, protons, alpha particles, and a few heavier atom nuclei) that bombard the earth at high speed from outer space. On entering our upper atmosphere, they commonly collide with gas atoms and alter the atomic structures of those atoms.

Costa A thickening of the nexine/endexine bordering an endoaperture, or following the outline of an ectoaperture.

The following forms are used: costa ectocolpi, costa ectopori, costa endocinguli, costa endocolpi, costa endopori.

Costa equatorialis Synonym of costa endocinguli.

Coup. Among many plains indian groups, some act of valour such as touching an enemy in battle, by which prestige was conferred upon an individual. Counting coup. The announcing of one's coups publicly.

Cranium The part of the skull that protects the brain in vertebrates.

Crassi A prefix for thick.

Crassitude Synonym of a thickening.

Craton A part of the earth's crust that has attained stability and has been little deformed for a prolonged period.

Creation science An assortment of many different, non-scientific attempts to disprove evolutionary theory, and efforts to prove that the complexity of living things can be explained only by the action of an "intelligent designer."

Creationism The religious doctrine that all living things on Earth were created separately, in more or less their present form, by a supernatural creator, as stated in the Bible; the precise beliefs of different creationist groups vary widely.

Cremation Destruction of the bodily remains of the deceased by burning. This mode of postmortem treatment may be favored for many reasons; to prevent the return of the dead, to protect the deceased from scavengers, or to prevent the transformation of the dead into a harmful entity. Treatment of the ashes is highly variable from one group to another. Cremation seems to have been particularly popular with Paleo-Indians and this is one of the reasons that skeletal remains dating to this period are so rare.

Crescentic Describing a pollen grain in (polar view with a very thick exine in the medium of the intercolpium, gradually thinning towards the colpi. Example: *Valerianella* (Valerianaceae>.

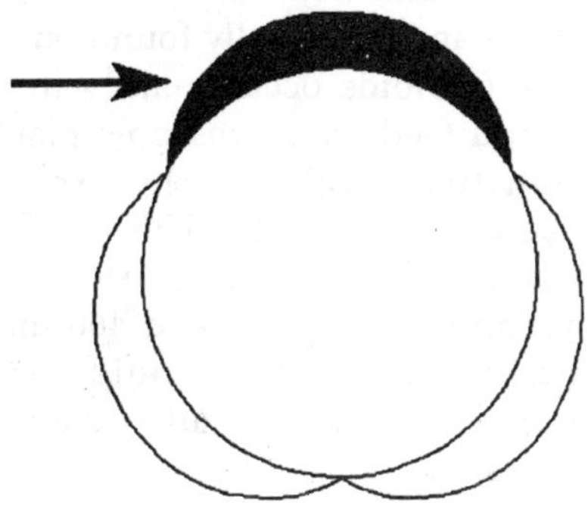

Fig. Crescentic

Cretaceous A period of time stretching from 142 - 65 million years ago. The end of this period (the KT boundary) was marked by the mass extinction of the dinosaurs.

Cretaceous period A period 144 to 65 million years ago, characterized by the growth of the first flowering plants and the height of the era of the dinosaurs. It ended with the complete extinction of the dinosaurs 65 million years ago.

Cretaceous western interior seaway The epicontinental sea that formed as marine waters from the north spread over North America from around 130 to 70 million years ago (Ma), at its peak in the Middle Cretaceous (~ 90 Ma) it extended from present-day Utah to the Appalachians and from the Arctic to the Gulf of Mexico; also refered to as the Western Interior Seaway.

Criminology Palynology is sometimes used to determine the past location of items or persons based on the pollen and spores on or in them

Crinoid A marine invertebrate animal belonging to a class (Crinoidea; about 700 species) of echinoderms, including sea lilies and feather stars. They have a small cup-shaped body covered with hard plates and five radiating pairs of feathery flexible arms surrounding the mouth at the top. Sea lilies, most of which are extinct, are fixed to the sea bottom or some other surface such as a reef by a stalk. Feather stars are free-swimming and are usually found on rocky bottoms. Crinoids occur mainly in deep waters and feed on microscopic plankton and detritus caught by the arms and conveyed to the mouth. The larvae are sedentary. They arose in the Lower Ordovician (between 500 and 460 million years ago), and fossil crinoids are an important constituent of Palaeozoic limestones.

Crinoids A group of marine 'sea-lilies' which attached themselves to sea-floor. Crinoids typically include a long thin stem and end with feather like arms arranged in a cup formation.

Crista A crest-like ornamentation element, taller than it is wide, characterized by a narrowly curved base and a sharp upper edge. Example: *Cristatisporites*.

Crop mark Differential vegetational growth as a result of buried features. Some species of plants are particularly sensitive to various subsurface conditions. For example cereals will not achieve normal height and will ripen sooner over wall foundations, while over ditches, or trenches they will grow taller and remain green longer. Study of these differences, particularly with the aid of aerial photography, can reveal such features in remarkable detail.

Cross dating A relative dating technique which attributes similar ages to two strata, components or sites on the basis of the recovery of similar artifacts from each; the use of an artifact whose age is known elsewhere, to date a new site.

Cross tetrad Synonym of decussate tetrad.

Crossing over The process during meiosis in which the chromosome of a diploid pair exchange genetic material, visible in the light microscope. At a genetic level, it produces recombination.

Croton pattern A characteristic type of ornamentation comprising rings of five or six (sometimes more) raised, often triangular, sexine elements arranged around a circular area, usually formed by capitate columellae (pila). Example: *Croton, Jatropha* (Euphorbiaceae), *Callitriche antartica* (Callitrichaceae), *Pimelea arenaria* (Thy-melaceae).

Crown clade A clade whose definition is restricted to the most recent common ancestor of all living members of the clade and all its descendants. For example, the most recent common ancestor of all modern birds and all its descendants is the crown clade Neornithes. Note that all crown clades are node-based clades, but not all node-based clades are crown clades.

Crustacean A group of marine invertebrates with exoskeletons and several pairs of legs. They include shrimp, lobsters, crabs, amphipods (commonly known as "sand fleas"), and many more.

Crustate Describing an aperture membrane that is thickly covered with coarse granules.

Cryptoaperture An endoaperture which is not apparent in surface view, because there is no ectoaperture.

Examples of such apertures have been described in *Phaleria* (Thymelaeaceae) (Erdtman, 1952), *Tetracera* (Dilleniaceae) (Kubitzki and Baretta-Kuipers, 1969) and *Phyllanthus* (Euphorbiaceae) (Meewis and Punt, 1980).

Cryptopolar Describing a spore in which the distal and proximal faces have dissimmilar sculpturing and lacks tetrad mark. Example: *Calobryum dentatum, Haplomitrium hookeri*.

Cryptospore Non-marine alete sporomorph with well differentiated contact areas but without (haptotypic) features such as laesurae.

Cryptospores Spores and other cells of ancient organisms of unknown affinity. Not strictly palynomorphs because the cell walls are destroyed by routine palynological processing. Cryptospores are generally small (10 - 20 µm) cylinders and spheres, smooth to ornamented. They are the primary indication of life during the Archean (3.5 - 2.5 GY) and Proterozoic (2.5 - 0.5 GY). However similar remains can be recovered by similar processes in Cenozoic sediments and in meteorites.

Crystal lattice The three-dimensional pattern of a crystalline solid. There are characteristic lattice patterns identified for different minerals.

Cucurbit The plant family which includes pumpkins, squash, gourds and cucumbers and which occurs in tropical and subtropical regions. Some members of this family were domesticated by Native North Americans.

Cultigen An initially wild plant which has undergone sufficient genetic changes due to nurturing (or conscious selection), so as to be entirely dependent upon man for its survival; a domesticated plant.

Cultivar A wild plant that is nurtured by humans. Cultivars may thus be found thriving outside of their normal habitats due to irrigation, fertilization or weeding.

Cultivation Preparation and use of land for the production of food.

Cult-statue A statue of a divinity found in a shrine dedicated to that divinity.

Cultural anthropology That branch of anthropology that concerns itself with homanity's non-biological adaptations. Occasionally it is used synonymously (but incorrectly) with social anthropology.

Cultural dating The process of comparing objects archaeologists find with information they already have; comparing cultural attributes.

Cultural dynamics The study of population movements and stability or cultural change and continuity. Cultural dynamics thus includes such phenomena as migration, diffusion, re-adaptation, population increases and expansions, etc. and attempts to identify the reasons for their occurrence.

Cultural ecology The study of the ways a society adapts to its environment.

Cultural evolution The study of how and why human adaptive systems have changed over time.

Cultural formation process Human activities responsible for forming and modifying the archaeological record.

Cultural resource management A branch of archaeology that is concerned with developing policies and action in regard to the preservation and use of cultural resources. Often called simply CRM.

Cultural resources Sites, structures, landscapes, and objects of some importance to a culture or community for scientific, traditional, religious, or other reasons.

Culture 1. The artifacts, behaviours and beliefs possessed by a particular people.

2. Our non-biological, non-"instinctual" means of adaptation; the shared pattern of ideas, beliefs and knowledge which form the basis for social interaction and which are learned by each generation.

3. The term is frequently (and incorrectly) used as a synonym for "society".

Culture area A geographic region within which the occupants are more similar to one another (particularly in terms of material culture) than to those beyond its limits. These rather frequently correspond to natural, environmental areas, thus reflecting a shared mode of adaptation to a similar environment. In practice, a culture area is defined on the basis of its center. The peripheries often share more traits with neighboring culture areas.

Culture hero In mythology, an animal, person or god(ess) who may be seen as the protector of a people, and/or as being the originator of their culture and circumstance. In Native North American folklore, he/she is frequently also a trickster.

Culture history The placement of the material remains of the culture(s) of a region into proper chronological order and the subsequent study of their development.

Culture process The underlying factors which bring about change in a culture. Processual archaeology attempt to identify such causes, and tests hypotheses thus generated against other archaeological data.

Cultus temple Temple dedicated to the worship of one or more deities.

Cuneiform The wedge-shaped characters of many ancient Near Eastern languages.

Cuneus A structure formed by the termination of the inner wall layer at the endogerminal which gives rise to a shallow V-shaped atrium that points towards the centre of the grain. Example: *Pseudotrudopollis*.

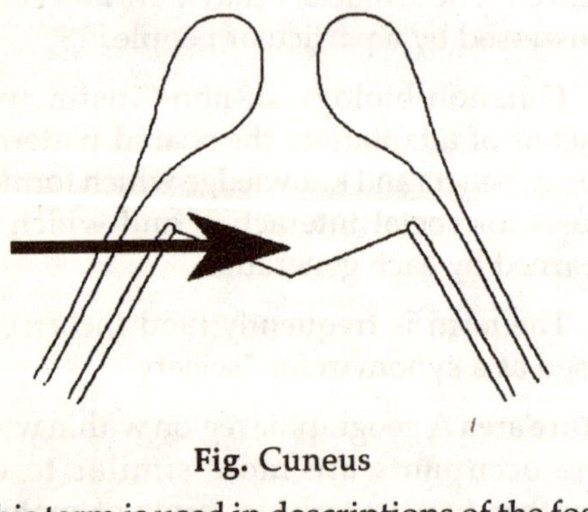

Fig. Cuneus

This term is used in descriptions of the fossil Normapolles group.

Cuniculus A space between the margin of the spore body and the inner face of the equatorially expanded spore wall, located at the equator.

Cupid Roman god of love. Knows as Eros by the Greeks.

Cupulate Synonym of foveolate.

Currie, Cameron A Canadian ecologist and recipient of the 2001 Natural Sciences and Engineering Research Council Doctoral Prize for his research on the complex symbiotic relationship of fungus-growing ants, the fungi they cultivate, mutualistic bacteria that the ants carry on their bodies, and pathogens that attack the fungi.

Currie, Philip J. Philip J. Currie is a Canadian paleontologist from the Royal Tyrrell Museum of Paleontology in Drumheller, Alberta. He has worked extensively in Canada and Asia, recently excavating feathered dinosaurs from China. He named: Caenagnathasia (1993, with Godfrey & Nessov), Callovosaurus (1980), Monolophosaurus (with Zhao, 1994), Ricardoestesia (with others, 1990), Sinraptor (with Zhao, 1993).

Cursorial Describes an animal built for rapid terrestrial locomotion.

Curvatura A line in trilete spores, extending from the extremities of the ends of the radii of the laesura and thus delimiting the contact areas. Examples: *Laevigatisporites glabratus, Divisisporites divisus*.

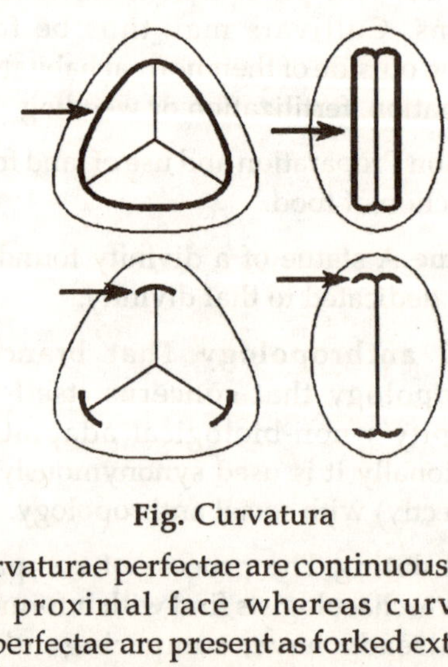

Fig. Curvatura

Curvaturae perfectae are continuous around the proximal face whereas curvaturae imperfectae are present as forked extensions that do not join.

Curvimurate Describing pollen grains with curved muri.

Cuvier, Georges Baron Georges Cuvier (1769-1832) was a French vertebrate zoologist who revolutionized biology by developing a natural system of classifying animals based on comparative anatomy. Cuvier studied fossils and founded the science of paleontology. Although he believed that catastrophic events caused regional extinction, he also believed in fixed species (as opposed to evolving species). Cuvier named many taxonomic groups of mammals, birds, reptiles, and fish, including the phylum Mollusca (1798), class Cephalopoda (1797), class Gastropoda (1797), and the genera: Pterodactylus (1809), Campylodon (1832, with Valenciennes), Cynodon vulpinus (1829), Palaeotherium (1825), Anchitherium (1825), Notidanus (1816), Spinax (1817), Hydrocyon (1819), Chaetopterus (1827), Orcynus (1817), Mydaus meliceps (1821), Xyrichtys (1799), Megalaima flavifrons (1816), Esacus (1829), Myliobatis(1817), Uropeltis (1829), Ziphius cavirostris= Cuvier's Beaked Whale (1823), Stenella frontalis = Atlantic Spotted Dolphin (1838), Grampus griseus = Risso's Dolphin (1828), Adapis parisiensis (1822), Paleosuchus palpebrosus = Cuvier's dwarf caiman (1807), Euselenops luniceps (1817), Jorunna tomentosa (1804), etc.

Cycads A fossil plant from Permian to Recent times. Cycads had short, squat trunks with a crown of large, palm-like leaves

Cycladic Dealing with the islands called the Cyclades, found between Greece and Turkey.

Cytoplasm The region of a eukaryotic cell outside the nucleus.

D

Daeschler, Ted Paleontologist and associate research curator at the Academy of Natural Sciences. Discoverer of late Devonian limbed fossils *Hynerpeton bassetti* and *Designathus rowei* (tetrapods) and *Sauripterus taylorii and Hyneria* (lobed-finned fishes), all early examples of animals exploiting both land and water environments. Author of two books on paleontology for young people.

Darius Persian king Darius III was defeated by Alexander the Great. Darius was killed by his own men.

Dark ages A period of chaos, destruction, and rebuilding that lasted from the fall of Rome in 476 C.E. to the emergence of stable Germanic kingdoms in the ninth century. Specifically, the Dark Ages are often said to have ended in 800 C.E., when Charlemagne was crowned the first Holy Roman Emperor.

Dart A (usually) large, arrow-like projectile propelled by either an atlatl or a blowgun.

Dart, Raymond Australian-born South African anatomist and anthropologist (1893-1988). In 1924 he described a fossil skull collected near Taung in South Africa, naming it *Australopithecus africanus*. Dart asserted that the skull was intermediate between the apes and humans, a controversial claim at the time, though later work made it clear that the Taung child, as it came to be known, was indeed a hominid.

Darwin charles R. Charles Robert Darwin (February 12, 1809-April 19, 1882) was an English naturalist who revolutionized scientific thought with the theory of evolution and natural selection. After studying at Cambridge University, Darwin served as naturalist on the ship *HMS Beagle* (captained by Robert Fitzroy) during its five-year voyage around the world traveling west (December 27, 1831-October 2, 1836). On this trip, Darwin studied a tremendous variety of plant and animal life. Darwin's book called "On the Origin of the Species by Means of Natural Selection, or the Preservation of Favoured Races in the Struggle of Life" was published in November 24, 1859 (and sold out in one day). This world-shaking book outlined the gradual change in a species from generation to generation through natural selectio Alfred Lord Wallace independently proposed the theory of evolution the same time as Darwi In 1871, Darwin published "The Descent of Man," which focused on the origins of people. The theory of evolution has been debated since it was proposed, but most scientists accept evolution and natural selection as the method of species formation and the foundation of biology.

Darwin, Charles The 19th-century naturalist considered the father of evolution. His landmark work, *On the Origin of Species*, published in 1859, presented a wealth of facts supporting the idea of evolution and proposed a viable theory for how evolution occurs — via the mechanism Darwin called "natural selection." In addition to his prolific work in biology, Darwin also published important works on coral reefs and on the geology of the Andes, and a popular travelogue of his five-year voyage aboard HMS *Beagle*.

Darwin, Erasmus The name shared by Charles Darwin's grandfather and brother, each important in his life and work. Charles's grandfather Erasmus (1731-1802) was a glorious polymath — physician, author, and botanist. His impact is reflected throughout a wide range of disciplines from the poetry to the technology of his day. Author of *The Loves of the Plants*, a 2,000-line poem detailing their sexual reproduction, and *Zoonomia, or the Theory of Generations*, whose themes echo throughout his grandson's work. Charles's older brother Erasmus (1805-1881), known as "Ras," used his network of social and scientific contacts to advance the theories of his shyer, more retiring sibling.

Darwinian evolution Evolution by the process of natural selection acting on random variation.

Darwinism Darwin's theory that species originated by evolution from other species and that evolution is mainly driven by natural selection. Differs from neo-Darwinism mainly in that Darwin did not know about Mendelian inheritance.

Data Information; the known facts; a series of measurements or observations.

Dating The process of determining the antiquity of an object or event. absolute dating. the determination of the age of an object relative to the present (eg. 1000 years ago or 43 B.C.). relative dating. the determination of the age of an object relative to others of unknown age (eg. B is older than A but younger than C). Relative dating can thus be used to establish a chronology or sequence whereas absolute dating is required to anchor the events firmly in time.

Datum point usually An arbitrarily-defined spot on or near an archaeological site which is used as a point of reference for the mapping of the site and for the plotting of the distribution of the artifacts which are recovered from it.

Datum Something used as the basis for reasoning, inference, calculating, or measuring. A reference point or origin for mapping.

Daub Un-tempered clay smeared over some rigid structure such as the woven frame (wattle) of a hut in order to keep out draughts. Daub is rarely preserved unless accidentally fired.

Dawkins, Richard An evolutionary biologist who has taught zoology and is the author of several books on evolution and science, including *The Selfish Gene* (1976) and *The Blind Watchmaker* (1986). He is known for his popularization of Darwinian ideas, as well as for original thinking on evolutionary theory.

De facto In fact; in reality. Something which exists or occurs de facto is not the result of a law, but because of circumstances.

Dead sea scrolls A collection of more than 800 manuscripts written on parchment, papyrus, and copper over 2,000 years ago. The Dead Sea Scrolls were found in 1947 in what is now Israel. The Scrolls contain many different kinds of texts, including the oldest known portions of the Old Testament of the Bible.

Debitage Debris; waste products or by products manufacturing process. Lithic debitage would thus include unused flakes, exhausted cores and broken artifacts.

Decans The (Greek) name given to the period of time (10 days) during which the Egyptians observed that certain constellations were visible on the horizon. There were 36 decans in the Egyptian year of 360 days, and tables were drawn up recording them so that the Egyptians were able to tell the time at night: a given constellation would be at a particular point in the sky depending on what hour it was.

Decapoda A group of Arthropods with ten legs, including shrimp, lobsters and crabs.

Decipher Cracking the code; figuring out something's meaning, especially an ancient language (for example, Egyptian hieroglyphics).

Decussate tetrad A multiplanar tetrad of pollen grains or spores arranged in two pairs lying across one another, the pairs (dyads) more or less at right angles to each other. Example: *Orophea* (Annonaceae).

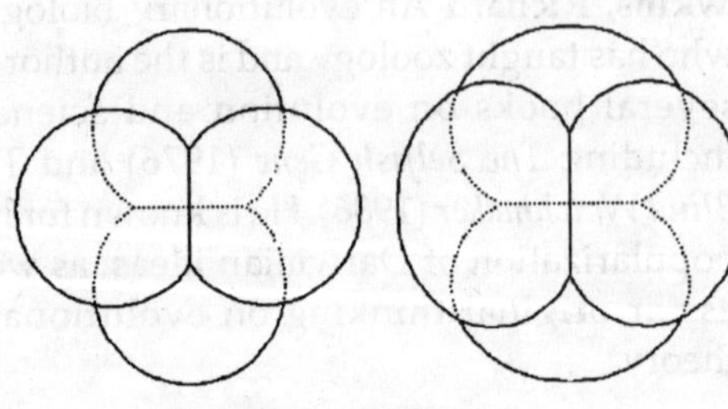

Fig. Decussate tetrad

Deduction A process of reasoning by which more specific consequences are inferred by rigorous argument from more general propositions.

Deep sea cores Cores drilled from the sea bed that provide the most coherent record of climate changes on a worldwide scale. The cores contain shells of microscopic marine organisms (foraminifera) laid down on the ocean floor through the continuous process of sedimentation. Variations in the ratio of two oxygen isotopes in the calcium carbonate of these shells give a sensitive indicator of sea temperature at the time the organisms were alive.

De-facto-refuse Artifacts - often still useable - left behind when an activity area, dwelling, or settlement is abandoned.

Deflation The removal of surficial deposits of soil, sand or fine gravel by wind action. Blowouts are formed as a result of deflation.

Deflesher A chisel-shaped, often toothed implement of bone, stone or metal used to remove the fat and flesh from the inner surface of a freshly skinned hide.

Degradation The wearing away or weathering of a surface by erosion.

Delphi Greek sanctuary of the god Apollo.

Delphinidae Family of cetaceans also known as "ocean dolphins." Includes Bottlenose

Delta A triangular-shaped body of land formed of alluvium at the mouth of a river.

Dembski, William A mathematician and philosopher who has written on intelligent design, attempting to establish the legitimacy and fruitfulness of design within biology.

Deme The local breeding population; the smallest reproductive population.

Demicolpus An ectocolpus divided into two parts. Example: *Amylotheca* (Loranthaceae).

Demographic transition A rapid increase in a society's population with the onset of industrialization, followed by a leveling off of the growth rate due to reduced fertility.

Demography The study of population statistics (population size, number of births and deaths, causes of death, diseases, age distribution, etc.), particularly as a means of making statements of living conditions.

Demotic Script used on Egyptian business documents (and whatever) from about 70 BCE. onwards. By the Graco-Roman period it had become the ordinary writing of everyday life. Word derives from Greek demoticos meaning popular.

Dendrochronology The scientific study of the annular growth of trees. Trees produce rings of various thickness annually in response to rainfall. Tree-rings therefore, can be used to reconstruct fluctuations in rainfall in the past, reflecting past climatic conditions.

Dennett, Daniel Philosopher and director of the Center for Cognitive Studies at Tufts University, whose work unites neuroscience, computer science, and evolutionary biology. Dennett sees no basic distinction between human and machine intelligence, advocating a mechanical explanation of consciousness. He is the author of *Brainchildren Essays on Designing Minds* and *Darwin's Dangerous Idea: Evolution and the Meanings of Life*, among many other books and publications.

Dentate A form of pottery decoration produced by impressing a toothed object of (usually) bone, wood or stone into the wet clay thus creating rows of small, square depressions.

Denticulate 1. Having a toothed edge.

2. A toothed stone tool which could have served as a saw.

Deposition (cultural) The laying down of deposits by human activities that move artifacts from systemic context to archaeological contest.

Deposition (environmental) The laying down of sediments by environmental agents such as wind and water.

Depositional environment The surroundings of artifacts in archaeological context.

Depositional history The order in which strata were laid down.

Depositional stratum A separable layer of material at a site.

Derived Describes a character or group of characters that has changed from its primitive state; advanced. The anatomy of a bird's forelimb (wing) is derived relative to the anatomy of a reptilian forelimb.

Derived characteristic Evolutionarily advanced character state (synonym of apomorph).

Derived homology Homology that first evolved in the common ancestor of a set of species and is unique to those species. Compare with ancestral homology.

Descendant Describes a species or taxon that has evolved from another species or from within a taxon.

Desert archaic tradition A seminomadic, hunting and gathering way of life that people in the Southwest adopted around 7000 B.C. The tradition is also known in Arizona as the Cochise, Amargosan, or Desert culture. The Desert Archaic lifeway was widespread, extending into the Great Basin of Utah and Nevada and the Mohave Desert of California. Although the Archaic lifeway gradually disappeared in southern Arizona as the Hohokam culture developed, the tradition was practiced into historic times by people such as the Great Basin Paiute.

Deshret The red crown This was the crown that represented Lower Egypt (northern).

Destructive archaeology Most methods used in archaeological research are destructive. We destroy what we are studying. Thus it's important to –

Deuterostome Clade of Bilateran animals that share multiple HOX genes for the posterior. Means "second mouth", so called because the mouth develops from the second opening into the embryonic gut. The first opening (the blastopore) forms the future anus only, and the mouth forms later. The body cavity (coelom) develops from buds off the gut. Deuterostomes belong to a larger group within the Animalia called the Bilateria, because they are bilaterally symmetrical with a left and a right side to their bodies. Some d euterostomes such as echinoderms like starfish exhibit radially symmetry. They are nonetheless in the group because during embryonic starfish develop, they begin bilaterally, but switch to radial symmetry during development.

Developmental cycle The stages passed through by individuals, behavioral components, artifacts, and artifact types.

Devil's claw Also called the unicorn plant. A coarse, low-growing annual with large, shallowly lobed leaves. All parts of the plant are covered with coarse, sticky hairs. The seedpods turn black at maturity and each is characterized by a long, slightly curved extension of its tip. The plants grows at elevations of 1000 to 5000 feet and ranges from western Texas to southern Nevada.

Devils lake-sourisford burial complex A late prehistoric period complex known from burial mounds in southwestern manitoba, southeastern saskatchewan and central north dakota. Some of the distinctive artifacts recovered from these sites include small, smoothed mortuary vessels, artifacts made of conch shell from the gulf of mexico, and tubular smoking pipes. The raw materials used and the style of their artwork indicates that these people participated in a far-flung trade network and were heavily influenced by mississippian cultures to the south. Although direct evidence is lacking, it is probable that these people made their living as nomadic bison hunters (syms 1979).

Devonian A period of the Paleozoic era, spanning the time between 410 and 360 million years ago. It is named after Devonshire, England, where rocks of this age were first studied.

Di A prefix for two.

Examples: dicolpate, dicolporate, diporate, disulcate, disulculate, diulculate.

Diadem Band worn on the head signifying royal power.

Diagenesis The physical, chemical or biological alteration of sediments into sedimentary rock at relatively low temperatures and pressures that can result in changes to the rock's original mineralogy and texture. After deposition, sediments are compacted as they are buried beneath successive layers of sediment and cemented by minerals that precipitate from solution. Grains of sediment, rock fragments and fossils can be replaced by other minerals during diagenesis. Porosity usually decreases during diagenesis, except in rare cases such as dissolution of minerals and dolomitization.

Diagnostic artifact An item that is indicative of a particular time period and/or cultural group.

Diagnostic Significant; distinctive or characteristic; of or pertaining to any artifact, feature or attribute which can provide useful information.

Diamicton Unsorted and unstratified sedimentary deposit of any origin, for example: till, landslide debris, mudflows, turbidites, *etc.*

Diapsida ("two holes") Clade of animals, partly distinguished by two temporal fenestrae, that includes the Archosauromporpha, Lepidosauromorpha, and several other closely related groups of animals –diapsid.

Diatom These single-celled algae are common among the marine phytoplankton. Their glassy, two-part shells have intricate patterns and fit together like the two parts of a shirt box.

Diatoms Microscopic photosynthetic organisms with siliceous secondary plant walls termed "frustules." Mostly 20 - 100 μm, and generally of two forms: pennate or centric.

Dichotomy principle Polytomies in a cladogram can have three different causes: Either a real polytomic speciation event (multiple split), or just a simple lack of knowledge about the synapomorphies that could resolve the polytomies into dichotomies, or the polytomies can even result as artefact of a "consensus-method" which is often applied in case of equally supported (equally parsimonious) cladograms of different topology (conflicting evidence). Out of this reason the preference for fully dichotomic trees must be understood as a heuristic principle, that requires that the search for further synapomorphies shall not be discontinued prematurely. The "principle of dichotomy" does by no means imply that polytomic speciations do not at all or do only rarely occur in evolution. This principle is not even essential or necessary for the application of the method of Phylogenetic Systematics. Nevertheless it seems quite plausible that from most polytomic speciation events of the past, not more than one or two species have survived till now, since extinction is a much too frequent phenomenon. Because of the fragmentary fossil record these species will often represent the only known descendant species of the referring polytomic speciation event. Therefore, most reconstructed speciation events will indeed appear to be dichotomic.

Dicolpate, dicolporate, diporate Describing pollen grains with two ectocolpi, two compound apertures or two pores.

The terms belong to the system of pollen classes introduced by Iversen and Troels-Smith.

Die Engraved stamp used to impress a design in softer material.

Diffusion The spread of cultural traits from one culture to another. direct diffusion. the spread of cultural traits by means of multiple hand-to-hand transmissions of adjacent groups rather than a migration of the original trait bearers. stimulus diffusion. the spread of the general idea of a culture trait which may subsequently manifest itself in the creation of the physical object or development of the custom by the recipient group.

Digitate With bacula or columellae which are branched distally into two or more parts. Example: *Polygonum bistorta*.

Digs Archaeological sites with on-going excavations.

Dinocyst A resting stage or reproductive stage in the life cycle of a dinoflagellate.

Dinoflagellate Possessing two tail-like extensions called flagella that are used for movement, these single-celled algae can live freely or in other organisms such as corals. When many dinoflagellates suddenly reproduce in great numbers, they create what are known as "red tides" by making the water appear red.

Dinoflagellates Unicellular aquatic organisms, motile and heterotrophic, parasitic, or photosynthetic. Generally with two flagella, one within an encircling sulcus, the second (longitudinal) oriented posteriorly. Life cycles various, but often with motile thecate stage having a cellulose cell wall, and a resting cyst (hypnozygote) with a chitinous or "dinosporinous" wall forming inside the theca. Form of the cyst ranges from similar to the theca (proximate) to bearing several trumpet-shaped processes (chorate). Theca and cyst often comprised of polygonal plates separated by sutures.

Dinosaur 1. In proper usage, describes any member of clade Dinosauria

2. In common usage, describes any member of clade Dinosauria that is not a "bird".

Dinosauria ("fearfully great lizards") Clade of animals, partly distinguished by a largely to fully open acetabulum, that first evolved in the latter third of the Triassic and was the dominant land animal from the early Jurassic to the end of the Cretaceous. Holtz and Padian (1995) define Dinosauria as the most recent common ancestor of *Triceratops* and modern birds, and all its descendants. Dinosaurs were wholly terrestrial, with no known aquatic species. Only one major clade of dinosaurs, Aves ("birds"), survives today.

Dinosaurs The general name given to describe the land-dwelling reptiles which lived between during the Triassic, Jurassic and Cretaceous period (248 - 65 million years ago).

Dionysos Greek god of wine. Known as Bacchus by the Romans.

Diorate Synonym of diploporate.

Dip net A fish net attached to a (usually) circular frame and often equipped with a handle.

Diploid Having two sets of genes and two sets of chromosomes (one from the mother, one from the father). Many common species, including humans, are diploid. Compare with haploid and polyploid.

Diploporate Describing an ectocolpus with two endoapertures.

Example: *Didymeles* (Didy-melaceae).

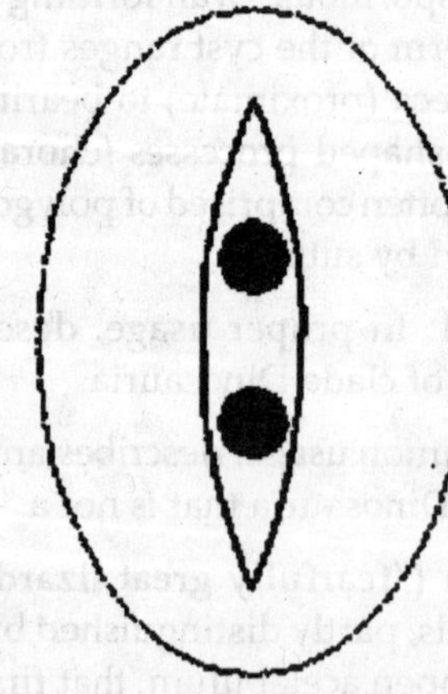

Fig. Diploporate

Diploxylonoid Describing bisaccate pollen grains in which the outline of the sacci in polar view is discontinuous with the outline of the corpus so that the grains seem to consist of three distinct, more or less oval parts.

Directional selection Natural selection where species survive by evolving new characters or modifying old characters. Genotypes that produce new characteristics have great reproductive success, and consequently the gene frequencies for these characters shift. Directional selection results in evolutionary changes; it is, however, a relatively short-lived process, because once a new adaptive state is reached the population then enters stabilizing selection

Discordant pattern A pattern in a tectate pollen grain in which the arrangement of the columellae is different from that of the elements on the tectum. Example: *Geranium* (Geraniaceae).

Antonym: concordant pattern.

Dispersal unit The morphological unit in which mature pollen grains or spores are shed, which may range from individuals (monads), to pairs (dyads), groups of four (tetrads), or groups of more than four (polyads). Larger, indeterminate numbers of pollen grains or spores may also be dispersed as pollinia or massulae.

Disruptive selection Natural selection where subpopulations of a single species within the same habitat exhibit different adaptations. Disruptive selection occurs when a habitat contains distinctly different types of soil conditions, food sources, or other factors, and developing subpopulations adapt to particular features of the area. Disruptive selection results in evolutionary changes and new species; once this occurs, however, the subpopulations enter into stabilizing selection.

Dissections Rounded to elongated cavities in a cingulum or zona. Example: *Vallatisporites ciliaris.*

Synonym of vacuoles.

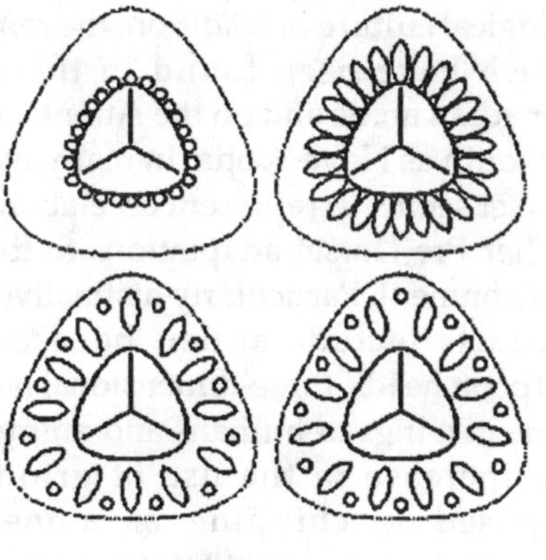

Fig. Dissections

Distal 1. The end of a bone furthest from the mid-line of the body when the individual assumes a normal standing posture.

2. The end of an artifact furthest from the user or observer; the end of a lithic artifact opposite the striking platform.

Distal face That part of a palynomorph that faces outwards the centre of the tetrad, between equator and distal pole.

Antonym: proximal face.

Distal pole The centre of the surface of the distal face.

Antonym: proximal pole.

Distance In taxonomy, referring to the quantitatively measured difference between the phenetic appearance of two groups of individuals, such as populations or species (phenetic distance), or the difference in their gene frequencies (genetic distance).

Disturbance Movement and damage of artifacts in archaeological context as the by-product of other activities.

Disulcate Describing pollen grains with sulci arranged in pairs. Two types of disulcate pollen are distinguished: equatorial disulcate, with opposing, eaquatorially arranged sulci and distal disulcate with paired sulci lying parallel to the long axis of the pollen grain on the distal face. Examples: *Metroxylon salomonense* (Palmae) (equatorial disulcate), *Chamaerops humilis* (Palmae) (distal disulcate).

Diversity (or genetic diversity) A measure of the possible choices of different information at a gene. For example, whether it codes for brown or blue eyes.

Diving bell Diving bells are described in Italy already in the 1530s. In the 17th century divers worked in very large bells. In the top was an air pocket, and below was free workspace towards the wreck. A step toward the earliest beginnings of marine archaeology.

Djed column Pillar symbol meaning "stability." Thought to represent the backbone of Osiris.

Dna (deoxyribonucleic acid) A large organic molecule that stores the genetic code for the synthesis of proteins. Each chromosome consists mostly of a DNA molecule. DNA is composed of sugars, phosphates and bases arranged in a double helix shaped molecular structure. Segments of DNA correspond to specific genes.

DNA Anucleic acid that constitutes the genetic material of all cellular organisms and the DNA viruses; DNA replicates and controls through messenger RNA the inheritable characteristics of all organisms. A molecule of DNA is made up of two parallel twisted chains of alternating units of phosphoric acid and deoxyribose, linked by crosspieces of the purine bases and the pyrimidine bases, resulting in a right-handed helical structure, that carries genetic information encoded in the sequence of the bases. Also, deoxyribonucleic acid.

Dobzhansky, theodosius A geneticist and zoologist best known for his research in population genetics using the fruit fly. His study of the evolution of races led to the discovery of genetic diversity within species, and confirmed his belief that genetic variation leads to better adaptability.

Dolichocephalic Long-headed; having a cephalic index of less than 75.

Domesticated plant A plant whose genetic characteristics are altered from their natural state by human propagation efforts.

Dominance (genetic) An allele (A) is dominant if the phenotype of the heterozygote (Aa) is the same as the homozygote (AA). The allele (a) does not influence the heterozygote's phenotype and is called recessive. An allele may be partly, rather than fully, dominant; in that case, the heterozygous phenotype is nearer to, rather than identical with, the homozygote of the dominant allele.

Dong zhiming Dong Zhiming is a Chinese paleontologist who named the following Chinese dinosaurs: Alxasaurus (with Russell, 1993), Archaeoceratops (with Azuma, 1998), Bellusaurus (1987), Chungkingosaurus (1983), Datousaurus (1984), Gasosaurus (1985), Gongbusaurus (1983), the familty Homalocephalidae (1978), Huayangosaurus (1982), Hudiesaurus (1998), Kelmayisaurus (1973), Microhadrosaurus (1979), Micropa-chycephalosaurus (1978), Nans-hiungosaurus (1979), Shanshanosaurus (1977), Shunosaurus (1983), Siluosaurus (1998), Sinornithoides (with Russell, 1994), Tianchiasaurus (1993), Tugulusaurus (1973), Tuojiangosaurus (1977), Wuerhosaurus (1973), Xiaosaurus (1983), Xuanhanosaurus (1984), Yangchuanosaurus (1978), and Zizhongosaurus (1983). He worked extensively with Dong Zhiming, and also worked with Li, Tang Zilu, Zhang, and Zhou Shiwu, who were co-namers of many of the above-listed dinosaurs.

Dorsal 1. L. The convex (excurvate) face of an artifact. 2. That face of an artifact which was furthest from the centre of the core from which it was manufactured. 3. Of or pertaining to the back or spinal part of the body.

Dorset A Middle Prehistoric Period archaeological culture or tradition, the remains of which have been found in the eastern Canadian arctic and on the Atlantic coast as far south as Nova Scotia. In many ways, the Dorset seems to represent an elaboration of earlier Pre-Dorset adaptations to the arctic environment. Particularly distinctive Dorset artifacts include antler, bone or ivory harpoon heads, three-dimensional ivory and bone carvings of humans and animals, and the increase in the use of grinding as opposed to chipping as a means of manufacturing projectile points and knives. These people pursued a seasonal round which involved the taking of sea mammals in the spring, fishing and caribou hunting inland during the latter part of the summer and fall, and a winter occupation on the sea ice subsisting largely on seals. Dorset culture appears to have disappeared rather suddenly and mysteriously about 1000 years ago with the expansion of the Thule people from Alaska.

Douglas fir A tall evergreen found at elevations of 6500 to 10,000 feet. Used by the Hohokam for timber, this tree can be found in the higher mountain ranges in Arizona and Nevada and at lower elevations in more northern regions.

Douglass, Earl Earl Douglass (1862-1931) was a US fossil hunter who in 1909 found the Carnegie Quarry (now known as Dinosaur National Monument) in Utah, USA. Douglass found many specimens (350 tons of fossils!), including Allosaurus, Apatosaurus, Camarasaurus, Camp-tosaurus, Diplodocus, Hoplosuchus, Othnielia, and Stegosaurus, for the Carnegie Museum in Pittsburgh, Pennsylvania, USA.

Drachma Coin that was the currency of Athens.

Dragged stamp A kind of pottery decoration found on some Laurel vessels produced by dragging a toothed (dentating) instrument across the wet clay, often in a zigzag fashion. The dragged stamp method is also known as push-pull.

Drawknife A woodworking tool consisting of a blade with perpendicularly-oriented handles at either end. This implement is also sometimes known as a spokeshave.

Drift copper Pieces of native copper which have been transported from their natural place of origin by glaciers. Drift copper may be found on the ground surface and undoubtedly was used by prehistoric peoples for the manufacturing of ornaments and tools.

Drift General term refering to any kind of glacially derived deposits. Comes from early idea that most drift was dropped from melting icebergs drifting in a glacial sea.

Drill A stone bit attached to a shaft and used to perforate dense materials.

Drumlin Elongated, streamlined hill molded by glacial deposition or erosion. Usually found in fields containing hundreds of individualdrumlins, located behind major end moraines and parallel to ice flow direction.

Dry farming Method where rainfall runoff is diverted or trapped to provide water for crops. Dry-farming systems include terraces, check dams, and small ditches.

Duat The Egyptian land of the dead. It Lies under the earth and is entered through the western horizon.

Duck bay A bay on the west shore of lake winnipegosis, manitoba, which has lent its name to a distinctive ceramic ware. Vessels are globular in shape with sharply angled necks and shoulders. Surfaces are fabric-impressed or roughened and decoration consists of rows of punctates (duck bay punctate type) or varying combinations of interior notches, punctates and cord-wrapped stick impressions on and near the lip (duck bay decorated lip type). This late woodland ware appears most frequently in the manitoba lowlands (snortlund-coles 1979).

Duncan A stemmed projectile point style of the Middle Prehistoric Period in the Northern Plains. Duncan points are included within the McKean Complex (Wheeler 1954).

Duplibaculate Synonym of duplicolumellate.

Duplication The occurrence of a second copy of a particular sequence of DNA. The duplicate sequence may appear next to the original or be copied elsewhere into the genome. When the duplicated sequence is a gene, the event is called gene duplication.

Duplicolumellate With columellae in two rows under each murus. Example: *Polygonum persicaria* (Polygonaceae).

Dyad General term for two microspores (pollen grains or spores) united as a dispersal unit.

Dynasty A family that retains political power over several generations.

E

Eared Used in the description of projectile points the ear-shaped or "tab-like" projections at the basal corners produced by the combination of a concave base and deep, wide side-notches.

Early man In the New World this term refers to the oldest known human occupants - i.e. prior to ca. 8,000 B.P.

Early prehistoric period The most ancient of the periods of human occupation in North America and closely equivalent in meaning to Paleo-Indian. In Europe, the Mideast and elsewhere, the time preceding the beginnings of agriculture.

Early side-notched point tradition A grouping of early Middle Prehistoric Period complexes characterized by side-notched projectile point styles (generally the first side-notched styles in the region) which have been given a number of different names: Mummy Cave, Bitterroot, Salmon River, Logan Creek, Simonsen, etc. These are by-and-large restricted to the Northern Plains and neighboring regions and coincide with the Altithermal or Atlantic Climatic Episode. A number of radiocarbon dates in excess of 8000 years clearly indicate that the authors of this tradition were contemporary with the Palaeo-Indians. Where preservation is good, these materials tend to be associated with the remains of now-extinct species of bison.

Earthenware A type of pottery made from common clay and fired at a temperature of less than 1000 C. The resulting vessel is soft and porous and requires a glaze to render it waterproof.

Earthwork A fortification, burial mound or other construction fashioned by the excavation and/or heaping of earth.

Easter island An island in the South Pacific (Polynesia), which has also been called Rapa Nui since 1860, where giant statues called "moai" are found.

Eastern cross timbers A relatively narrow, north-south strip of forested land that divides the Grand Prairie to the west from the Blackland Prairie to the east in North Central Texas. The Eastern Cross Timbers are associated with the sandier soils of the Woodbine geologic formation.

Eastern triangular. A late woodland projectile point style. The point is isosceles-triangular in outline, approximately 24 mm in length and 15 mm wide. In form it is very similar to (and therefore readily confused with) the many other triangular point styles of the late prehistoric period across north america.

Eccentric flint Flint cut in a pattern.

Ecclesiastical cult A highly complex religious system headed by a full-time priest.

Ecdysozoans Clade of animals that: grow by periodically moulting - shedding their skin or exoskeleton (timed by steroid hormone signals); share a unique pattern of HOX genes, lack cilia; have separate sexes that copulate to achieve egg fertilization.

Echinate Describing pollen and spores with an ornamentation comprising spines longer than 1μm.

Echinoderm Echinoderms, whose name means "spiny skin," are a group of marine invertebrates that includes starfish, britt-lestars, basket stars, sea cucumbers, sand dollars, sea urchins, and others. They live in environments from shallow coastal waters to deep-sea trenches, from the tropics to the poles.

Echinoderms (crinoids, blastoids, starfishes, sea lilies, and sea urchins) Sea animals covered with calcite plates or spines. They could be free-swimming or found attached to the sea bottom. They usually have a five-fold radial symmetry.

Echinoids A group of marine invertebrates which lived on and and beneath the sea bed. Some species are characterised by the presence of spines protruding from a spherical shell, others lived beneath the sea floor, feeding on organic materials trapped within the sediment.

Echinolophate Describing a lophate pollen grain with echinate ridges.

Echo-sounding An acoustic underwater-survey technique, used to trace the topography of submerged coastal plains and other buried land surfaces.

Ecofact An object or substance found in a site which is of natural origin but which nonetheless provides information pertinent to archaeology. Examples might include fauna, flora, pollen and soil.

Ecological determinism A form of explanation in which it is implicit that changes in the environment determine changes in human society.

Ecological genetics The study of evolution in action in nature, by a combination of field work and laboratory genetics.

Ecological niche The function or position of an organism in a community of organisms.

Ecological species concept A concept of species, according to which a species is a set of organisms adapted to a particular, discrete set of resources (or "niche") in the environment. Compare with biological species concept, cladistic species concept, phenetic species concept, and recognition species concept.

Ecology The study of organisms and their interactions with their environment. Ecosystem - A system made up of a community of animals, plants, and bacteria and their interrelated physical and chemical environment.

Economic anthropology A subdiscipline of anthropology that attempts to understand how the schedule of wants and demands of a society is balanced against the supply of goods and services available, withthe recognition that economic processes cannot be interpreted without culturally defining the demands and understanding the conventions that dictate how and when they are satisfied.

Economic class A group that is defined by the economic position of its members in relation to the means of production in the society—the wealth and relative eocnomic control they may command.

Economic system The ideas and institutions that people draw upon and the behaviors in which they engage in order to secure resources to satisfy their needs and desires.

Economy 1. The full range of human activities dealing with the acquisition, distribution and consumption of subsistence-related commodities.

2. The way in which these resources are managed.

Ecosystem A group of organisms with specific relationships between themselves and a particular environment.

Ectexine The outer part of the exine, which stains positively with basic fuchsin in optical microscopy and has higher electron density in conventionally prepared TEM sections. Orthographical variant: ektexine.

Ectexine includes the foot layer (nexine 1), if present. Erdtman introduced the term in 1943, but used sexine in his later publications.

Ecto A prefix for outer.

Ectoaperture An aperture in the outer layer of the sporoderm.

Examples: ectocolpus, ectopore.

Ectointine Synonym of exintine.

Ectotherm An animal that derives much of its body heat from external heat sources.

Eden A town in wyoming which has lent its name to a plano projectile point style of the northwestern plains. Eden points are 8 to 12 cm in length and approximately 1.5 cm wide. They are collaterally and horizontally flaked and thus display a diamond-shaped cross-section. Most have stems that are only slightly narrower than the blade — in fact in some cases it is produced solely by lateral grinding. Because they are sometimes found in association with alberta and scottsbluff points, they are included in the cody complex.

Effigy A likeness of a person or animal.

Effigy bowl A vessel crafted in the shape of a person or animal.

Effigy mound An earthwork constructed in the shape of an animal or bird. Those of the upper Mississippi are the most well-known in North America. Of probable Hopewell affiliation, these may only be a metre or two in height but may be over 100 m. in length. Occasionally, human burials have been found within them.

Effigy pipe An aboriginal smoking pipe shaped to resemble a human or animal form.

Effigy vessel A container fashioned in the likeness of a human or animal.

Egalitarian society A society that recognizes few differences in wealth, power, prestige, or status.

Ektannulus An abrupt thickening of the outer wall layer in the region of the exogerminal. Example: *Krutzschipollis.*

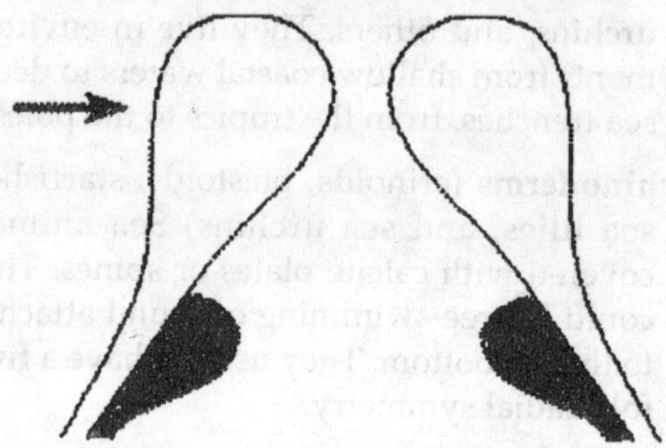

Fig. Ektannulus

This term is used in descriptions of the fossil Normapolles group for part of an annulus.

Ektexine Orthographic variant of ectexine.

The original spelling was soon superseded by the latinized spelling "ectexine" in the publications of Erdtman and others.

Ekto Orthographical variant of the prefix ecto-.

El dorado A Spanish phrase, meaning "the Gilded One" or "the Golden Man," which refers to a legendary South American king who covered himself in gold. European adventurers (mostly Spanish) became obsessed with finding the legendary treasures of El Dorado, and searched for a city filled with gold throughout the 16th- and 17th-centuries.

Elater A spirally twisted, filamentous band attached to certain spores and some fossil pollen grains Example: *Equisetum* (Equisetaceae).

Eldredge, Niles A paleontologist and evolutionary biologist with the American Museum of Natural History, Eldredge, together with Stephen Jay Gould, proposed the theory of punctuated equilibria, providing paleontologists with an explana-tion for the patterns which they find in the fossil record. He has written several books for a general audience, including *Time Frames: The Evolution of Punctuated Equilibria* and *Life in the Balance: Humanity and the Biodiversity Crisis*.

Electrical resistivity Electrolysis A standard cleaning process in archaeological conservation. Artifacts are placed in a chemical solution, and by passing a weak current between them and a surrounding metal grill, the corrosive salts move from the cathode (object) to the anode (grill), removing any accumulated deposit and leaving the artifact clean.

Electron probe microanalysis Used in the analysis of artifact composition, this technique is similar to XRF (X-ray fluorescence spectrometry), and is useful for studying small changes in composition within the body of an artifact.

Electron spin resonance A chronometric dating technique based upon the behavior of electrons in crystals exposed to naturally occurring radioactivity; used to date limestone, coral, shell, teeth, and other materials. Enables trapped electrons within bone and shell to be measured without the heating that thermoluminescence requires.

Electron spin resonance dating (ESR) A radiometric dating method based on the fact that background radiation causes electrons to separate from their atoms and become trapped in the crystal lattice of minerals. When odd numbers of electrons are separated, there is a measurable change in the magnetic field (or spin) of the atoms. Since the magnetic field progressively changes with time in a predictable way as a result of this process, it provides an atomic clock, or calendar, that can be used for dating purposes. This technique is employed primarily to chronometrically date calcium carbonate in limestone, coral, teeth, mollusks, and egg shells. It can also date quartz and flint.

Electrophoresis The method of distinguishing entities according to their motility in an electric field. In evolutionary biology, it has been mainly used to distinguish different forms of proteins. The electrophoretic motility of a molecule is influenced by its size and electric charge.

Electrum A mixture of gold and silver.

Element A substance made up of only one kind of atom.

Elevation A measurement of vertical distance in mapping.

Eluviation The process by which fine particles of soil are moved downward through the soil profile by rain water thus leaving the coarser particles nearer the surface.

Embalm Process of preserving a dead body.

Embryo An early stage of animal development that begins after division of the zygote (the earliest stage, in which joined egg and sperm have not yet divided).

Embryonic Related to an embryo, or being in the state of an embryo.

Embryophyte spores A wide variety of microscopic unicellular reproductive cells developing into gametophytes. Examples include liverworts, club mosses, horse tails, ferns, and many extinct forms. Typically with a triangular or oval outline and various scultpure.

Emic A perspective in ethnography that uses the concepts and categories that are relevant and meaningful to the culture under analysis.

Emigration The movement of organisms out of an area.

Emlen, Stephen A world authority on the social behavior of animals, particularly birds. Emlen's interests center on evolutionary or adaptive aspects of animal behavior. The goal of his research is to better understand the social interactions that occur between individuals, especially cooperation and conflict.

Empathetic method The use of personal intuition (in German Einfuhlung to seek to understand the inner lives of other people, using the assumption that there is a common structure to human experience. The assumption that the study of the inner experience of humans provides a handle for interpreting prehistory and history is made by idealist thinkers such as B. Croce, R.G. Collingwood and members of the "postprocessual" school of thought.

Emperical Received through the senses (sight, touch, smell, hearing, taste), either directly or through extensions.

Empericism Reliance on observable and quantifiable data.

Empire The behavioral component of regional systems tied together by trade and political and military activities.

Empirical Determined by experimentation.

Emulation One of the most frequent features accompanying competition, where customs, buildings, and artifacts in one society may be adopted by neighboring ones through a process of imitation which is often competitive in nature.

Enculturation The process by which the young learn their own culture.

Endannulus An annulus formed by the endexine of a pollen grain. Example: *Atlantopollis*.

This term is used in description of the fossil Normapolles group.

Endexine The inner part of the exine which remains relatively unstained with basic fuchsin in optical microscopy and has a lower electron density in conventionally prepared TEM sections.

Synonym of nexine 2 (Erdtman, 1960b).

Originally endexine was defined simply as the inner of the two main layers of the exine (Erdtman, 1943) to which Erdtman later applied the term nexine. However, as more recently defined, endexine does not include the foot layer (nexine 1) which is considered part of the ectexine.

Endintine The inner, cellulosic zone of the intine which is adjacent to the cytoplasm and in fluorescence microscopy stains positively with PAS calcofluor.

Antonym: exintine.

Orthographical variant: endointine.

Endler, John A zoologist and professor with interests in evolution and how it affects geographic variation. His current research focuses on guppies *(Poecilia reticulata)* in their natural habitat, and how visual signs and vision dictate their behavior.

Endo A prefix for inner.

Endoaperture An aperture in the inner layer of the sporoderm, often the inner aperture of a compound aperture.

Examples are: endocingulum, endocolpus, endopore.

Endoaperture area The region of the nexine of a single compound aperture which encloses a number of endoapertures.

The term is used where the number of endoapertures is more than two and also where the endoapertures are not sharply delimited. Examples: *Deplanchea* (Bigno-niaceae), *Phyllanthus* spp. (Eupho-rbiaceae).

Endocingulum A ring-shaped endoaperture continuous around a pollen grain and lying in the equatorial plane. Example: *Polygonum convolvulus* (Polygonaceae).

Synonym of colpus equatorialis and zonorate.

Endocrack An irregular groove occurring in the inner surface of the nexine/endexine and readily apparent in acetolysed pollen. Example: *Anemone nemorosa* (Ranunculaceae).

Endogamy The restriction of choice of spouses to members of one's own group.

Endogerminal Describing an aperture in the inner wall layer.

This term, used in descriptions of the fossil Normapolles group, is essentially synonymous with endoaperture.

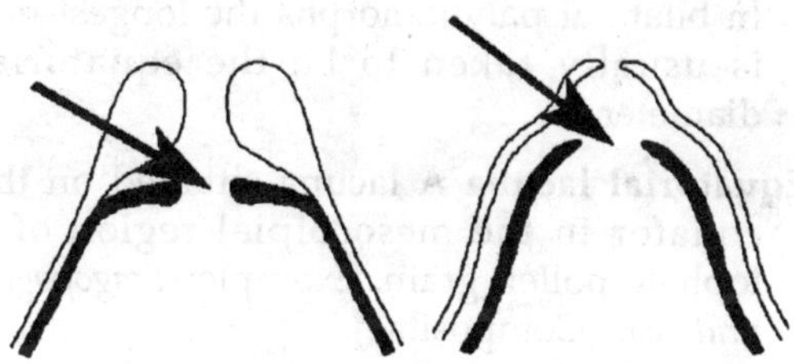

Fig. Endogerminal

Endointine Orthographic variant of endintine.

Endoplica A fold or swelling of the inner wall layer.

A term originally used for features in fossil pollen of the genus *Triatriopollenites*, and later used in the fossil Normapolles group where endoplicae are present as Y-shaped structures centered over the poles.

Endosculpture Sculpturing Occurring on the inner surface of the nexine/endexine.

Endospore The innermost layer of a spore wall

It is probably homologous with the intine of a pollen grain. Also used as a synonym for the corpus in camerate spores. Mycologists use the term for spores formed on the inside of a sporangium.

Endosymbiosis The relationship between organisms which live one within another (symbiont within host) in a mutually beneficial relationship. This process is believed to have accurred in the the evolution of eukaryotes from prokaryotes involved the symbiotic union of several previously independent ancestors. According to the theory, these ancestors included a host cell, an ancestor of mitochondria, an ancestor of chloroplasts, and, more controversially, a prokaryote that brought with it the structures that today provide cellular motion.

Endotherm An animal whose body heat is regulated by internal physiological mechanisms.

End scraper A stone tool formed by chipping the end of a flake of stone which can then be used to scrape animal hides and wood.

Engineers level An optical surveying instrument designed to obtain accurate level lines of sight and turn.

Ennead A group of 9 Egyptian deities that are associated with a major cult center. The best known is the great ennead of Heliopolis, It consists of Atum, Shu, Tefnut, Geb, Nut, Osiris, Isis, Seth and Nephthys.

Entrepreneurs Individuals who are willing to take risks and break with traditional practices in order to make a profit.

Entrepreneurship Economic innovation and risk taking.

Environment The sum of the external conditions and influences which surround an object or organism – particularly the ecological and social settings in which people work and live.

Environmental archaeology A field in which inter-disciplinary research, involving archaeologists and natural scientists, is directed at the reconstruction of human use of plants and animals, and how past societies adapted to changing environmental conditions.

Environmental circumscription An explanation for the origins of the state propounded by Robert Carneiro that emphasizes the fundamental role exerted by environmental constraints and by territorial limitations.

Environmental formation process Biological, chemical, and physical aspects of the environment responsible for forming and modifying the archaeological record.

Environmental palynology The use of palynomorphs, their identification, distribution, and abundance to determine past changes in the biota, climate, or surficial geology of an area. Includes Quaternary palynology and Archaeological palynology . Exclusive of stratigraphic palynology,, but the study of sedimentary sequences often includes both stratigraphic and environmental palynology.

Environmental variance Within a population, the measure of how much of the variation of a particular phenotype is due to environmental factors (as opposed to variations in genotype - see genetic variance). An example might be the height of a plant as determined by such factors as nutrition or damage during development.

Environmental zone Regional plant-animal associations that are climatically determined.

Enzyme A protein that acts as a catalyst for chemical reactions.

Eocene An epoch of the lower Tertiary period, spanning the time between 55.5 and 33.7 million years ago. Its name is from the Greek words "eos" (dawn) and "ceno" (new).

Eoliths Very crudely chipped or shaped stones that were first thought to be tools. Some scholars have thought they were natural pieces of stone, not shaped by humans.

Eon The largest formal unit of geologic time. There are three eons: the Archean, Proterozoic, and Phanerozoic.

Epidermis The outermost layer of the skin.

Epistasis An interaction between the genes at two or more loci, such that the phenotype differs from at would be expected if the loci were expressed independently.

Epoch A division of geologic time less than a period and greater than an age.

Equator The dividing line between the distal and proximal faces of a pollen grain or spore.

Equatorial axis Often misappropriately used as a synonym of equatorial diameter.

Equatorial bridge Synonym of bridge.

Equatorial diameter A line, lying in the equatorial plane, perpendicular to the polar axis and passing through it.

In bilateral palynomorphs the longest line is usually taken to be the equatorial diameter.

Equatorial lacuna A lacuna situated on the equator in the mesocolpial region of a lophate pollen grain. Example: *Tragopogon pratensis* (Compositae).

Equatorial limb Synonym of equatorial outline.

Equatorial outline General description of the equator when a pollen grain is seen in polar view.

Synonym of limb.

Equatorial plane The plane perpendicular to the polar axis and lying midway between the poles.

Equatorial ridge A ridge lying along the equator in the mesocolpial region, usually applied to lophate pollen grains. Example: *Cichorium intybus* (Compositae).

Equatorial view The view of a pollen grain or spore where the equatorial plane is directed towards the observer.

Antonym: polar view.

Equiaxe Pollen Grains with a polar axis equal to the equatorial diameter.

Erect Synonym of prolate.

Era An interval of geologic time shorter than an eon. An era includes two or more periods. There are four major eras: Precambrian, Paleozoic, Mesozoic, and Cenozoic.

Erikson, leif Leader of the Vikings (also called Norsemen) who came from Europe and discovered North America in A.D. 1000, 492 years before Columbus.

Eros Greek god of love. Known as Cupid by the Romans.

Erosion The term given to describe the action of weathering followed by the transportation of the eroded materials by wind, water, gravity or ice.

Erratic Any kind of rock material that has been transported by ice or water a great distance from its origin and now rests on bedrock of different character, also called exotics.

Erwin, Douglas Dr. Erwin is a paleobiologist with the National Museum of Natural History in the Smithsonian Institution. His research is concerned with aspects of major evolutionary novelties, particularly the Metazoan radiation and post-mass extinction recoveries. Recent work has involved the developmental events associated with the Cambrian along with their environmental context. He also works on the rate, causes and consequences of the end-Permian mass extinction.

Esker Long, narrow, high ridge of sand and gravel representing ice-contact deposits of some kind of melt-water stream or delta; generally normal to ice margin, behind major end moraines.

Eskimo 1. The Native inhabitants of Alaska, northern Canada and Greenland. Traditionally, the Eskimo lived in small bands and followed a seasonal round of activity. These people are known for their remarkable technological and behavioral adaptations to one of the world's most trying environments. 2. The language of these people.

Etectate Describing pollen grains interpreted as having lost their tectum during evolutionary development.

Ethnic cleansing The systematic elimination of a particular ethnic group from a region or society, by means including deportation, forced emigration, or genocide.

Ethnic group The behavioral components of groups of households that recognize a very general common ancestry.

Ethno_archaeology The approach to archaeology which makes use of observations of the "archaeological" remains of living non-literate peoples in an attempt to gain understanding of the nature of the material evidence associated with specific activities. This contrasts with the study of prehistoric activity areas and the derivation of plausible inferences to define them.

Ethnocentrism The judging of other cultures by the standards of one's own culture. This common human tendency almost inevitably leads to the conclusion that other cultures are inferior to one's own.

Ethnographic information Information obtained from the anthropological study of living peoples.

Ethnography The by-and-large descriptive, non-interpretive, non-comparable study of another culture.

Ethnohistory The study of the development of a culture by means of archaeological, anthropological and documentary evidence.

Ethnology The scientific, interpretive, comparative study of other cultures.

Etruscans People who ruled Etruia (modern Tuscany), prior to being absorbed by the Romans.

Eu A prefix for true.

When applied to patterns of structuring, such as eurugulate, eustriate and eureticulate, the prefix indicates the arrangement of the columellae. This contrasts with patterns on the tectum which are indicated by the prefix supra-.

Eugenics The science or practice of altering a population, especially of humans, by controlled breeding for desirable inherited characteristics. The term was coined in 1883 by Francis Galton, who was an advocate of "improving" the human race by modifying the fertility of different categories of people. Eugenics fell into disfavour after the perversion of its doctrines by the Nazis.

Euintine Synonym of endintine.

Eukaryote Any organism made up of eukaryotic cells. Eukaryotes are generally larger and have more DNA than prokaryotes (whose cells do not have a nucleus to contain their DNA). Almost all multicellular organisms are eukaryotes.

Eukaryotic cell A cell with a distinct nucleus.

Eumetazoa The primary clade of Kindom Animalia that can be considered a subkingdom. The cells of Eumetazoans truly co-operate to form unmistakable tissues and organs. Nutrients and signals (e.g., hormones) flow efficiently betwwen cells, such that only specialized cells are involved in the acquisition of food.

Euramerica A supercontinent that existed in the Late Silurian through Devonian, formed by the collision of Baltica, Laurentia, and Avalonia; included modern North America, Greenland, Scandinavia, and Europe; also called the "Old Red Continent" for the red color of its oxidized deposits.

Eurypalynous Describing plant taxa characterized by possession of a great diversity of palynomorphs.

Antonym: stenopalynous.

Eustasy Global position and changes in position of sea level.

Eutectum Describing a pollen grain with a continuous tectum.

Synonym of pertectate, tectum imperforatum and tectum solidum.

Antonym: semitectum.

Evidence Data proving a point or contributing to a solution.

Evolution Darwin defined this term as "descent with modification." It is the change in a lineage of populations between generations. In general terms, biological evolution is the process of change by which new species develop from preexisting species over time; in genetic terms, evolution can be defined as any change in the frequency of alleles in populations of organisms from generation to generation.

Evolutionary classification Method of classification using both cladistic and phenetic classificatory principles. To be exact, it permits paraphyletic groups (which are allowed in phenetic but not in cladistic classification) and monophyletic groups (which are allowed in both cladistic and phenetic classification) but excludes polyphyletic groups (which are banned from cladistic classification but permitted in phenetic classification).

Evolutionary systematics The so-called "Evolutionary Systematics" is a syncretistic approach to biological systematics, which was especially endorsed by ernst mayr and simpson. Although the proponents of this approach mostly accepted the Hennigian methodology as adequate technique for the reconstruction of phylogenetic trees, they strongly objected against a strictly cladistic classification, since they wanted to use paraphyletic groups in their classifications. These paraphyla (e.g. "Reptilia") were a direct corollary of their desire to assign a higher categorial rank. (e.g. class) to a monophyletic taxon with numerous autapomorphies (e.g. birds), than to its sistergroup (e.g. crocodiles), if the latter has retained numerous symplesiomorphies with other groups (e.g. lizards). This desire was justified by the greater "evolutionary divergence" compared to the common ancestor, and the possession of a new "adaptational level". Except the difficulty to define and measure "evolutionary divergence" and "adaptational level", the main problem of this approach is the extreme arbitrariness in the classification and delimitation of paraphyla: Should animals been divided in "Protozoa" (paraphyletic) versus Metazoa (monophyletic), or rather in "Invertebrata" (paraphyletic) versus Vertebrata (monophyletic), or better in "Protostomia" (paraphyletic) versus Deuterostomia (monophyletic); should one likewise divide Vertebrata in "Pisces" (paraphyletic) versus Tetrapoda (monophyletic), or in "Anamnia" (paraphyletic) versus Amniota (monophyletic), or maybe even in non-mammals (paraphyletic) and Mammalia (monophyletic), which curiously was never proposed yet. Even the exclusion of man from the kingdom of animals as separate kingdom (regnum) "Psychozoa" has been proposed and would be absolutely compatible with the principles of "Evolutionary Systematics". The grouping of crocodiles and birds as sistergroups in a monophyletic taxon Archosauria has been dismissed as absurd by evolutionary systematists, while they accepted a group like Deuterostomia without any protest, although it is including such divergent organisms as sea-cucumbers and man. A further critique against Phylogenetic Systematic was the mere conjecture that it shall completely neglect the evidence from symplesiomorphic characters, although they are important homologies, too. This statement is of course nonsense, since all homologous characters are recognized and used in Phylogenetic Systematics, and symplesiomorphic characters are therefore not neglected at all, but are recognized on that hierarchical level on which they do represent a synapomorphy (compare apomorphic). A more general problem of "Evolutionary Systematics" is the circumstance that several very different criteria are used for the construction of a classification (phylogeny, divergence, adaptational level), but in the resulting system it is not recognizable which criteria have been used for a particular taxon. Because of the mentioned problems and the theoretical, practical and heuristic superiority of Phylogenetic Systematics, the number of proponents of "Evolutionary Systematics" has strongly declined in the past decades.

Ewald, paul Professor of biology at Amherst College, specializing in hummingbird and flower coevolution and the evolution of infectious diseases. His research on disease focuses on the evolutionary effects of various public health interventions.

Exaptation (formerly *preadaptation*) Adoption of a character that had one use in an ancestral form into a new, different use in a descendant form. For example, three bones in the jaws of the ancestors of mammals were exapted into the hammer, stirrup and anvil, the bones of the middle ear -exaptation.

Excavation Systematic digging and recording of an archaeological site.

Exclusive anchor taxon Describes the anchor taxon not included in the taxon being defined. For example, in the stem-based clade definition of Carnosauria, *Allosaurus* and all taxa sharing a more recent common ancestor with *Allosaurus* than with "birds," "birds" is the exclusive anchor taxon.

Exine The outer layer of the wall of a palynomorph, which is highly resistant to strong acids and bases, and is composed primarily of sporopollenin.

Exintine An outer (pectic) layer of the intine, lying below the nexine, and which stains positively with alcian blue.

Synonym of ectointine.

Antonym: endintine.

Exitus The site of exit of the pollen tube from the sporoderm.

Exoexine Synonym of sexine.

Exogerminal Describing an aperture formed in the outer wall layer of the exine.

This term, used in descriptions of pollen of the fossil Normapolles group, is essentially equivalent to ectoaperture.

Exon The nucleotide sequences of some genes consist of parts that code for amino acids, with other parts that do not code for amino acids interspersed among them. The coding parts, which are translated, are called exons; the interspersed non-coding parts are called introns.

Exoskeleton In contrast to an endoskeleton, is an external anatomical feature that supports and protects an animal's body. Many invertebrate animals such as insects, spiders, crustaceans and shellfish have exoskeletons. Lobsters, for example, have tough outer shell systems which provide rigidity and shape to their bodies. An exoskeleton may interfere with an animal's growth. To overcome this, arthropods go through a process called moulting in which they shed their exoskeleton and replace it with a new, larger one.

Exospore The outer layer of a spore wall.

A perispore, if present, lies outside the exospore, and is not part of it. Probably homologous with exine in pollen. Mycologists use the term for spores formed on the outside of a sporangium.

Exosporium Synonym of exine.

Experimental archaeology Scientific studies designed to discover processes that produced and/or modified artifacts and sites.

Extant Describes types or species of animals that are currently living. Not extinct

Extended family. The nuclear family (husband, wife, children) together with other relatives. Included among the latter are either the wife's married daughters, their husbands and children, or the unmarried children and wives and children of married sons. In general usage, the term may simply imply a group of relatives beyond the nuclear family.

Extended mckean-oxbow complex. The term applied to the hypothesized complex consisting of a blending of the mckean and oxbow cultures and its persistence until the tenth century a.d.

Extinct When all members of a species have died out.

Extinction The resulting term given to mark the end of a species or group of creatures. Extinction usually results from a long-term change in the environment, which subsequently triggers a period of decline and eventual end. The dinosaurs are perhaps the best example of a group of reptiles which faced mass extinction at the end of the Cretaceous period.

Extralimital No longer living in the local area.

Extrusion Something (rock, fossil, etc.) that is forced, pressed, or pushed out.

Extrusive igneous rock Rocks formed from volcanic processes - i.e. those that are erupted onto the surface of the earth (i.e. basalt, andesite, rhyolite).

F

Fabric A material woven of plant or animal fibers.

Fabric-impressed A term used to describe the surface finish of many of the late woodland pottery types of manitoba. Under ideal conditions, it is possible to see the individual woven strands of the fabric which was wrapped around the paddle or which comprised the mould within which the pot was formed.

Facies Lateral or vertical changes in the character of a sediment or rock unit. Changes in mineralogy, color, texture, fossil content, *etc.*

Facies, metamorphic A set of metamorphic mineral assemblages, repeatedly associated in space and time, such that there is a constant and therefore predictable relationship between mineral composition and chemical composition. That relationship is a consequence of conditions of temperature and pressure under which the assemblages are stable.

Fact A natural phenomenon repeatedly confirmed by observation.

Factor analysis A multivariate statistical technique which assesses the degree of variation between artifact types, and is based on a matrix of correlation coefficients which measure the relative association between any two variables.

Faculative biped Describes an animal that walks or otherwise moves about on its two hindlimbs, but can and does spend some time locomoting on all four limbs

Faculative quadruped Describes an animal that walks or otherwise moves about on all four limbs, but can and does spend some time locomoting on only its hindlimbs.

Faience Quartz sand heated in crucible with soda until the quartz melts then solidifies into glaze. Many colors were done, but most popular was green or turquoise made by added copper filings before heating. An art found mostly in ancient Egypt.

Fall off analysis The study of regularities in the way in which quantities of traded items found in the archaeological record decline as the distance from the source increases. This may be plotted as a falloff curve, with the quantities of material (y-axis) plotted against distance from source (X-axis).

False door A carved or painted door in an Egyptian tomb which was an entry and exit point for the soul of the deceased.

Family The category of taxonomic classification between order and genus . Organisms within a family share a close similarity; for example, the cat family, Felidae, which includes lions and domestic cats.

Family unit Among chimpanzees, a small group consisting of a mother with some or all of her offspring.

Fastigium Cavity in a colporate grain, appearing as a separation of the inner part of the exine from the domed sexine in the region of the endoaperture. Example: *Hypericum* (hypericaceae).

Synonym of caverna.

The term describes a feature morphologically similar to the vestibulum; the latter is restricted to porate pollen grains.

Faulting A fracture in the earth's crust (fault) that causes the land on one side of the fault to be raised above the land on the other side.

Fauna A Latin term which refers to animals remains, as opposed to flora which refers to plant remains.

Faunal analysis In archaeology, the scientific study of animal remains. As different species are adapted to different environments, the kinds of animal bone found in an archaeological site can reveal information about local conditions. For example, the dominance of bison in the faunal record might indicate proximity to grasslands at the time that the site was occupied. Because many species bear young only in a certain season, and since an expert can accurately determine an animal's age at the time of death, faunal analysis can also yield information of the time of year in which a site was occupied. The presence of seasonally migratory species may lend additional support to such conclusions. Finally, because most faunal material in sites are the remains of feasts, analysis can reveal information on the diet of the site's occupants and allow estimates of the number of people who may have resided there.

Faunal dating A method of relative dating based on observing the evolutionary changes in particular species of mammals, so as to form a rough chronological sequence.

Faunal remains The (usually) hard tissues of birds, fish and animals which survive in the archaeological record.

Feature Something distinctive encountered on the ground surface or during the course of excavations which is not artifactual in the usual sense. Its significance may lie not in the object or objects which constitute the feature, but rather in the relationship of the objects to each other. Thus while a cobble, fleck of ash or fragment of burned bone would mean little if found in isolation, a concentration of bone and ash surrounded by a circle of cobbles would suggest a cooking area, and this patterning would constitute the feature. Other examples of features could include post moulds, storage pits, a garbage dump, a cache of tools, a flint knapping area, a collapsed dwelling or a burial.

Feldspar A group of rock-forming minerals all of which consist of aluminum silicates and which may contain potassium, sodium, calcium or barium. Feldspars are the chief elements of igneous rock.

Fenestrate Describing a class of pollen grains characterized by large, window-like spaces lacking a tectum.

The term is accepted as a category in the classification system of Iversen and Troels-Smith (1950) and was especially established to include lophate pollen grains of the Compositae. Although useful for defining a class of pollen grains the term is not recommended in descriptions. Many such grains can be described as lophate.

Fennoscandian ice sheet Pleistocene ice sheet centered on Norway, Sweden, and Finland, and extended at times into the northern European lowlands.

Fermentation A series of reactions occurring under anaerobic conditions (lacking oxygen) in certain microorganisms (particularly yeasts) in which organic compounds such as glucose are converted into simpler substances with the release of energy. Fermentation is involved in bread making where the carbon dioxide produced by the yeast causes dough to rise.

Fertile crescent A region of the Middle East arching across the northern part of the Syrian Desert and extending from the Nile Valley to the Tigris and Euphrates rivers. The ancient civilizations of Egypt, Phoenicia, Assyria, and Babylonia developed in this area, which was also the site of numerous migrations and invasions.

Fetus The embryo of a mammal that has reached a stage of development in the uterus in which most of the adult features are recognizable. Specifically in humans it refers to the stage of development after the appearance of bone cells, a process occurring 7 to 8 weeks after fertilization.

Fictive kin Persons such as godparents, and old family friends whom children call "aunt" and "uncle".

Field data forms Printed forms used to record archaeological survey or excavation information. Special forms are frequently used to record artifact proveniences; features and burials; site locations and descriptions; and level-notes.

Field dependence The tendency to see the field of vision as a single unit, with separate objects existing only as part of the whole.

Field independence The tendency to see the objects in one's field of vision as discrete units, distinct from the field as a whole.

Field notes Archaeologists keep a notebook with them when they are digging so they can note when they change levels and what kinds of things they find. They need to keep another record in case the profile or floor plan they drew wasn't very clear. Later, in the lab, archaeologists might question the context of an object. If they have notes to go back and look at, it makes it easier to figure out what was going on.

Fieldwork The firsthand observation of human societies.

Filigree Fine open metalwork using wires and soldering, first developed in the Near East.

Fill Sand, earth or other material which is contained within a feature or overlying a site.

Fimbria Long, hair-like appendages

The term has been used in the descriptions of fossil spores. Example: *Radiatisporites radiatus*.

Synonym of capillus.

Findspot The location in which an artifact is found.

Fire drill A fire-making device consisting of a wooden shaft, the tip of which is twirled against another piece of wood until the friction creates a spark.

Fire spall A flake detached by exposure to intense heat.

Fire-broken rock Stone which has been fractured by exposure to heat. Generally, a fire fracture is difficult to distinguish from other forms of breakage such as that due to freezing. For that reason other evidence (such as scorching) is required to identify the cause of fracturing.

Fired Hardened (as in ceramics) by exposure to intense heat.

Firn Old snow which has survived at least one summer melt season, density in the range of 0.4 to 0.5.

Fischer's law/rule The name given to the widespread arrangement in developmental tetrads of tri-aperturate pollen grains whereby the apertures form in pairs at six points in the tetrad. Example: Ericaceae .

Fishing camp A site used primarily for the acquisition of fish and possibly the taking of other aquatic species.

Fission track dating A method of dating an object that counts the number of tracks made by the breakdown of radiocarbon elements. The older an object is the more tracks it leaves. This method is used mostly on rocks, pottery, and glass.

Fission-fusion society A constantly changing form of social organization whereby large groups undergo fission into smaller units and small units fuse into larger units in response to the activity of the group and the season of the year.

Fissura A sharp, straight split that appears during germination in certain inaperturate pollen grains. Examples: *Taxodium* (Taxodiaceae), *Cupressus* (Cupressaceae).

Fitness The success of an individual (or allele or genotype in a population) in surviving and reproducing, measured by that individual's (or allele's or genotype's) genetic contribution to the next generation and subsequent generations.

Fixation A gene has achieved fixation when its frequency has reached 100 percent in the population.

Fixed 1. In population genetics, a gene is "fixed" when it has a frequency of 100 percent.

2. In creationism, species are described as "fixed" in the sense that they are believed not to change their form, or appearance, through time.

Fjord Deep, long, steep-sided glacial trough that extends into the sea.

Flagging Brightly colored plastic ribbon used to mark features, sites, surveyed stakes etc., to aid in their relocation.

Flake A thin chip of stone detached from either a larger flake or a core by the application of pressure or a blow (Percussion). Characteristically, manufactured flakes have a bulb of percussion, a bulbar scar and compression rings radiating outward from the point of impact on the ventral face, and the remnant of the striking platform. channel flake. a long, thin flake detached in such a way as to produce a "groove" on the finished artifact. Such artifacts are said to be fluted. decortification flake. a flake which serves to remove part of the weathered outer surface (cortex) of a core. lamellar flake. a flake with parallel edges. Prismatic flake. a parallel-sided flake, either triangular or quadrilateral in cross-section, produced from a specially prepared core. Prismatic flakes are also known as blades (sense 4). waste flake. one which is produced as a by-product of the manufacture of something else; a discarded unused flake.

Flaker An implement of bone, antler, stone or other material, used to remove flakes from a core or preform.

Flaking Knapping; chipping; the act of removing flakes from a core or preform. alternate flaking. the process of removing flakes from alternate faces along the edge of a tool, thus producing a wavy or sinuous edge. collateral flaking. a kind of flaking produced by removal of flakes from the face of a blade which begin at either edge and terminate at the midline. This kind of flaking commonly produces a diamond-shaped cross-section. horizontal flaking. a kind of flaking in which the flake scars are at right angles to the long axis of the blade. oblique flaking. a kind of flaking in which the flake scars are at an angle to the long axis of the blade. parallel flaking. a kind of flaking in which the flake scars are parallel to one another. percussion flaking. the removal of flakes by striking. pressure flaking. the removal of flakes by the application of pressure. retouch flaking. a form of secondary flaking, always accomplished by pressure, which is used to sharpen or straighten an edge. ripple-flaking. a fine form of parallel flaking which gives a surface the appearance of ripples. secondary flaking. a fine form of flaking intended to remove surface irregularities, or to sharpen or straighten an edge. transverse flaking. a kind of flaking in which the flake scars run across the full width of the blade.

Flammer, Larry A retired high school biology teacher and co-founder of the Santa Clara County Biotechnology-Education Partnership, which provides teacher training and lab equipment for local schools. He is a current member and Web writer for the evolution and nature of science institute (ENSI).

Flange A general term, used to describe equatorial extensions of spores

This term is widely used in Palaeozoic spores, but is not precisely defined.

Flexed burial A human interment where the body is placed in a semi-fetal Position with the knees drawn up against the chest and hands near the chin.

Flint A kind of hard, rock with a glassy opaque appearance. It was a favorite raw material for many prehistoric humans for the manufacture of tools such as knives and scraping implements.

Flint knapping The flaking of stone for the purpose of manufacturing tools regardless of whether the stone is in fact flint.

Floodwater farming The practice of planting crops in areas that are flooded every year in the rainy season, the floodwaters thus providing natural irrigation.

Floor plan Archaeologists draw a floor plan of the unit they are digging in at the bottom of every level, or when they find a feature such as a fire pit. A floor plan shows how something looks from above.

Flora Plants of a given region or period of geologic time.

Floral remains Pollen, seeds, wood charcoal and other plant parts which may be preserved in the archaeological record. Analysis of these can provide information on past environments and subsistence patterns.

Flotation A method of obtaining seeds and other organic materials from soil by using liquids.

Fluorine analysis dating A **relative dating** method based on the fact that bones buried in the ground progressively lose nitrogen and gain fluorine and other trace elements. The rate at which these changes occur depends on the local environment. If two bones from the same site have markedly different amounts of nitrogen and fluorine, it is a strong indication that they did not come from the same time period. The bone with the least amount of nitrogen and the greatest amount of fluorine is most likely the oldest.

Fluorine dating. A relative dating technique which uses the amount of fluorine present in bones or teeth as an indication of their age. Living bone has a known and more-or-less constant fluorine content with some minor regional fluctuations due to differences in drinking water. After death, bone begins to absorb fluorine from ground water in the soil in which it is buried. The process continues until a theoretical maximum is reached. Comparison of the fluorine content of two bones, therefore, will allow the archaeologist to determine which is older, or if the two are of approximately the same age. Because the rate of fluorine uptake varies from place to place, all of the samples to be dated must come from the same site or at least from similar situations. The most famous application of fluorine dating was its contribution to exposing the piltdown hoax in england. It was found that the human skull was much more recent than the bones of the extinct animals among which it had been planted. In manitoba, fluorine analysis has been used primarily as a means of verifying series of radiocarbon dates or as a means of helping estimations of age when radiocarbon dating is not possible.

Fluorine test A dating method that measures the amount of fluorine, nitrogen, and uranium in bones. Older bones have more fluorine and uranium and less nitrogen. But because decomposition happens at different speeds in different places, it's not possible to compare bones from different sites.

Flute A long, narrow flake removed from a spear point to aid in the binding of the point to the spear shaft.

Fluvium Any river-deposited sediment.

Focus. In the midwestern taxonomic method devised by mckern (1939) a focus was defined as a group of components with very similar traits.

Folk taxonomy The classification of phenomena on the basis of cultural tradition.

Folktales Traditional stories found in a culture (generally transmitted orally) that may or may not be based on fact.

Folsom A town in New Mexico which has given its name to a distinctive fluted projectile point and to the Palaeo-Indian Complex or culture of which it is a part. The Folsom site is of particular significance to the history of American archaeology because it was here that the discoveries were made (1926-28) that conclusively demonstrated the contemporaneity of man with now-extinct species of animals in the New World. The projectile points of the Folsom Complex are among the finest examples of the flint knapper's art found anywhere in the world. Ranging in length from 2 to 7.5 cm, Folsom points are either lanceolate or paralle-sided in outline, and are deeply concave at the base which may give the basal edges an "eared" appearance. Occasionally a small "nipple" or projection may be present at the center of the base. This is a remnant of the striking platform created to enable the removal of the channel flakes which often extend the full length of each face. Associated artifacts include a variety of scraping tools, gravers, knives, grinding stones, hammerstones and gaming pieces. Where preservation is good the predominant faunal association is bison, thus marking a change from the earlier Clovis peoples' focus upon the mammoth. Folsom points occur over a fairly broad area, but excavated sites cluster between Montana and Texas. Folsom points are nearly as rare in Manitoba as Clovis and for much the same reason; Lake Agassiz covered much of the province and the southwestern corner of the province which was available for occupation did not support the kind of vegetation suitable for the animals which Clovis and Folsom peoples hunted. Folsom radiocarbon dates range from approximately 8000 to 9000 B.C.

Food chain A sequence of organisms in a community where each member of the chain feeds on the member below it. The chain moves from plants that make their own food to herbivores to carnivores.

Food web How plants, predators and prey interact. The interaction of all food chains found in an ecological community.

Foot layer The inner layer of the ectexine. Synonym of nexine 1, pedium and sole.

Foraging Collecting wild plants and hunting wild animals for subsistence.

Foramen Synonym of pore.

This term is sometimes used in its adjectival form forate (e.g. Walker and Doyle, 1975) which is a synonym of pantoporate. A special form of the term is internal foramen, which refers to features in the exine for example of certain Compositae and Acanthaceae pollen grains.

Foramen magnum The large opening at the base of the skull through which the spinal cord passes.

Foraminifer Protozoans that belong to the subclass Sarcodina, order Foraminifera, that have a test of one to many chambers composed of secreted calcite or agglutinated particles.

Foraminifera These invertebrates are very common in the global ocean, and their distinctive, chambered shells are common in the fossil record as far back as 550 million years. Although very few today exceed 9 mm in diameter, fossils have been found that measure 15 cm across.

Forensic anthropology The application of the techniques of osteology and skeletal identification to legal problems.

Foreshaft In a compound dart or spear, a shaft to which is attached the projectile point and which in turn fastens to the main or backshaft. The latter falls away after the foreshaft penetrates the prey and thus may be retrieved.

Forge 1. A place used for working metal by heating and hammering; a furnace or hearth used for heating metal.

2. To shape metal by heating and hammering.

Formal dimension The physical properties of artifacts.

Formal interview An interview that consists of questions designed to elicit specific facts, attitudes, and opinions.

Formal organization A group that restricts membership and makes use of officially designated positions and roles, formal rules and regulations, and a bureaucratic structure.

Formalism A school of economic anthropology which argues that if the concepts of formal economic theory are broadened, they can serve as analytic tools for the study of any economic system.

Formation A mappable, lithologically distinct unit of rock.

Formation process Those processes affecting the way in which archaeological materials came to be buried, and their subsequent history afterwards. Cultural formation processes include the deliberate or accidental activities of humans; natural formation processes refer to natural or environmental events which govern the burial and survival of the archaeological record.

Fossaperturate Refers to an equatorially aperturate, lobate pollen grain with the apertures in the indentations between the lobes.

The term belongs in the system of ambs as described by Erdtman (1952).

Fossil A remnant or impression of plant or animal life which is preserved. In strictest terms, fossilization refers only to the loss of fats and gelatine from bone and not necessarily the subsequent replacement of these by minerals (mineralization). In its most general (and most incorrect) usage, a fossil may be anything dug from the ground.

Fossil beach A lake or ocean beach developed when the water-level was significantly different from that of the present. Most commonly these will be "raised beaches", or old strandline features and sediments found above the modern shoreline.

Fossil cuticles The outermost protective layer of the skin of leaves or blades of grass, made of cutin, a very resistant material that survives in the archaeological record often in feces. Cuticular analysis is a useful adjunct to palynology in environmental reconstruction.

Fossil ice wedges Soil features caused when the ground freezes and contracts, opening up fissures in the permafrost that fill with wedges of ice. The fossil wedges are proof of past cooling of climate and of the depth of permafrost.

Fossilization All the processes that involve the burial of a plant or animal in sediment and the eventual preservation of all, part, or a trace of it.

Fossula A feature of ornamentation consisting of an elongated, irregular groove in the surface.

Foulke, William parker William Parker Foulke was a US scientist and dinosaur artist who found the first American dinosaur skeleton, Hadrosaurus, a duck-billed dinosaur. The fossilized bones were found by workmen in a Cretaceous marl (a crumbly type of soil) pit on the John E. Hopkins farm in Haddonfield, New Jersey beginning in 1838. Foulke heard of the discovery and recognized its importance in 1858. The dinosaur was excavated and named by US anatomist Joseph Leidy who named it Hadrosaurus foulkii (meaning "Foulke's big lizard").

Founder effect The loss of genetic variation when a new colony is formed by a very small number of individuals from a larger population.

Founder principle The situation in which a founding population does not represent a random sample of the original population; a form of sampling error.

Foveola A feature of ornamentation consisting of more or less rounded depressions or lumina more than 1μm in diameter. The distance between foveolae is greater than their breadth.

Freehold Private ownership of property.

Frequency dimension The number of occurrences of an artifact type.

Frequency-dependent selection Selection in which the fitness of a genotype (or phenotype) depends on its frequency in the population.

Fresco A type of wall painting. Paint was applied to freshly spread plaster before it dried. This method locked in the colors, creating vivid and beautiful art.

Friable Easily crumbled, as in the case of rock or pottery.

Frieze Ornamental band.

Frost action The process by which objects buried in the ground are moved about by the freezing and expansion of water.

Frustillum Synonym of areola.

Functionalism The theory that all elements of a culture are functional in that they serve to satisfy culturally defined needs of the people in that society or requirements of the society as a whole.

Funerary cones Clay cones inserted above an Egyptian tombs entrance with the name and title of the deceased.

Fungal spores Unicellular or multicellular, reproductive or distributional cells developing into a number of different phases of the complex life cycles of the fungi.

Fungi A group of organisms comprising the kingdom Fungi, which includes molds and mushrooms. They can exist either as single cells or make up a multicellular body called a mycelium. Fungi lack chlorophyll and secrete digestive enzymes that decompose other biological tissues.

Furrow A common word for an elongate aperture.

Examples: colpus, sulcus, tenuitas.

Furrow membrane Synonym of colpus membrane.

G

Gaff 1. A barbed fishing spear.

2. A shaft with a hook attached intended to aid the landing of large fish.

Galea A relatively large element of the outer wall of a spore consisting of a sharply tapering spine and a broad bulbous base.

Galena A lead ore, crystals and fragments of which were widely traded and used for ornamental purposes in Native North America.

Galton, Francis A cousin of Charles Darwin, Galton was a British explorer and anthropologist. He was known for his studies of human intelligence and later for his work in eugenics (a term he coined), the "science" of improving human hereditary characteristics. Known for his efforts at various sorts of measurement (he developed fingerprinting and was a pioneer in statistics), he was knighted in 1909.

Galton, Peter M. Peter M. Galton is a British paleontologist working in the USA. He named: Aliwalia (1985), Blikanasaurus (with J. van Heerden, 1985), Bugenasaura (1995), Callovosaurus (1980), Camelotia (1985), Dracopelta (1980), Gravitholus (with W. P. Wall, 1979), Lesothosaurus (1978), Ornatotholus (with H. Sues,1983), Othnielia, (1977), Stygimoloch (with H. Sues,1983), Torvosaurus (with J.A. Jensen, 1979), Valdosaurus (1977), and Yaverlandia (1971). He named the dinosaur families: Blikanasauridae (with J. van Heerden, 1985), Fabrosauridae (1972), Staurikosauridae (1977), and Ruehleia (2001). He named the order Herrerasauria (1985). He also championed the cladistic theory that birds are modern-day dinosaurs (with R. Bakker, 1974), showed that Hypsilophodon was not arboreal (did not live in trees), that hadrosaurs did not drag their tails but used the tail as a counterbalance for the head, and that the Pachycephalosaurs butted heads like rams.

Gamete Haploid reproductive cells that combine at fertilization to form the zygote, called sperm (or pollen) in males and eggs in females.

Gaming piece An artifact often of stone, bone, or pottery which served as a token or counter in a game of chance or skill.

Gardener A horticulturalist; one who relies only upon manpower for the nurture and harvesting of crops.

Gastropod (snails) Most gastropods have coiled shells, either flat (like a wheel) or spiral (like a garden snail).

Gastropod Meaning "stomach foot," this name refers to the class of mollusks that contains the most species. Gastropods include snails and slugs that are marine, freshwater, and terrestrial.

Gastropoda An order of molluscs more commonly known as snails.

Gathering Among chimpanzees, the largest observed group within the community.

Gauthier, Jacques A. Jacques A. Gauthier is a US paleontologist and Professor of Geology & Geophysics at Yale University. He has worked extensively on the classification of dinosaurs, birds, and all saurians (including lizards, crocodylians, and rhynchocephalians). In 1986, J. A. Gauthier looked at over 100 characteristics of birds and dinosaurs and showed that birds belonged to the clade of coelurosaurian dinosaurs.

Gehring, Walter J. Dr. Gehring and his research group discovered the homeobox, a DNA segment characteristic for homeotic genes which is not only present in arthropods and their ancestors, but also in vertebrates up to humans. Their work on the "master control gene" for eye development sheds light on how the mechanism for building eyes may have evolved long ago in the ancestor of what are now very different types of organisms.

Geminicolpate Describing pollen grains with colpi arranged in pairs.

Gemma A sexine element which is constricted at its base, higher than 1μm, and that has approximately the same width as its height.

Gender A cultural construct consisting of the set of distinguishable characteristics associated with each sex.

Gene A sequence of nucleotides coding for a protein (or, in some cases, part of a protein); a unit of heredity.

Gene expression The degree to which a gene is active in a certain tissue of the body, measured by the amount of mRNA in the tissue.

Gene family A set of related genes occupying various loci in the DNA, almost certainly formed by duplication of an ancestral gene and having a recognizably similar sequence. Members of a gene family may be functionally very similar or differ widely. The globin gene family is an example.

Gene flow The movement of genes into or through a population by interbreeding or by migration and interbreeding.

Gene frequency The frequency in the population of a particular gene relative to other genes at its locus. Expressed as a proportion (between 0 and 1) or percentage (between 0 and 100 percent).

Gene pool All the genes in a population at a particular time.

Generalized reciprocity Informal gift giving for which no accounts are kept and no immediate or specific return is expected.

Generalized species Species that can survive in a variety of ecological niches.

Generalized trait A trait used for many functions.

Generative cell The cell in a pollen grain which divides to form male gametes (Jackson, 1928).

Genetic code The code relating nucleotide triplets in the mRNA (or DNA) to amino acids in the proteins.

Genetic determinism The idea that all behavior, including very specific behavior, is biologically based, in contrast to cultural determinism.

Genetic drift Changes in the frequencies of alleles in a population that occur by chance, rather than because of natural selection.

Genetic engineering Removing genes from the DNA of one species and splicing them into the DNA of another species using the techniques of molecular biology.

Genetic load A reduction in the average fitness of the members of a population because of the deleterious genes, or gene combinations, in the population. It has many particular forms, such as "mutational load," "segregational load," and "recombinational load."

Genetic Related to genes. A gene is a sequence of nucleotides coding for a protein (or, in some cases, part of a protein); a unit of heredity.

Genetic variance Within a population, the measure of how much of the variation of a particular phenotype is due to genotypic variation. An example might be the height of a tree as determined by genes inherited from the parents.

Genetics The study of the mechanisms of heredity and biological variation.

Geniculum A bulge in the equatorial exine of the colpus, often associated with a separation of the sexine from the nexine and the rupturing of the latter. Examples: *Fraxinus* (Olea ceae), *Quercus* (Faga- ceae).

If the separation forms a cavity the term fastigium should be applied.

Genome The full set of DNA in a cell or organism.

Genomics The science dealing with analysis of the full range of genes in an organism.

Genotype A specific combination of genes that determine a character. For example; in humans each character is determined by the combination of two genes specific to that character. Let B and b represent the two genes that determine height (a capital letter denotes a dominant gene, a lowercase letter denotes a recessive gene). Each human will have one of the height genotypes BB, Bb or bb. If two humans produce an offspring, the offspring will receive one of its height genes from the mother and the other from the father. If the father has the height genotype BB and the mother bb, then the offspring will have the height genotype Bb (B from the father, b from the mother). Parents with height genotypes BB and BB can only produce offspring with height genotype BB (similarly bb and bb can only produce bb). BB and Bb can produce offspring with genotypes BB or Bb; Bb and Bb can produce offspring with genotypes BB, Bb or bb. In mixed genotypes (in our example Bb) the dominant gene determines the character (if B is tall and b is short, the offspring with genotype Bb will be tall)

Genus The second-to-lowest category in taxonomic classification. The phrase "species name" generally refers to the genus and species together, as in the Latin name for humans, Homo sapiens.

Geochemical analysis The investigative technique which involves taking soil samples at regular intervals from the surface of a site, and measuring their phosphate content and other chemical properties.

Geochemistry The science that deals with chemical changes in and composition of the earth's crust.

Geochronology A relative dating method based on the association of early human sites with changing features of the land, such as the advance and retreat of glaciers or the rise and fall of sea levels. When these events are well dated, geochronology could be considered a reliable calibrated relative dating technique. Associated evidence may be changes in the frequency of plant species measured by pollen count and other kinds of paleoecology evidence.

Geographic coordinates The world-wide system of latitude and longitude used to define the location of any point on the earth's surface.

Geographical isolation A form of reproductive isolation in which members of a population become separated from another population by geographical barriers that prevent the interchange of genes between the separated populations.

Geographical race A major division of humankind into large geographical areas wherein people resemble one another more closely than they resemble people in different geographical areas.

Geologic column A column representing the eras and periods of time with their major events and changing life forms.

Geologic time The period of time extending from the formation of the earth to the present.

Geologic time scale An arbitrary chronologic sequence of geologic events, used as a measure of the age of any part of geologic time, usually presented in the form of a chart showing the names of the various rock-stratigraphic, time-stratigraphic, or geologic-time units.

Geological unconformities Breaks in the original stratigraphic sequence of a geological deposit. Understanding all of the unconformities in an area is a prerequisite for using **stratigraphy** for **relative dating** purposes.

Geologist A scientist who studies rocks, including how and when they were formed, the minerals they contain, and how they have changed through time.

Geology The study of the planet Earth - the materials of which it is made, the processes that act on these materials, the products formed, and the history of the planet and its life forms since its origin.

Geomagnetic reversals An aspect of archaeomagnetism relevant to the dating of the Lower Paleolithic, involving complete reversals in the earth's magnetic field.

Geometric period Period of cultural development in Greece after the end of the Mycenean kingdom. Named after the geometric decoration typical of the time.

Geomorphology A subdiscipline of geography, concerned with the study of the form and development of the landscape, it includes such specializations as sedimentology.

Georadar A technique used in ground reconnaissance, similar to soil-sounding radar, but with a much larger antenna and more extensive coverage.

Germ plasm The reproductive cells in an organism, or the cells that produce the gametes. All cells in an organism can be divided into the soma (the cells that ultimately die) and the germ cells (the cells that are perpetuated by reproduction).

Germinal aperture A hole in the furrow membrane through which the pollen tube emerges.

Examples: germ pore (synonym of pore), germinal furrow (synonym of ectocolpus). Synonym of exitus.

Germination The initial stages in the growth of a seed to form a seedling. The embryonic shoot (plumule) and embryonic root (radicle) emerge and grow upward and downward, respectively. Food reserves for germination come from tissue within the seed and/or from the seed leaves (cotyledons).

Gesso Mixture of glue and chalk or plaster applied to walls and objects producing a smooth finish.

Gestation The period in animals bearing live young (especially mammals) from the fertilization of the egg and its implantation into the wall of the uterus until the birth of the young (parturition), during which the young develops in the uterus. In humans gestation is known as pregnancy and takes about nine months (40 weeks).

Giant ground sloth With a body the size of a cow, they pulled leaves and twigs to their mouths with their long-clawed hands and long tongues. Found from the Pliocene to Pleistocene epoch.

Gill net A kind of net suspended vertically in the water which snares fish by the gills as they attempt to swim through it.

Gill pouches Structures that form in the early human embryo and that are thought to be homologous to the gill slits of other chordates.

Gilmore, Charles W. Charles Whitney Gilmore (1874-1945) was a scientist who studied North American and Asian dinosaurs (including those in the Gobi Desert), and fossil lizards. He named Alamosaurus (1922), Alectrosaurus (1933), Archaeornithomimus (1920), Bactrosaurus (1933), Brachyceratops (1914), Chirostenotes (1924), Mongolosaurus (1933), Parrosaurus (1945), Pinacosaurus (1933), Thescelosaurus (1913), and the family Troodontidae (1924). Gilmoreosaurus (Brett-Surman, 1979) was named to honor Gilmore.

Gingerich, Philip Gingerich is interested in evolutionary change documented in the fossil record and how this relates to the kinds of changes observable in the field or laboratory on the scale of a few generations. His ongoing fieldwork in Wyoming, Egypt, and Pakistan is concerned with the origin of modern orders of mammals, especially primates and whales.

Glacial lake A lake formed of ponded glacial meltwater, or by the damming of a drainage system by glacial activity. A "pro-glacial" lake has at least one margin formed by glacial ice.

Glacial maximum The position and period of greatest advance of a glacier.

Glaciation The expansion of the polar ice caps and the covering of portions of the earth with large masses of ice and snow. At least four major periods of glaciation are recognized within the Pleistocene epoch: the Nebraskan, the Kansan, the Illinoian and most recently, the Wisconsinan.

Glacier accumulation/ablation Gain of ice mass mainly through snowfall by a glacier; ablation is the loss of mass mainly by melting or calving.

Glacier ice Ice formed from the accumulation and recrystallization of snow, which exhibits past or present flow. Typically has a bluish-green color due to tiny gas bubbles or may be dirty with included sediment. Density 0.5-0.9 g/cc.

Glacier surge Rapid, wavelike movement of ice down a glacier which results in sudden advance of the ice margin. Mass transfer of material from the accumulation zone to the ablation zone with high ice velocities. Occurs cyclically in certain glaciers, but never occurs in others.

Glaze Glossy finish or covering on a surface. Glaze can be applied to pottery or can form as a coating on glass like materials during the heating process.

Gleysol A soil type which develops under boggy conditions. It is characterized by a grayish coloration, mottled appearance and a reduced concentration of iron and other elements.

Globular Shaped like a globe or sphere. When used in reference to pottery, it implies a generally rounded vessel shape.

Gloger's rule A rule which states that within the same species of endotherms, more heavily pigmented forms tend to be found near the equator and lighter forms away from the equator.

Glottochronology A technique in linguistics used to date the amount of time elapsed since speakers of a once-common languge separated from one another. If a group of people becomes divided into two due to a migration or by some other means, their modes of speech will change differentially and after one or two thousand years they will be incapable of understanding each other. Since the rate of change in basic vocabulary is known and fairly constant (about 18% every thousand years) a comparison of the vocabularies of related languages can allow the calculation of the time of divergence.

Glyph A character, symbol or picture carved in stone, metal or some other substance.

Goldfarb, Alex A Russian-born microbiologist now at the Public Health Research Institute in New York City, Dr. Goldfarb is piloting a program in the Russian prison system to combat the further evolution of drug-resistant strains of tuberculosis, which have infected at least 30,000 inmates.

Gondwana A supercontinent that existed from Cambrian to Jurassic time, mainly composed of South America, Africa, Madagascar, India, Antarctica, and Australia.

Goniotreme Synonym of angulaperturate.

Gorge A small tool, often of bone, pointed at either end and equipped with a hole near its center for attachment to a line. It is baited and imbeds itself crosswise in the throat of fish or small game when the line is pulled.

Gorget A plaque-like or crescent-shaped object of stone, wood, shell, etc., often elaborately decorated, and perforated for suspension around the neck.

Gouge A chisel- with a broad groove ground into one face thus giving the bit a U-shape. These are often interpreted as woodworking tools.

Gould, Stephen Jay A professor of geology and zoology at Harvard University since 1967. A paleontologist and an evolutionary biologist, he teaches geology and the history of science, as well. With others, he has advanced the concept that major evolutionary changes can occur in sudden bursts rather than through the slow, gradual process proposed by the traditional view of evolution. In addition to his scholarly works, Gould has published numerous popular books on paleoanthropology, Darwinian theory, and evolutionary biology.

Grammar The formal structure of a language, comprising phonology, morphology, and syntax.

Grammatical structure The rules for organizing elements of a language into meaningful utterances.

Granite Coarse-grained, light-colored igneous rock composed of quartz, feldspar and biotite or hornblende.

Grant, Peter and Rosemary Biologists whose long-term research focuses on finches in the Galapagos Islands, and the evolutionary impact of climatic and environmental changes on their populations. They live part of the year in the Islands, and have received honors for their work since they began in 1973.

Granular exine A type of exine stratification in which the infratectal layer is composed of more or less rounded, granules rather than of columellae or other structures.

Granulate granulatus With granules.

Granulation The soldering of grains of metal to a background, usually of the same metal, and much used by the Etruscans.

Granule General word for a small, rounded element.

Granulum A very small and rounded element of the **sexine/ectexine** that is less than 1μm in all directions.

A granulum is the diminutive for granum.

Granum The combined elements gemmae and verrucae.

Graphic arts Those forms of art such as painting and drawing.

Graptolite A small, colonial, often planktonic marine animal that was very abundant in the oceans 300 to 500 million years ago; now extinct.

Graptolites A group of small marine creatures, that existed during the Paleozoic period (545 - 248 Million years ago). These thin, tubular organisms often resemble pencil markings.

Grass river The name of a tributary of the Nelson which lent its name to a late woodland pottery type and the phase within which it was believed to occur. Known only from a handful of sites, grass river vessels were globular in shape and were impressed with a ribbed-fabric covered paddle. Decoration usually consisted of a single row of circular punctates around the rim. Virtually no other diagnostic artifacts were found in association with the ceramics. There is now a consensus that the grass river ceramics are in fact Selkirk ware and as a consequence the grass river phase concept has fallen into disuse.

Grassland An environment, such as prairie, upon which grasses dominate, either naturally or due to human intervention.

Grave A place where the dead are buried.

Grave escort An individual who is killed and placed in the grave with a person of higher status so that he may accompany the latter to the afterlife and serve him.

Grave goods Objects (tools, weapons, ornaments, etc.) placed in the grave with the deceased so that he may use them in the afterlife or simply as an expression of affection for the individual.

Gravel A mixture of sand, pebbles, and small cobbles which range in diameter from 2 to 200 mm.

Graver A small, sharp-pointed tool used for engraving or incising bone, antler, ivory, wood, etc.

Graviportal Describes an animal built for slow terrestrial locomotion as a result of its comparatively heavy body weight.

Great apes The orangutan from Asia and the common chimpanzee, bonobo (pygmy chimpanzee), and gorilla from Africa.

Great English vowel shift A linguistic change during the Middle English period, when speakers of English began to alter the sounds of vowels, eventually changing all vowel sounds in the language.

Greaves Bronze shin guards worn by Greek soldiers.

Greene, Mott A historian of science who has written extensively about the development of geological thought during the 19th and early 20th centuries, including the development of the theory of continental drift.

Greenhouse gases Gases that absorb and reradiate infrared radiation. When present in the atmosphere, these gases contribute to the greenhouse effect, trapping heat near the surface of the planet. On Earth, the main greenhouse gases are carbon dioxide, water vapor, methane, nitrous oxide, ozone, and some halocarbon compounds.

Gregorian calendar The most commonly used calendar system in the world today. It is based on the older Julian calendar of the late Roman Empire but more accurately corresponds to the solar year of 365.2422 days. The Julian calendar year is 365.25 days. This is inaccurate by approximately 11 minutes a year. The result is that the Julian calendar is out of sync by one day every 131 years. The Gregorian calendar fixed this problem by changing the rules for leap years. In the Julian system, a day is added to the end of February every 4 years. In the Gregorian system a year is a leap year if it is divisible by 4 unless it is divisible by 100 but not by 400. This eliminates leap years periodically in order to keep the calendar more synchronized with the solar year. The Gregorian calendar was named after Pope Gregory XIII who officially approved it in

1582 A.D. England and its colonies did not adopt it until 1752.

Grid A network of uniformly spaced squares that divides a site into units; used to measure and record an object's position in space.

Griffin Mythological beast with the head, forelegs and wings of an eagle and the body and hind legs of a lion.

Grinding stone Abrader; shaft smoother; whetstone; any coarse-grained stone used to sharpen, dull, shape or polish other tools by abrasion.

Grinding The shaping of an object or the dulling of an edge by means of abrasion with another object or substance. basal grinding. the smoothing of the proximal end of a tool (especially a projectile point) so that it will not cut through its bindings after hafting. lateral grinding. intentional smoothing of the blade edges of a tool (especially a projectile point) so that it will not cut through its bindings after being hafted.

Grit and grog tempered pottery Sand (grit) and crushed pottery sherds (grog) mixed in the unfired clay to make ceramic vessels stronger. These inclusions prevented the rapid expansion of the paste as the clay's water content was boiled away when the pottery was fired.

Grit tempered pottery Pottery to which small particles of sand or stone have been added to the clay. Temper lends strength to the vessel and helps prevent it from breaking when fired.

Grooming In primates, the activity of going through the fur with hand or teeth to remove insects, dirt, twigs, dead skin, etc.; also acts as display of affection.

Grooming cluster A small group of closely related females that engage in a high degree of grooming.

Groove A general descriptive word.

Ground reconnaissance A collective name for a wide variety of methods for identifying individual archaeological sites, including consultation of documentary sources, place-name evidence, local folklore, and legend, but primarily actual fieldwork.

Ground running and walking A form of quadrupedalism in which the animal walks on the ground using the hands and the feet; the palms of the hand are flat on the ground.

Ground stone Stone artifacts shaped by sawing, grinding, and/or polishing with abrasive materials (e.g. "ground slate knives", "polished soapstone pendants" etc.).

Ground water Fresh water found under the surface of the Earth (usually in aquifers) that often supplies wells and springs. Also referred to as underground water and subsurface water.

Groundplan The complete set of character states of the last common stem-species of a monophyletic group has been termed the groundplan of this group. This definition should be supplemented by the phrase "... at the time of its splitting into descendant species", to take into account the case of a phyletic change within the stem-species. The groundplan includes all the plesiomorphic states, as well as all the apomorphic states of this stem-species. The apomorphic groundplan characters are the autapomorphies of the referring mono-phyletic group. The term groundplan is one of the most important terms of Phylogenetic Systematics, although it was never clearly defined by hennig himself. The groundplan concept is completely different from the typological concept of the "Bauplan". Out of this reason AX (1984) suggested the new (synonymous) term "Grundmuster" (= ground-pattern). Of course a groundplan cannot be observed (unless one has a time-travel apparatus), but only reconstructed on the basis of a phylogenetic hypothesis. Thus a groundplan is not evidence FOR a phylogenetic reconstruction, but a conclusion FROM a phylogenetic reconstruction.

Group A number of individuals who interact on a regular basis and have a sense of collective identity.

Group selection The selection operating between groups of individuals rather than between individuals. It would produce attributes beneficial to a group in competition with other groups rather than attributes beneficial to individuals.

Gula A rather ornate projecting, neck-like, extension on the proximal face of a trilete spore. Example: *Lagenicula horrida*.

This term is mainly used in the description of some fossil megaspores.

Gulf of california Also known as the Sea of Cortez, it is the body of water separating Baja, California from the rest of Mexico. The Colorado River empties into the gulf.

Gumbotill Till which is highly weathered and becomes sticky and plastic when wet.

Gun flint A square blade-segment of flint used to ignite the powder charge of a flint-lock gun. Often mistaken for an aboriginal artifact

Gyttja A kind of deposit consisting largely of only partially decomposed organic matter.

H

H endoaperture An elaborate endoaperture, consisting of a central part which connects two lateral, longitudinal elongations, forming an "H" shape. Examples: *Cornus* (Cornacaea), *Centaurium* (Gentianaceae).

Haab One of the three Maya calendars, and the one which corresponds most closely to ours in length. The Haab is also known as the "Vague Year" by archaeologists, since it is 365 days in length, or about a quarter day short of the actual solar year.

Habitat The place or type of site where a plant or animal naturally or normally lives and grows.

Habitation area A generalized term for a house or tent floor, or the remains of any other type of aboriginal shelter.

Habitation site An archaeological site which served as a residence. Evidence pointing toward this function might include remains of dwelling structures and cooking areas.

Habitus As defined by Bourdieu, a culturally specific way not only of doing and speaking, but also of seeing, thinking and categorising. Habitus tends to be"naturalized" in that it is taken for granted or assimilated into the unconscious so that habitus is a necessary condition of action and shared understanding.

Hadean The earliest subdivision of the Precambrian, spanning the time between the formation of the Earth, about 4.5 billion years ago, and the start of the Archaean era, 3.8 billion years ago. This interval predates the period of true geologic time since no rocks of this age are known on Earth, with the exception of a few meteorites.

Hades The underworld and also the name of the Greek god of the underworld. The Romans called the god Hades, Pluto.

Hadith A collection of sayings and acts of Muhammad and the first Muslims. They are the record of the Prophet's Sunna, which is second only to the Quran in authority for Muslims.

Hadrian Roman emperor from 117 to 138 CE.

Haeckel, Ernst A German biologist who lived from 1834-1919, Haeckel was the first to divide animals into protozoan (unicellular) and metazoan (multicellular) forms. His notion of recapitulation is no longer accepted.

Haft The handle of a knife, the shaft of a spear, etc.; to equip with a haft.

Hafting The process of equipping a blade (sense 2) with a handle; the handle itself together with the bindings.

Haile selassie, Yohannes A paleoanthropologist who, while doing field work in Ethiopia for his doctoral dissertation at the University of California, Berkeley, discovered *Ardipithecus ramidus kadabba*, a bipedal hominid dated at 5.2 million years old.

Half life The amount of time for half of the atoms of a radioactive element in a sample to decay, or fission. The reduction in the number of atoms follows a geometric scale—at the end of 2 half-lives, there are 1/4 of the atoms left; at the end of 3 half-lives, there are 1/8 remaining; etc.

Halo A clear zone around a well defined feature such as a spine or an aperture. Examples: *Ranunculus acris* (Ranunculaceae), *Valeriana* (Valerianaceae).

Hamilton, W.D. A naturalist, explorer, and zoologist who worked in the world of mathematical models, including "Hamilton's Rule," about the spread through a population of a gene for altruistic self sacrifice. He was also interested in the evolutionary impact of parasites as the key to many outstanding problems left by Darwin, including the baffling riddle of the evolution of sex. This led him to extensive work in computer simulations.

Hammerstone A rounded cobble, sometimes equipped with a groove to facilitate hafting. Signs of use may include pecking facets or battering at the working end.

Hamulate Describing a form of rugulate ornamentation consisting of irregularly arranged, winding, or angular rounded muri of varying thickness, which do not form a distinct reticulum, but rather a maze-like pattern. Example: *Lycopodiella inundata* (Lycopodiaceae).

Mostly used in spore terminology.

Hand axe A core tool flaked on both sides, generally fist-size, and commonly manufactured in the later portion of the Lower Paleolithic and Middle Paleolithic periods.

Hand level A small, simple, hand-held surveying instrument for establishing horizontal lines-of-sight over short distances.

Hand-maul A carefully manufactured unhafted stone hammer.

Hanging gardens of babylon Huge step pyramids covered with lush vegetation, built about 600 B.C. by King Nebuchadnezzar for his wife in Mesopotamia, which is now modern Iraq. One of the Seven Wonders of the Ancient World, the Hanging Gardens no longer exist.

Hanna A corner-notched, expanding-stemmed, concave-based projectile point style of the Middle Prehistoric Period in the Northern Plains. Hanna points are included within the McKean Complex (Wheeler 1954).

Haploid The condition of having only one set of genes or chromosomes. In normally diploid organisms such as humans, only the gametes are haploid.

Haplotype A set of genes at more than one locus inherited by an individual from one of its parents. It is the multi-locus analog of an allele.

Haploxylon type Bisaccate pollen in which the outline of the sacci in polar view is more or less continuous with the outline of the corpus, so that the grains appear a more or less smooth ellipsoidal form. Examples: *Pinus cembra*, *Picea* (Pinaceae).

Haploxylonoid Describing bisaccate pollen in which the outline of the sacci in polar view is more or less continuous with the outline of the corpus, so that the grains appear a more or less smooth ellipsoidal form.

Hardness The quality of being resistant to scratching or deformation. Hardness is considered to be a diagnostic attribute of pottery and is measured by means of the Moh Scale. Aboriginal pottery in Manitoba ranges from 2 to approximately 3.5 on this scale of 10.

Hardy Weinberg principle In population genetics, the idea that if a population experienced no selection, no mutation, no migration, no genetic drift, and random mating, then the frequency of each allele and the frequencies of genotype in the population would remain the same from one generation to the next.

Hardy Weinberg ratio The ratio of genotype frequencies that evolve when mating is random and neither selection nor drift are operating. For two alleles (A and a) with frequencies p and q, there are three genotypes: AA, Aa, and aa. The Hardy-Weinberg ratio for the three is: p^2AA : 2pqAa : q^2aa. It is the starting point for much of the theory of population genetics.

Harem A subunit of a larger social group consisting of a male associated with two or more females.

Harmomegathy The process by which pollen grains and spores change in shape to accommodate variations in the volume of the cytoplasm caused by changing hydration.

Harpoon A spear with a (frequently) barbed, detachable head. Upon striking the prey, the main shaft falls away and the harpoon head, with a line attached, remains in the animal.

Harpoon head The arming tip of a harpoon. generally classifiable into 2 main forms - toggling and barbed - each of which may be composite or single-piece, and may or may not carry additional cutting-blades or side-blades. Always have line-guards or other means of line attachment.

Harvey, Ralph A geologist whose work includes the study of geological processes at a range of scales, from the smallest nanometer to broader-scale interpretations of the history experienced by geological materials.

Hasp A fastening device.

Hatshepsut Queen of New Kingdom Egypt from 1479 to 1458 B.C., first alongside her half-brother, Thutmosis II (who became her husband), and then alongside her nephew, Thutmosis III. A great builder and leader, she is often thought of as Egypt's greatest woman ruler.

Hauterivian European stage of the Lower Cretaceous, spanning the time between 122 and 118 million years ago.

Hawkins, Benjamin W. The first dinosaur models were made by Benjamin Waterhouse Hawkins of England in 1854. He made and sold plaster-cast dinosaurs through the Ward's catalogue of scientific supplies. His original models included Igauanodon, Hylaeosaurus, Megalosaurus, Plesiosaurus and Ichthyosurus. The first dinosaur used for adult amusement was a life-size model of an Iguanodon (made of concrete by Hawkins) that was used to house a dinner party for scientists (including Richard Owen, who coined the term dinosaur) at a major exhibition in London, England, in 1854. The invitations to the party were sent on fake pterodactyl wings.

Hearth A fireplace; an open area within which a fire has been deliberately kindled for cooking, light or warmth. Such a feature may be defined on the basis of ash, charcoal, blackened earth, an encircling ring of cobbles, fire-broken rock, burned bones or a baked clay floor.

Heat treatment An aboriginal process by which the flaking properties of a rock were improved by controlled heating in a fire.

Heavy metals Metals with a high relative atomic mass, such as lead, copper, zinc and mercury. Many of them are toxic.

Hedjet A white crown. This was the crown of Upper Egypt (southern).

Heel toe stride A method of progression characteristic of humans in which the heel strikes the ground first and the person pushes off on the big toe.

Hegemony Preponderant influence or authority of one individual or social group over another.

Heliocentric A sun-centered model of the universe.

Hell gap The name of a valley in eastern wyoming which has given its name to a plano projectile point style as well as to the complex of which it is a part. In some respects, the hell gap projectile point is similar to the agate basin type and the close relationship between the two complexes has often been commented upon. The former style, however, exhibits a base which is constricted to such a degree as to render the point a stemmed appearance. Lateral grinding is heavy along the "stem". The length of these specimens ranges from 5 to 16 cm. Flaking may be parallel or more irregular. Other items of the complex include a variety of side- and end-scrapers, a number of knife styles including a single-shouldered variety, burins, spokeshaves, gravers, perforators and denticulates. Hell gap sites are most common in the northern plains and adjacent regions. Where preservation is good, the fauna found in association is predominantly bison of now-extinct species. At one site, it was suggested that natural features such as parabolic sand dunes were used to help in the entrapment of a number of animals at once. Radiocarbon dates range from 8300 to 7600 b.c. (frison 1978). A few hell gap points have been found in manitoba but unfortunately, all are surface finds.

Hellenistic period Historic period that begins after the death of Alexander the Great. His empire was broken up and ruled by dynastics mostly descended from his generals and were strongly influenced by Greek culture.

Hematite A red to black oxide of iron having variable hardness and commonly used as a point by Native North Americans. Hematite is also known as red ochre, particularly in its softer forms or after having been

Hemoglobin A protein that carries oxygen from the lungs throughout the body.

Hendrickson, Sue Sue Hendrickson (December 2, 1949 -) is a self-taught fossil hunter (specializing in fossil inclusions in amber), marine archaeologist, adventurer and explorer. In South Dakota in 1990, Hendrickson found the remarkable T. rex fossil that is now known as Sue. This T. rex fossil is the largest and most complete T. rex found to date. Sue (the fossil) is now displayed at the Field Museum in Chicago, Illinois, USA.

Hennig's auxiliary principle According to this principle a shared derived similarity of a group of organisms has to be regarded as their synapomorphy (thus as a homology), unless the paraphyly of the group can be demonstrated with other characters, that are conflicting to the putative synapomorphy but regarded as stronger evidence by quantity or quality. In cases of structural similarity of a sufficient degree, homology must be assumed, while convergence should never been assumed a priori, but only be postulated on the basis of the total evidence of all available characters (a posteriori). Basically this principle can be regarded as a special case of the principle of parsimony.

Hephaistos Greek god of fire and blacksmithing, known as Vulcan by the Romans.

Hera Wife of Zeus and Greek goddess of family and marriage. The Romans called her Juno.

Heraea games The female version of the Greek Olympic Games. The games honored the goddess Hera, wife of Zeus, and featured foot races for three different age groups along a 525-foot track. The women ran in a tiny tunic (a chiton), which was cut to expose one breast. Like men in the Olympic Games, winners of the Heraea Games received trophies.

Herakles Greek mythological hero, son of the god Zeus and the mortal Alkmene. Known for courage and great strength. Called Hercules by the Romans.

Herbivore An animal that only eats plants.

Herbivores The collective name for animals which feed on grass and other plant life. During the Jurassic period the mighty Sauropods were among the largest herbivores to walk the earth.

Herculaneum Less famous than Pompeii but much better preserved, Herculaneum was also buried by the A.D. 79 eruption of Mt. Vesuvius. The city lies two miles northwest of Pompeii and was home to about 5,000 people. The city has been difficult for excavators to uncover, because it lies under a substance harder than concrete and because much of the modern Italian city of Ercolano was built right on top of it.

Hercules The legendary hero of Greek and Roman mythology forced to perform Twelve Labors in order to make the Gods happy and pay for the crime of killing his wife and children. Hercules is one of the most commonly shown figures in ancient art. Called Herakles by the Greeks.

Heredity The process by which characteristics are passed from one generation to the next.

Heritability Broadly, the proportion of variation (more strictly, variance) in a phenotypic character in a population that is due to individual differences in genotypes. Narrowly, it is defined as the proportion of variation (more strictly, variance) in a phenotypic character in a population that is due to individual genetic differences that will be inherited in the offspring.

Heritable Partly or wholly determined by genes; capable of being passed from an individual to its offspring.

Herm Sculpted stone pillar topped with a head.

Hermeneutics Principles of interpretation; the term is often used with reference to the study of Jewish and Christian scriptures.

Hermes Greek god who was herald and messenger of the other gods. Known as Mercury to the Romans.

Herrnstein, Richard J. A professor of psychology and an author of notable books on intelligence and crime. He has primarily done research on human and animal motivational and learning processes. His books include *Psychology and I. Q. in the Meritocracy*, and he coauthored (with Charles Murray) *The Bell Curve: Intelligence and Class Structure in American Life* (1994).

Hesperides Nymphs who guarded the golden apples, the stealing of which was the last of Hercules labors.

Heterobathmy The simple fact that plesiomorphic and apomorphic character states occur in different number and combinations in different groups, because of their different age of origin, was called the "heterobathmy of characters" by willi hennig. Ernst mayr called the same phenomenon "mosaic evolution", a term that is frequently misunderstood. It is only because of the heterobathmy of characters that there is the possibility to reconstruct the past phylogeny with the present pattern of characters. This heterobathmy is caused by an evolutionary process that has been called "additive typogenesis": derived characters did not originate all at the same time but successively one after the other, so that groups that originated from successive splitting events will of course have a different set of these characters (fewer apomorphies if they originated early and more apomorphies if they originated later).

Heterobrochate Describing a reticulum with brochi of different sizes. Examples: *Adoxa Moschatelina* (Adoxaceae), *Forsythia europaea* (Oleaceae).

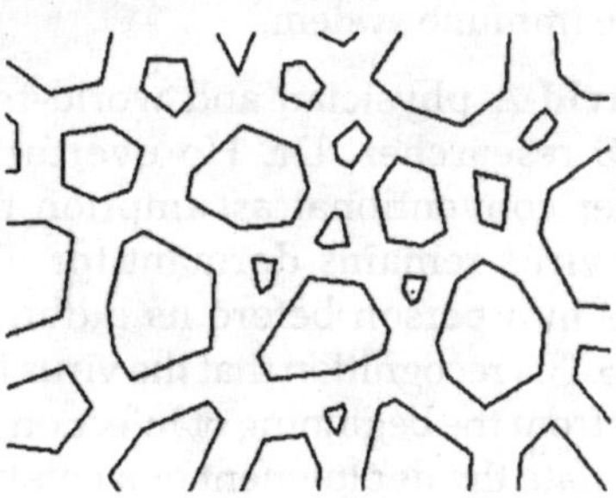

Fig. Heterobrochate

Heterocolpate Describing pollen grains with both simple and compound colpi present. Examples: *Lythrum*, *Peplis* (Lythraceae), *Myosotis* Boraginaceae). Heterogametic The sex with two different sex chromosomes (males in mammals, because they are XY). Compare with homogametic.

Heteropolar Describing pollen or spores in which the distal and proximal faces of the exine are different, either in shape, ornamentation or apertural system. Example: *Echium vulgare* (Boraginaceae).

Antonym: isopolar.

Heterosporous Describing plants producing both microspores and megaspores
Antonym: homosporous.

Heterotrophic Comprise organisms that are not self-sustaining; that is, they derive energy from the oxidation of organic compounds either by consumption or absorption of other organisms.

Heterozygosity (for most purposes) The proportion of individuals in a population that are heterozygotes.

Heterozygote An individual having two different alleles at a genetic locus. Compare with homozygote.

Heterozygote advantage A condition in which the fitness of a heterozygote is higher than the fitness of either homozygote.

Heterozygous Having two different alleles for a particular trait.

Hewn Wood shaped by heavy cutting or chopping blows struck by hand tools such as axes or adzes.

Hieratic Cursive Egyptian script derived from hieroglyphs. Word comes from Greek hieratikos (priestly).

Hieroglyphs The earliest Egyptian script, introduced about 3000 B.C. and used until about A.D. 4. It was used for important inscriptions, although it was also often painted or written on papyrus rather that carved.

High altitude mountain sickness A condition that includes shortness of breath, physical and mental fatigue, rapid pulse rate, headaches; occurs in persons not acclimatized to high altitudes.

High priest The head of the local priesthood.

Higher taxa Taxa above the species level, such as family, order, class, phylum, and kingdom

Hill, Andrew A paleontologist and professor at Yale University. His work with Mary Leakey's team at Laeotoli, Tanzania, in the 1970s helped lead to the discovery of the fossilized footprints of early hominids and other mammals. His current research interests include hominid evolution, paleoecology and taphonomy.

Hilum Circular, indistinctly delimited, irregular aperture or thinning in spores. Examples: *Aequitriradites verrucosus*, *Couperisporites tabulatus*.

This kind of feature is present in certain bryophytes and fungi and may occur on the proximal or distal face.

Hinduism The third largest religion in the world (behind Christianity and Islam).

Hinduism is native to and predominantly practiced in India. Followers of the Hindu religion are known as Hindus. Hinduism is a diverse body of religion, philosophy, and cultural practice. Unlike Christianity or Islam, Hinduism has no founder. Hindus believe in a supreme soul called Brahman, who has no real shape or form. Different aspects or characteristics of Brahman are represented by the gods and goddesses of Hinduism. Hinduism's three main gods are Brahma, the creator; Vishnu, the protector; and Shiva, the destroyer.

Hinge feature A kind of cleavage produced by a flake which does not run its full length. The point of termination is abrupt, and the flake is rounded on its ventral face and sharp on its dorsal face at the distal end.

Historic archaeology The archaeology of a people for whom there are written records. In North America therefore historic archaeology is the archaeology of people after European contact.

Historical linguistics The study of how languages change over time.

Historical particularism A detailed descriptive approach to anthropology associated with Franz Boas and his students, and designed as an alternative to the broad generalizing approach favored by anthropologists such as Morgan and Tylor.

Historical record Artifacts, particularly documents and photographs, retained within an adaptive system.

Historiographic approach A form of explanation based primarily on traditional descriptive historical frameworks.

Hitchcock, Edward B. Edward B. Hitchcock (1793-1864) was a US clergyman and geologist who found the first large dinosaur trackways (in Connecticut, USA). Hitchcock collected over 20,000 dinosaur fossil footprints; he thought that the trackways had been made by huge, extinct birds (and was essentially correct).

HIV (human immunodeficiency virus) The virus causes AIDS by inactivating the T cells of the immune system.

Ho, David A physician and world-renowed AIDS researcher. Dr. Ho overturned an earlier conventional assumption that the HIV virus remains dormant for up to 10 years in a person before its outbreak into AIDS. His recognition that the virus is active right from the beginning of infection led him to initiate the deployment of a combination of drugs to overpower the virus.

Hoards Deliberately buried groups of valuables or prized possessions, often in times of conflict or war, and which, for one reason or another, have not been reclaimed. Metal hoards are a primary source of evidence for the European Bronze Age.

Holism The philosophical view that no complex entity can be considered to be only the sum of its parts; as a principle of anthropology, the assumption that any given aspect of human life is to be studied with an eye to its relation to other aspects of human life.

Holmes, Arthur Arthur Holmes (1890 - 1965) was a British geologist who first proposed the idea of a geologic time scale in 1913. This was soon after the discovery of radioactivity, and using radio-isotope dating, Holmes estimated that the Earth was about 4 billion years old - this was much greater than previously believed.

Holocene (recent) epoch Last and still continuing portion of the Quaternary Period. Characterized by wholly modern fossils (all fossils are of modern organisms). Began 10,000 radiocarbon years BP with the end of the last major glaciation.

Holocene The most recent epoch of the Quaternary; that period of time since the end of the Pleistocene or "Ice Age" (10,000 or

12,000 years ago) until the present. According to some, the Holocene is not a separate epoch at all — it is simply a brief warm spell within the Pleistocene. The Holocene is also known as the Recent Epoch.

Holomorph The entire set of all morphological, physiological, chemical / molecular, and ethological features of a studied specimen, has been termed the holomorph of this semaphoront by willi hennig. The "holomorph of a biospecies" (new term!) includes the holomorphs of all its semaphoronts, thus all inheritable features of all ontogenetic instars of both sexes, and in case all morphs (e.g. castes) and all different generations of a heterophasic cycle of generations, including all populations of all generations that existed between the origin and the end of this species.

Holotype Specimen that serves as the standard bearer of a species or subspecies name.

Holtz, Thomas Thomas R. Holtz, Jr. (1965-) is an American vertebrate paleontologist and author. Holtz is a lecturer at the Department of Geology, University of Maryland, College Park. He was born in Los Angeles and spent his early childhood near Houston, Texas. He was an undergraduate at the Johns Hopkins University, and received his Ph. D. at Yale University. Holtz's main research is on the evolution, anatomy, and ecology of theropod dinosaurs (especially tyrannosaurids). Holtz is the coauthor of The World of Dinosaurs (with Michael Brett-Surman, illus. by James Gurney - Greenwich Workshop Books, 1998), contributor to The Complete Dinosaur [Dinosaur hunters of the Southern Continents and other sections] (ed. James O. Farlow and Michael Brett-Surman - Indiana University Press, 1997), coauthor of the Dinosaur Field Guide (with Michael Brett-Surman - Random House, 2001), author of the Little Giant Book of Dinosaurs (2001, Sterling Press), author of the Tyrannosaur sections of the Univ. of Arizona's Tree of Life classification project on the Web (2000), and has many other publications.

Homalozoa (informally carpoids) Extinct subphylum (Phylum Enchinodermata) whose members have no trace of radial symmetry. The theca is depressed and asymmetrical. Homalozoa are known from Middle Cambrian to Middle Devonian rocks.

Home base A location to which males and females return in human societies.

Home range The area occupied by an animal or animal group.

Homeobox genes A set of genes that are important in developmental patterns. These establish segments in an embryo that may become specific organs or tissue types. In general, "homeotic" genes are genes that control the development of organisms, and "homeogenes" or "homeobox genes" are the subset of homeotic genes that contain "homeoboxes". "Hox" genes are a subset of homeogenes that determine positional cell differentiation and development. Mutations in Hox genes result in the conversion of one body part into another: for example, in the fruit fly *Drosophila*, a specific Hox mutation results in a leg developing where an antenna would normally be.

Homeobox Homeoboxes are relatively short, very similar or identical sequences of DNA, characteristic of homeotic genes (which play a central role in controlling body development) and shared by almost all eukaryotic species. Homeoboxes encode a protein "homeodomain", a protein domain that binds to DNA. The DNA-binding homeodomain consists of approximately 60 amino acids, and these homeodomain motifs are involved in orchestrating the development of a wide range of organisms.

Homeostasis (developmental) A self-regulating process in development, such that the organism grows up to have much the same form independent of the external influences it experiences while growing up.

Homeotic mutation A mutation causing one structure of an organism to grow in the place appropriate to another. For example, in the mutation called "antennapedia" in the fruit fly, a foot grows in the antennal socket.

Homer Greek poet who composed the Iliad and the Odyssey.

Hominid A species of the family Hominidae. Humans are the only hominids today, however, there were others in the past 4-5 million years.

Hominid human Part of the family of Hominidae, which includes both extinct and modern forms of humans.

Hominidae Family of the superfamily Hominoidea that includes humans.

Hominids Members of the family Hominidae, which includes only modern humans and their ancestors since the human lineage split from the apes.

Hominoid A member of the superfamily Hominoidea, which includes apes and humans.

Hominoidea Superfamily of the suborder Anthropoidea that includes the apes and humans.

Homo erectus Meaning "upright man." This Early Stone Age hunter colonized new habitats throughout Africa, Europe, and southern Asia. Bigger and brainier than Homo Habilis or Australopithicus, Homo Erectus had the thickest skull of any member of the human species. Its strong muscles joining its neck to the rear skull bump stopped its heavy head from sagging forward. Average height was between 5 and 6 ft., and weight, 88 to 160 lbs.

Homo habilis Meaning "handy man." This was the first known species of our genus, Homo. Homo Habilis had a larger brain, but a smaller face and jaw and a more rounded head, than Australopithicus. Homo Habilis was about 5 feet tall, weighed about 110 pounds, and had a brow ridge, flat nose, and projecting jaw. This species lived about 1.52 million years ago, perhaps longer. Its hand and foot bones suggest that it was bipedal (walked upright) and had a strong yet sensitive grip.

Homo neanderthalensis A species of hominid that lived between 150,000 and 30,000 years ago in Europe and Western Asia, originally thought to be a geographic variant of *Homo sapiens* but now generally accepted to be a distinct species.

Homo sapiens The biological class to which all living human beings belong as well as their immediate ancestors including the "Cro-Magnons". Some include the Neanderthals within this species also.

Homobrochate Describing a reticulum with brochi of the same sizes. Example: *Armeria maritima* (Plumbaginaceae).

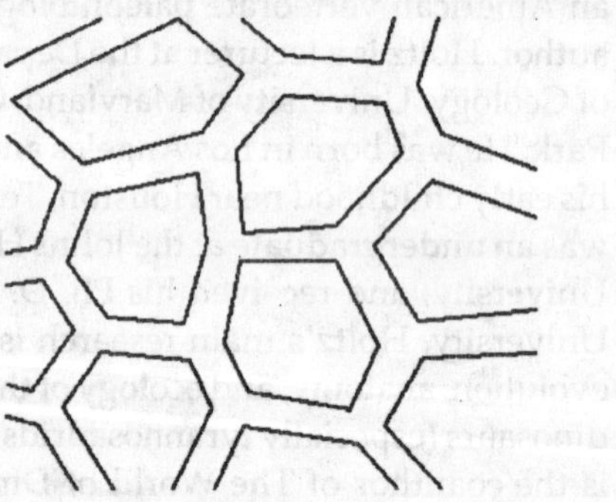

Fig. Homobrochate

Homogametic The sex with two of the same kind of chromosomes (females in mammals, because they are XX). Compare with heterogametic.

Homologous structures The structures shared by a set of related species because they have been inherited, with or without modification, from their common ancestor. For example, the bones that support a bat's wing are similar to those of a human arm.

Homology A character shared by a set of species and present in their common ancestor. Compare with analogy. (Some molecular biologists, when comparing two sequences, call the corresponding sites "homologous" if they have the same nucleotide, regardless of whether the similarity is evolutionarily shared from a common ancestor or convergent. They likewise talk about percent homology between the two sequences. Homology in this context simply means similarity. This usage is frowned upon by many evolutionary biologists, but is established in much of the molecular literature.).

Homonomy (= serial homology) Homonomous organs are organs with an identical or at least similar construction ("Bauplan") within a single organism, e.g. the fore leg and hind leg of tetrapods, or the segments of annelid worms. This phenomenon is often termed "serial homology" to indicate that the similarity might be based on a common genetic origin. In molecular biology there is a similar phenomenon because of the frequent existence of multiple copies of genes in a string of DNA. These copies are so to speak "homonomous genes", but have be called paralogous genes, while homologous (s. str.) genes in different organisms have been called orthologous genes.

Homoplasy 1. A character state present in two or more species of organisms that is not present in their common ancestor

2. In cladistics, a character present in at least two clades that is absent in the common ancestor of the two clades.

Homosporous Describing plants producing only one kind of spores

Synonym of isosporous.

Antonym: heterosporous.

Homozygote An individual having two copies of the same allele at a genetic locus. Also sometimes applied to larger genetic entities, such as a whole chromosome; a homozygote is then an individual having two copies of the same chromosome.

Homozygous Having identical alleles for a particular trait.

Hopewell A Middle Woodland Period culture (or cult) which occupied (or "influenced") much of eastern North America. Closely related to Adena, this Ohio/Illinois-centered "complex" consisted of log tombs within burial mounds, grit-tempered utilitarian pottery, effigy vessels for inclusion with the dead, elaborate ceramic figurines, stone platform pipes which often incorporated human and animal likenesses, sheet copper ornaments, earspools, finely made ceremonial knives of obsidian and effigy mounds – the latter in the shape of panthers, bears, birds, turtles, and other animals. Hopewell artifacts which are believed to have been ceremonial in function are very similar over vast areas while utilitarian objects vary regionally. This has led some to believe that Hopewell is simply a religion, cult or belief system shared by a number of groups with different languages and subsistence modes. The sheer ambitiousness in some of the Hopewell earthworks and the fineness of their artworks have suggested to some the ranked society, division of labor, and occupational specialization usually associated with farming societies. Direct evidence, however, is not overly convincing. It seems more probable that Hopewell people made their living by hunting, fishing and gathering a wide variety of resources within a rich environment.

Hoplite A citizen of a Greek city state rich enough to afford body armor and required to serve in a time of crisis.

Horizon 1. In pedology, a horizontal layer within a soil profile which is chemically, physically and/or biologically distinct from other layers

2. In reference to a specific site, a certain cultural level; for example, the agate basin horizon of the hell gap site.

3. An artifact class or style, an art style or other distinguishing cultural trait(s) which occurs over a broad area in a relatively brief period of time. The presence of such objects or styles (horizon "markers") are thus useful as a means of dating a site.

Horizontal angle In mapping, the angle of sight measured on the level or horizontal plane.

Horizontal circle With major surveying instruments, the graduated horizontal table around which the sighting telescope revolves; used to measure the horizontal angle.

Horizontal datum A base measuring point ("0.0 point") used as the origin of rectangular coordinate systems for mapping or for maintaining excavation provenience.

Horizontal distance The measurement of distance on a true level plane.

Horizontal migration A nomadic pattern characterized by regular movement over a large area in search of grass; also called plains migration.

Horizontal provinience The location of an object on a two-dimensional plane surface.

Horn An elongated part of an endoaperture, which is curved towards one of the poles. Example: *Anagallis arvensis* (Primulaceae).

Horner, John R. John R. (Jack) Horner is a US paleontologist (born on June 15, 1946 in Shelby, Montana) who named: Maiasaura (with Makela, 1979) and Orodromeus (with D.B. Weishampel, 1988). Horner discovered the first egg clutches in the Americas (Maiasaura) and the first evidence of parental care from dinosaurs (also Maiasaura). Horner is the Curator of Paleontology at the Museum of the Rockies in Bozeman, Montana, USA. Horner attended the University of Montana, majoring in geology and zoology. Horner is the author of: "Digging Dinosaurs" (Workman Pub., 1988), "Complete T-rex" (with D, Lessem, Simon and Schuster,1993), "Maia, A Dinosaur Grows Up" (Museum of the Rockies, Montana State University, 1985), "Digging Up Tyrannosaurus rex" (with D, Lessem, 1992), "Dinosaur Eggs and Babies" (Cambridge University Press, 1994), and "Dinosaur Lives" (HarperCollins, 1997). Horner was a technical advisor for the movies Jurassic Park and The Lost World.

Horner An archaeological site near the town of cody in northwestern wyoming which produced eden and scottsbluff projectile points and cody knives. The site, radiocarbon dated between 7500 and 6500 b.c., is of significance because it appears to represent the communal hunting and kill of approximately 200 bison. As used by irwin (1971), the term also refers to a phase which he dates between 7500 and 6500 b.c. and within which he includes the cody complex as well as alberta points.

Horsetail A seedless plant related to ferns. Twenty-five species of only one genus, *Equisteum*, remain today, whereas many different species, some the size of modern trees, were abundant in ancient swamps. Along with lycophytes and ferns, horsetails were among the first terrestrial plants to appear.

Horticulture The science and art of growing fruit, flowers, ornamental plants, and vegetables in small gardens.

Horus Egyptian god associated with the living pharaoh and often portrayed with the head of a falcon, son of Osiris and Isis, opponent of the god Seth.

Horus name A Pharaoh's name. It identifies the pharaoh with a form of the god Horus.

House pit An aboriginally excavated house floor.

Household A domestic residential group whose members live together in intimate contact, rear children, share the proceeds of labor and other resources held in common, and in general cooperate on a day-to-day basis.

HOX genes (hox cluster) Genes (transcription factors) paramount in development and found to be highly conserved in evolution. Occuring in clusters they act as master switches for other genes in that express and time developmental processes. The genomes of all animals that have been sequenced have at least one Hox cluster that show strong homology to the genes in Drosophila (the friut fly). For example, mice and humans have 4 Hox clusters (a total of 39 genes in humans) located on four different chromosomes.

Humus A dark brown or black soil material consisting of partially decomposed animal and vegetal matter.

Hunter gatherers People or societies that are dependent on wild food resources. Hunter-gatherers are usually from technologically simple societies that are highly mobile. Also sometimes called "foragers."

Huxley, Thomas H. Thomas H. Huxley (1825-1895) was a British scientist and friend of Charles Darwi He was the first scientist to notice the similarity between birds and dinosaurs. He named: Acanthopholis (1865), the family Archaeopteryglidae (1871), Euskelosaurus (1866), Hypsilophodon (1869), and the family Megalosauridae (1869).

Huxley, Thomas Henry British intellect, photographer, and contemporary of Darwin. He was the first to apply the theory of natural selection to humanity to explain the course of human evolution.

Hybrid The offspring of a cross between two species.

Hydration Formation of a bond between water and some material.

Hydria A pot used to bring water to the table. It had two horizontal handles for carrying and a vertical handle for pouring.

Hydrogeology The science that deals with subsurface waters and geologi aspects of surface waters.

Hydrophyte A damp-loving plant which grows only under water or in very moist soil.

Hydrothermal vent A volcanic opening on the ocean floor, often located along the mid-ocean ridges. These vents are known to support rich marine organisms, independent of light.

Hyksos People from western Asia (Palestine) who conquered the Nile Delta and ruled Egypt during the Second Intermediate Period. Their expulsion by the Pharaoh Amose may have given rise to the Exodus story. The name Amose in Hebrew is Moses.

Hypostyle hall From the Greek word meaning; "bearing pillars". It is a term used to describe the grand, outermost halls of Egyptian temples. They are believed to represent a grove of trees.

Hypothesis An explanation of one or more phenomena in nature that can be tested by observations, experiments, or both. In order to be considered scientific, a hypothesis must be falsifiable, which means that it can be proven to be incorrect.

Hypothetico-deductive explanation A form of explanation based on the formulation of hypotheses and the establishment from them by deduction of consequences which can then be tested against the archaeological data.

Hypoxia Low oxygen pressure due to being at high altitude.

I

Iapetus ocean A relatively small ocean that existed between the continents of Laurentia, Baltica, and Avalonia from the Late Precambrian to the Devonia.

Ibu The tent of purification. This is the place where Egyptian mummification was performed.

Ice age A period of many thousands, even millions of years, during which time the Ice Caps extend towards formerly mild/warmer areas of the globe. An Ice Age may follow the dramatic climate change triggered by a large scale impact, such as an asteroid collision.

Ice cores Borings taken from the Arctic and Antarctic polar ice caps, containing layers of compacted ice useful for the reconstruction of paleoenvironments and as a method of absolute dating.

Ice lobes, fans, domes, etc Dynamic features within large ice sheets which reflect glacier accumulation and ablation, ice temperature, bedrock topography, ice topography, marginal seas, and ice flow:

Dome - high, central stable portion of ice sheet with low rates of accumulation and slow ice movement.

Saddle - depression or sag on ice sheet between domes.

Divide - line across central domes or saddles marking iceshed position.

Stream - narrow zone extending from dome toward margin in which concentrated, fast ice movement takes place, often developed over subglacial troughs or valleys. Also called *fleuve de glâce*.

Lobe - marginal extension of ice stream along broad bedrock trough or valley, may be 100s of km long.

Tongue - small marginal extension of ice sheet into narrow valleys, generally few km long.

Fan - coalescence of thin, marginal, ice lobes on land.

Shelf - portion of ice margin floating on sea or lake, ablation by calving of icebergs into water.

Ice man The name given to the 5,300-year-old body of a man found preserved in a glacier on the border between Italy and Austria in 1991.

Ice sheets Ice sheets or glaciers. covered 30 percent of the world, including much of North America, during the Ice Age.

Ice wedge A vertical wedge-shaped vein of ground ice found in permafrost areas. causes "polygonal ground" and may result in severe disturbance of archaeological sites.

Ice-creeper An Eskimo invention consisting of a spiked piece of ivory or bone which is attached to the soles of mukluks to facilitate movement over the ice.

Ichnocoenose An assemblage of trace fossils, such as an area containing dinosaur trackways.

Ichnofossils Trace fossils that are seen as preserved tracks or other signs of the behaviors of animals in the substrate. Ichnofossils can provide insights on the behavior of an extinct animal. Very rarely is the animal itself found in direct association with the ichnofossil it created.

Ichthyosaurus A group of marine predators which lived between the Triassic and Cretaceous period. These diverse reptiles were shaped like a dolphin and characterised by a long jaw, containing razor sharp teeth and large eyes.

Iconography An important component of cognitive archaeology, this involves the study of artistic representations which usually have an overt religious or ceremonial significance; e.g. individual deities may be distinguished, each with a special characteristic, such as corn with the corn god, or the sun with a sun goddess etc.

Idealism The philosophical theory that there are fundamental non-material "ideas," "plans," or "forms" underlying the phenomena we observe in nature. It has been historically influential in classification.

Idealist explanation A form of explanation that lays great stress on the search for insights into the historical circumstances leading up to the event under study in terms primarily of the ideas and motives of the individuals involved.

Idealize To present something as more perfect than it is.

Ideo-function The role an artifact has in a society's ideology.

Ideology A society's beliefs, attitudes, values, knowledge, and information.

Ieb This is the heart. The Egyptians believed the heart was the center of all consciousness, even the center of life itself. When someone died it was said that their "heart had departed." It was the only organ that was not removed from the body during mummification. In the Book of the dead, it was the heart that was weighed against the feather of Maat to see if an individual was worthy of joining Osiris in the afterlife.

Igloo An often dome-shaped Eskimo structure built by the piling of blocks of snow and intended as a shelter for either people or animals.

Igneous Formed from molten lava which has hardened on or below the surface of the earth.

Igneous rocks Rocks formed from igneous processes.

Imam There are many usages of this word. Generally and in lowercase, it refers to the leader of congregational prayers; it implies no ordination or special spiritual powers. For many Sunni Muslims it is a figurative term for the leader of the Islamic community. Among Shii Muslims the word has many complex meanings; generally it indicates that particular descendant of the House of Ali ibn Abu Talib, who is believed to have been God's designated repository of the spiritual authority inherent in that line. The identity of this individual and the means of determining his identity has caused divisions among Shii Muslims.

Immigration The movement of organisms into an area.

Immutability The ability to withstand change.

Impacts The changes that archaeological resources undergo as the result of some action.

Impression mark A mark on the proximal face of a pollen grain retained from the post-meiotic stage. This mark can be linear from tetragonal tetrads or Y-shaped from tetrahedral tetrads. Examples: *Nypa fruticans*, *Howea belmooreana* (Palmae).

In A prefix used to emphasis the absence of a feature, as for example in inaperturate.

The term should be used where apertures are completely absent. If ectoapertures are absent, but endoapertures are present, the pollen is cryptoaperturate.

In situ A latin phrase meaning literally "in place". In archaeology, it refers to an artifact or other object found in its original position.

Inca A culture that thrived in the 15th century A.D. Inca culture was centered in the town of Cuzco, in what is now Peru, although the Incas controlled a territory spanning 2,600 miles, from what is today the southern border of Colombia to central Chile. Inca is commonly spelled INKA.

Incensario A Mesoamerican incense burner.

Incest Sexual intercourse between closely related persons.

Incest taboo The prohibition of sexual intimacy between people defined as close relatives.

Incidence A structure formed by the termination of the inner wall layer at the endogerminal which gives rise to a deep V-shaped atrium that points towards the centre of the grain. Example: *Pseudotrudopollis*.

This term is used in descriptions of the fossil Normapolles group.

Incipient Beginning; in an initial stage; not fully developed or completed.

Incised A decoration found on pottery consisting of lines drawn into wet clay. When fired, the arrangement of lines leaves a permanent design on the vessel surface.

Incising Producing lines or patterns by cutting into or engraving a surface. As a means of pottery decoration, incising refers to the freehand etching of narrow deep lines on the vessel surface with a sharp instrument prior to firing.

Inclined sights In mapping, a vertically angled line of sight.

Inclusion An intentional cultural association, such as grave-goods with a burial.

Inclusive anchor taxon Describes the anchor taxon included in the taxon being defined. For example, in the stem-based clade definition of Carnosauria, *Allosaurus* and all taxa sharing a more recent common ancestor with *Allosaurus* than with "birds," *Allosaurus* is the inclusive anchor taxon.

Inclusive fitness An individual's own fitness plus his or her effect on the fitness of any relative.

Increment borer A hand-operated coring device for obtaining tree-ring samples.

Independent family household A single-family unit that resides by itself, apart from relatives or adults of other generations.

Independent variable The variable that can cause change in other variables.

Index A spirit-bubble leveling device on the vertical circle of major surveying instruments.

Index fossil Remains of a plant or animal of a species that is known to have lived only during a specific time period. The discovery of such a fossil in an archaeological site is circumstantial evidence of the approximate time period that it was occupied. Fossil bones of horse and elephant related species are often used to relatively date fossils of our ancestors. Index fossils are used for biostratigraphy.

Indigenous Native to, originating in or occurring naturally in a given place; aboriginal.

Indirect percussion A technique for flaking stone artifacts by interposing a bone or antler punch between the hammer and the raw materials. Allows greater control than direct percussion flaking.

Individualistic cult The least complex form of religious organization in which each person is his or her own religious specialist.

Induction A method of reasoning in which one proceeds by generalization from a series of specific observations so as to derive general conclusions.

Industrial age A cultural stage characterized by the first use of complex machinery, factories, urbanization, and other economic and general social changes from strictly agricultural societies.

Industrial society A society consisting of largely urban populations that engage in manufacturing, commerce, and services.

Industrialism A form of social organization in which the population's needs for food, manufactured products, transportation, and many services are met through the use of machines powered largely by fossil fuel.

Industry All of the artifacts of a single material (bone, stone, ceramic, etc.), made by members of a culture.

Infantile The period in an individual's life cycle from birth to the eruption of the first permanent teeth.

Inference Assumption based on an observation.

Informal interview An unstructured question-and-answer session in which the informant is encouraged to follow his or her own train of thought, wherever it may lead.

Informant A person who provides information about his or her culture to the ethnographic fieldworker.

Infra A prefix meaning below or beneath.

The term is often used for patterns underneath a complete or partial tectum. Examples: infrareticulate, infrarugulate, infrastriate, infrastructure, infratectate.

Infrared absorption spectroscopy A technique used in the characterization of raw materials, it has been particularly useful in distinguishing ambers from different sources: the organic compounds in the amber absorb different wavelengths of infrared radiation passed through them.

Infratectum A general term for the layer beneath the tectum, which may be alveolar, granular, columellar, or structureless.

Synonym of interstitium (Walker and Walker, 1981).

Infraturma An artificial grouping of form-genera of spores and pollen in the turma-system of Potonié.

Inheritance of acquired characters Historically influential but factually erroneous theory that an individual inherits characters that its parents acquired during their lifetimes.

Inhumation The placement of the dead in an excavated pit beneath ground surface.

Iniskim A fossil ammonite whose natural appearance resembles a bison.

Inka A culture that thrived in the 15th century A.D. Inka culture was centered in the town of Cuzco, in what is now Peru, although the Inkas controlled a territory spanning 2,600 miles, from what is today the southern border of Colombia to central Chile. Inka is commonly spelled INCA.

Inner tetrad mark A tetrad mark at the inner layer of a spore and which does not reach up to the margin when seen in polar view and always smaller than the outer tetrad mark.

Innovation The process of adopting a new thing, idea, or behavior pattern into a culture.

Inordinate Describing a pollen grain or spore with an arrangement of elements irregularly distributed.

Antonym: ordinate.

The elements can be of structural or sculptural origin. Examples: columellae under a tectum; scabrae on the tectum.

INQUA International union for quaternary research; iternational scientific organization for study of Quaternary geology, climate, biology, and anthropology, founded in 1928.

Insectivorous Feeding largely or exclusively on insects.

Instinct A genetically-determined pattern of behavior that is characteristic of a species and is often a response to specific internal or environmental stimuli.

Institutions A society's recurrent patterns of activity, such as religion, art, a kinship system, law, and family life.

Instrument A general term for major optical surveying equipment, including transits, alidades, and surveyor's levels.

Instrument height The elevation of the line-of-sight of a surveying instrument above the immediate ground surface.

Instrument position The location at which a surveying instrument is established to obtain a sighting.

Insula Synonym of areola.

Intaglio Design cut into the surface of a hard material.

Intectate Describing pollen grains without a tectum, but with sculpturing. Examples: *Viscum* (Loranthaceae), *Ilex* (Araliaceae).

Intelligent design The non-scientific argument that complex biological structures have been designed by an unidentified supernatural or extraterrestrial intelligence.

Inter A prefix for in between.

In palynology the prefix has been used in many different contexts, from indicating position of apertures to defining areas of the surface.

Interaspidium The area of exine between (aspides. Example: *Dorstenia* (Moraceae).

Intercolpium The meridional segment between two colpi and extending to the poles.

The terms mesocolpium and apocolpium are recommended because they enable a more precise subdivision of the pollen surface.

Interglacial A warm interlude between two glacial periods.

Interlacunar gaps Gaps in the short sexinous ridges which often divide the ectoapertures of lophate pollen grains into lacunae. Example: *Cichorium intybus* (Compositae).

Interlacunar ridges Ridges separating the lacunae in lophate pollen grains. Example: *Cichorium intybus* (Compositae).

Interloculum A space between the inner- and outer wall layers that is present around most or all of the pollen grain. Example: *Extratriporopollenites conjunctus.*

The term was originally used for features in pollen of the Triatrioaperturate category of the fossil Normapolles group.

Intermediate periods Times of confusion in Egyptian history when different parts of the country were ruled simultaneously by different kings.

Internal foramen Foramen present within the **sexineectexine.**

Internal tectum A more or less continuous layer within the outer sexineectexine composed of laterally connected parts of columellae. Example: *Nigella damascena* (Ranunculaceae).

Differs from infratectum and interstitium in being a single layer rather than referring to a number of elements under the tectum.

Interporal lacuna A lacuna occurring between the ends of two ectoapertures and adjacent to the polar area of a lophate pollen grain. Example: *Tragopogon pratensis* (Compositae).

Interporium The meridional segment between two pores extending to the poles.

The terms mesoporium and apoporium are recommended because they enable a more precise subdivision of the pollen surface.

Interstadial A brief, warm period within a glacial period.

Interstitium The layer of the exine situated between the nexine and the tectum.

Synonym of infratectum.

Intexine Synonym of nexine.

Intine The innermost of the major layers of the pollen grain wall underlying the exine and bordering the surface of the cytoplasm.

The intine is not acetolysis resistant and is therefore absent in conventionally prepared palynological material.

Intra A prefix meaning within.

Intrareticulum Synonym of infrareticulum.

Intron The nucleotide sequences of some genes consist of parts that code for amino acids, and other parts that do not code for amino acids interspersed among them. The interspersed non-coding parts, which are not translated, are called introns; the coding parts are called exons.

Intrusive An artifact or feature found within a feature, component or stratum of which it was not originally a part.

Intrusive igneous rock A body of rock that intrudes and melts its way into the crust, and solidifies before reaching the surface - i.e. the Sierra Nevada batholith.

Intrusive object An object that is believed to have entered the body of a person and to be the cause of illness and/or death. Such diagnoses are common among many pre-industrial cultures worldwide. The cure often consists of the removal of the object by a shaman or sorcerer. Sometimes this is accomplished by (apparently) sucking the object through a bone tube (sucking tube) and then displaying the object as proof of the cure.

Inuit 1. An Eskimo word meaning literally "The People".

2. The name by which the North American (and Greenland) Eskimo refer to themselves.

3. The language of these people, more properly termed "Inuktitut".

Inuksuk The Eskimo word for the likeness of a human made by piling stones. These are often erected in a long V-shaped formation bordering a drive lane. Caribou, mistaking the inuksiuk (plural) for people, can then be stampeded towards the narrow end of the V where a corral and/or hunters await them.

Inversion An event (or the product of the event) in which a sequence of nucleotides in the DNA is reversed, or inverted. Sometimes inversions are visible in the structure of the chromosomes.

Invertebrates The name given to creatures without a backbone (vertebrae).

IQ An abbreviation of "intelligence quotient," usually defined as the mental age of an individual (as measured by standardized tests) divided by his or her real age and multiplied by 100. This formulation establishes the average IQ as 100. The usefulness and reliability of IQ as a measure of intelligence has been questioned, in part because of the difficulty of devising standardized tests that are free of cultural biases.

Iron age The third age in Thomsen's three-age system, referring to the period when iron tools were manufactured.

Iron pyrite A metallic, yellow-to-brown sulfide of iron. This widely occurring mineral is also known as fool's gold. The mineral is cubic in crystal form.

Iroquoian A family of languages whose speakers originally occupied much of the St. Lawrence River valley and isolated portions of the east-central United States. The major languages within this family include Seneca,

Cayuga, Onondaga, Mohawk, Oneida, Wyandot (Huron), Cherokee and Tuscarora.

Isis Egyptian goddess venerated as ideal mother, wife of Osiris and mother of Horus.

Islamist One who seeks to make Islam a more prominent part of the political and social order, usually by implementing some version of Islamic law, or Sharia. Often used as a more accurate replacement for the term "fundamentalist."

Islamization The process of Islam's becoming a more prominent part of the political and social order. This may be through adoption of laws based on Islam, a stricter code of behavior for Muslims in a community, or significant portions of a population converting to Islam.

Isodiametric tetrad A general term for tetrads in which all members are more or less the same in size.

Antonym: anisodiametric tetrad.

Isolated find The recovery; usually from the surface, of a single artifact with no other artifacts in association.

Isolating mechanism Any mechanism, such as a difference between species in courtship behavior or breeding season, that results in reproductive isolation between the species.

Isolation Synonym for reproductive isolation.

Isopolar Describing a pollen grain or spore in which the proximal and distal faces of the exine are alike.

Antonym: heteropolar.

Isospore A spore of a plant producing only one kind of spore Isosporous Describing plants producing only one kind of spore

Synonym of homosporous.

Antonym: heterosporous.

Isostasy The rising of a land surface following the removal of the enormous weight of glacial ice. This phenomenon is of particular importance in Manitoba archaeology as the south-to-north tilting of the landscape due to the retreat of the Wisconsinan ice sheet influenced the shoreline of glacial Lake Agassiz and thus the areas which were available for human occupation.

Isotope Form of an element with a certain number of neutrons, for example carbon exists in three isotopes: C^{12}, C^{13}, and C^{14}. Some isotopes are naturally unstable and spontaneously decay at a fixed rate; other isotopes are stable.

Isotopic dating Radiometric dating; all methods of age determination based on nuclear decay of naturally occurring radioactive isotopes. Age in years for geologic materials are calculated by measuring the presence of a short-life radioactive element, e.g. carbon-14, or by measuring the presence of a long-life radioactive element plus its decay product, e.g. potassium-40/argon-40.

Ithyphallic From the Greek word meaning; "with erect penis". Various gods are represented in this form. Most notably Min and Amun.

J

Jablonski, David Paleontologist and professor in the Department of the Geophysical Sciences of the University of Chicago. His research emphasizes combining data from living and fossil organisms to study the origins and fates of lineages and adaptations, to develop an understanding of the underlying dynamics of speciation and extinction that could lead to a general theory of evolutionary novelty. He is interested in the way evolutionary patterns are shaped by the alternation of extinction regimes, with rare but influential mass extinctions driving unexpected evoutionary shifts.

Jade A general term which may refer to either true jade (nephrite) or jadeite. Most prehistoric jade objects in Canada originate in British Columbia but some artifacts particularly adzes were traded into Alberta. It was also a commonly used stone for religious purposes and decoration in Mesoamerica and Asia.

Janensch, Werner Werner Janensch was a German paleontologist and museum curator (the Natural History Museum of Berlin) who led an expedition (with Edwin Hennig) to the Tendaguru Beds in what is now Tanzania, Africa. That expedition found many late Jurassic period dinosaurs, including some Brachiosaurus. Janensch named Dicraeosaurus (1914) and Elaphrosaurus (1920).

Jasper A non-translucent microcrystalline quartz which occurs in a wide variety of colors and which was used for ornaments and chipped stone artifacts in prehistoric North America.

Jean-francois champollion French linguist and Egyptologist who is credited with using the writings of the Rosetta Stone to decipher Egyptian hieroglyphics, from 1808 to 1822.

Jensen, James A. James A. Jensen is a US paleontologist from Brigham Young University who discovered Supersaurus (1972) and Ultrasauros (1979). He named Cathetosaurus (1988), Dystylosaurus (1985), the family Torvosauridae (1985), and Torvosaurus (with P.M. Galton, 1979).

Johanson, Don A paleontologist and founder of the Institute for Human Origins. Johanson discovered Lucy (at that time the oldest, most complete hominid skeleton known) in 1974, and the following year unearthed the fossilized remains of 13 early hominids in Ethiopia. He is the author of several popular books on human origins.

Johnson, Jerry Johnson's research interests focus on the interactions and evolutionary relationships of amphibian and reptilian species of tropical American and Mexican desert ecosystems. Johnson specializes in field research in places such as Yucatan, Jalisco, Zacatecas, and Chiapas, Mexico. He has done research on the biochemical analysis of rattlesnake venom using immunological techniques, snake ecology, and lizard ecology.

Johnston, Victor Professor of biopsychology at New Mexico State University in Las Cruces. His research interests include the evolution of consciousness and perceptions of beauty. He is the author of *Why We Feel: the Science of Human Emotions.*

Jökulhlaup Icelandic term for massive outburst flood of glacial melt water, often associated with a subglacial volcanic eruption.

Judah Region in Palestine from about 922 to 587 BC named after one of the tribes of Israel.

Juncture The linkage or separation of syllables by pauses.

Juniper Small to medium-sized evergreen trees with scaly leaves. Trunks and branches of the trees are often twisted. Various species grow at elevations of 3000 to 8000 feet in southern Arizona.

Juno Roman goddess of marriage and wife of Jupiter. Called Hera by the Greeks.

Jupiter Roman king of the gods. Called Zeus by the Greeks.

Jurassic period Between 206 and 144 million years ago. Period characterized by the development of many new kinds of dinosaurs and the first birds.

Jurassic The middle period of the Mesozoic era, spanning the time between 213 and 145 million years ago. It is named after the Jura Mountains between France and Switzerland, where rocks of this age were first studied.

Juvenile ceramics Generally small and crudely made pottery vessels believed to be the product of a child learning the potter's art.

Juvenile The period in an individual's life cycle that lasts from the eruption of the first to the eruption of the last permanent teeth.

K

K selection Natural selection typically occurring in populations near the carrying capacity (K) of the environment. This tends to result in a species which survives by producing few young with substantial resource investment by the parents, who typically care for the young until they reach maturity —K-selected.

K T boundary The event that marked the end of the Cretaceous and the beginning of the Tertiary period, 65 million years ago. This period of time was characterised by a period of mass extinction, widely accepted to have resulted from dramatic climate change. Among the contributors to this change include a six mile wide asteroid which struck the earth, throwing billions of tones of dust into the atmosphere. The name originates from the German 'K' for Kreide, meaning Chalk and 'T' for Tertiary.

Ka & Ba Although it is difficult to find modern concepts which faithfully convey the Egyptian ideal of what the Ba & Ka were, the Ba is often described as the soul of a man, the Ka as his double.

Kame Hill or terrace deposit of ice-contact, stratified drift laid down in ice-marginal streams or deltas; generally parallel to ice margin.

Kame hills complex Woodland/protohistoric complex centred in the southern indian lake region in northern manitoba. Included in the artifact inventory are side- and corner-notched projectile points, triangular points, a variety of scraper and biface forms, whetstones, adzes, gravers and grooved hammerstones. The bone industry consisted of points, awls and beads and a harpoon head made of antler. It is the ceramics, however, which are the most distinctive and which provided the impetus to define the complex. Large and small pots and plates were fashioned from clay as were bowls, cups and smoking pipes. Although the majority of the pots were of the clearwater lake punctate type, variation in vessel shape and in the combination of decorative elements used, served to distinguish the southern indian lake pots from those from neighbouring areas. Dickson (1980) estimates the complex to date from a.d. 850 to 1750.

Kandile Footed bowl with perforated handles used as a lamp or possibly a storage vessel in the Cycladic civilization of the Greeks.

Kaolin A fine white clay consisting of decayed feldspar used to make porcelain; also known as china clay.

Katun A Mayan word meaning 20 years. In actuality, it is 20 times 360 days.

Kayak A double-pointed, decked, skin-covered canoe found among some Eskimo groups. In some varieties, a waterproof coat covers the cockpit and fastens tightly around the neck and wrists of the traveller, thus rendering the vehicle and clothing contiguous and the kayak virtually unsinkable. A skilled individual can travel in heavy water and breakers in such a craft.

Keewatin lanceolate A projectile point style defined by Harp (1961) on the basis of his investigations along the Thelon River in the Northwest Territories. As the name implies, these specimens are un-notched and in terms of most particulars, including size, shape, thickness, type of flaking and presence and extent of basal and lateral grinding, they are essentially identical to Agate Basin points. Wright (1972) considers Keewatin Lanceolate points and the complex within which they occur to represent an occupation by Plano peoples who were ancestral to the authors of his Shield Archaic Tradition.

Kegl, Judy A linguist who works on theoretical linguistics as it applies to signed and spoken languages. Among her research interests is a study of Nicaraguan Sign Language.

Kerameikos Area in which potters lived in Athens. Also used as a burial place.

Kettle hole A depression in the ground formed by the melting of a buried block of ice. The debris which formerly covered the ice is thus allowed to settle to the bottom of the kettle.

Khepresh The Egyptian blue crown was a ceremonial crown and/or war crown.

Khnum Egyptian god of creation represented by a ram.

Kill site An archaeological site which was used primarily for the slaughter and (usually at least the preliminary) butchering of game. As one would expect, the artifacts commonly found at such sites include projectile points and large knives. Faunal remains are often extremely numerous and an analysis of these may provide details on the method used to butcher the animals.

Killed artifact An artifact which has been purposely broken so as to release its spirit. Killed artifacts may be found as grave goods and the implication is that this was accomplished so that the spirit of the artifacts could accompany the spirit of the deceased to the afterlife.

Kiln A brick-lined oven used to fire ceramics.

Kilt Short skirt worn by Egyptian men.

Kimbel, Bill An anatomist, Kimbel worked with Don Johanson and assembled Lucy's skull fragments. In 1991, Kimbel and Yoel Rak found a 70 percent complete skeleton of *Australopithecus afarensis*.

Kin selection The process whereby an individual's genes are selected by virtue of that individual's increasing the chances that his or her kin's genes will be propagated into the next generation.

Kin terminology The terms that systematically designate distinctions between relatives of different categories.

Kindred A collection of bilateral kin.

Kingdom The second highest level of taxonomic classification of organisms (below domains). Classification schemes at the kingdom level have changed over time. Recent molecular data have generally reinforced the evolutionary significance of the kingdoms Animalia, Plantae, and Fungi. The single-celled eukaryotes once lumped into the kingdom Protista are now known to be very diverse, and not closely related to one another. The prokaryotic organisms once lumped into the kingdom Monera are now considered to belong to separate domains: Eubacteria and Archaea.

Kingdom animalia Comprises the animals. These are multicellular organisms with cells that lack a cell wall. Many are capable of movement, or movement of some of their body parts, at some time of their life. Animals are heterotrophic, that is, they cannot obtain energy from non-living forms and must eat other organisms, or their products, to obtain energy.

Kingdom fungi Comprise mushrooms, yeasts, and other fungi. These are eukaryotic multicellular organisms that are typically non-moving, have a cell surrounded by a cell wall, and are heterotrophic, that is, they cannot use energy from non-living forms and must consume other organisms, or their products, to obtain energy. Many are decomposers, that is, they obtain energy by breaking down molecules in dead, decaying organisms.

Kingdom monera Comprise bacteria and cyanobacteria. These are unicellular organisms that are prokaryotic, that is, do not have a nucleus surrounded by a nuclear membrane. The bacteria are mostly heterotrophic, that is, they cannot obtain energy from non-living forms and must eat other organisms, or their products, to obtain energy. The cyanobacteria are autotrophic, that is, they can convert energy from non-living forms into biologically useful energy (stored in the chemical bonds of biological molecules); cyanobacteria, like plants, accomplish this through the process of photosynthesis.

Kingdom plantae Comprise the plants. These are eukaryotic multicellular organisms with cells surrounded by a cell wall made of the carbohydrate cellulose. Plants are typically non-mobile. They are autotrophic, that is, they can convert energy from non-living forms into biologically useful energy (stored in the chemical bonds of biological molecules); plants accomplish this through the process of photosynthesis, synthesis of energy-containing biological compounds by trapping light energy.

Kingdom protista Comprise unicellular organisms that are eukaryotic, that is, have a nucleus separated from the cytoplasm of the cell by a nuclear membrane. Some are plant-like in that they are autotrophic, while others are animal-like in that they are heterotrophic.

Kinnikinnick A smoking mixture of the North American Indians consisting of the bark and leaves of certain plants occasionally mixed with tobacco. Some of the plants used produced a mild narcotic effect.

Kirchweger, Gina An Austrian biologist interested in the biological evolution of skin tone. Her essay, "the biology of skin color," concerns the evolution of race.

Kirkland, James I. James Ian Kirkland (1954-) is an American geologist and paleontologist who has studied dinosaurs from the southwestern USA for over 20 years, discovering many new and important genera. Kirkland named (or co-named) the dinosaurs: Animantarx (Carpenter, Kirkland, Burge, and Bird, 1999), "Eohadrosaurus" (Kirkland, 1997 [nomen nudum]), Eolambia (Kirkland, 1998), Gastonia (Kirkland, 1998), Mymoorapelta (Kirkland and Carpenter, 1994), Nedcolbertia (Kirkland, Britt, Whittle, S. K. Madsen, and Burge, 1998), Utahraptor (Kirkland, Burge, and Gaston, 1993), and Zuniceratops (Wolfe and Kirkland, 1998). Kirkland is an adjunct Professor of Geology at Mesa State College, Grand Junction, Colorado, a research Associate of the Denver Museum of Natural History, and a Utah State Paleontologist with the Utah Geologic Survey.

Kiva A ceremonial room, often circular and semi-subterranean, used by Pueblo Indians.

Kluger, Matthew A researcher whose work on lizards demonstrated that fever is beneficial and can improve the immune response to infection. The implication for humans is still being researched, but evidence indicates that mild fevers can have a number of important immunological functions that allow us to better fight bacterial and viral infections.

Knapping The production or shaping of stone artifacts by means of pressure and/or percussion flaking.

Knife river flint Brown chalcedony; a dark, brown translucent siliceous stone found in the Knife River area of North and South Dakota and particularly favored for its flaking properties.

Knowlton, Nancy Dr. Knowlton is Professor of Marine Biology at the Scripps Institution of Oceanography, University of California San Diego, and Staff Scientist at the Smithsonian Tropical Research Institute in Panama. Her primary research interests concern various facets of marine biodiversity. These include the nature of species boundaries in corals, elucidating biogeographic patterns in tropical seas, the ecology of coral-algal symbiosis, and threshold effects in coral reef ecosystems.

Knuckle walking Semierect quadrupedalism, found in chimpanzees and the gorilla, in which the upper parts of the body are supported by the knuckles rather than the palms.

Kondrashov, Alexey A population geneticist specializing in mathematical analysis who has studied the evolutionary role of slightly deleterious mutations. He has theorized that a primitive organism's strategy for protecting itself against damaging mutations may have been the first step in the evolution of sexual reproduction.

Kouros A statue of a nude Greek youth. A young man.

Krater Bowl for mixing wine and water.

Kreiswirth, Barry Director of the Public Health Research Institute TB Center in New York, Dr Kreiswirth uses DNA fingerprinting to study the evolution of antibiotic resistance in *Mycobacterium tuberculosis*, the pathogen that causes TB.

Kula ring A system of ceremonial, non-competitive, exchange practiced in Melanesia to establish and reinforce alliances. Malinowski's study of this system was influential in shaping the anthropological concept of reciprocity.

Kurdish Part of the Indo-European family of languages, Kurdish is the language of the Kurds and is spoken in Turkey, Iran, Iraq, Syria, and Russia.

Kurds A non-Arab Middle Eastern minority population that inhabits the region known as Kurdistan, an extensive plateau and mountain area extending across Turkey, Iraq, Iran, Syria, and Armenia. In the late '90s, estimates projected that the Kurds number more than 20 million. The majority of Kurds are devout Sunni Muslims.

Kylix A shallow footed bowl used for drinking.

Kyrtome A more or less arcuate fold or band in the interradial areas outside the (laesurae of trilete spores. Examples: *Ahrensisporites guerickei, Concavisporites rugulatus.*

Some palynologists prefer to use torus for separate interradial bands, and kyrtome for a connected feature.

L

Labret A pulley-shaped object of stone, bone or wood, inserted in a perforation of the lower lip as an ornament or status symbol by some aboriginal peoples.

Labrum The elevated and/or thickened part of a laesura between the commissure and the remainder of the proximal surface.

The term is mainly used in the descriptions of fossil spores. Because of possible confusion with the term labrum (sensu Thomson and Pflug (1953)) it is, however, suggested to use the term margo instead.

Labyrinth A building with intricate passages. Also the place where King Minos of Crete kept the monster Minotaur.

Lac A resinous deposit of an insect that lives on creosote bushes.

Lactation The production of milk by a female mammal

Lacuna A depressed area surrounded by ridges in lophate pollen grains.

Lacustrine deposits Lake sediments; usually fine laminated silts and clays.

Lacustrine Describes anything characteristic of lakes. In geology it describes deposits formed by lakes or preserving lake environments. Lacustrine deposits can be limestone, shale, mudstone or saline deposits. The lithographic limestones of Las Hoyas, Cuenca, Spain and the volcanic ash deposits of Liaoning, China, are lacustrine deposits.

Laesura The arm of a proximal fissura or scar of a spore.

A monolete spore has one laesura, a trilete spore three (although some palynologists consider a trilete spore to have a single triradiate laesura). A laesura comprises a commissure which may be bordered by a margo (labrum).

Laevigate A general term for smooth, as if polished

Synonym of psilate.

Orthographical variant: levigate.

Although the correct Latin spelling is levigate, the spelling laevigate is accepted throughout the botanical literature The term has mostly been used in the descriptions of fossil spores. Example: *Laevigatisporites*.

Lagerstatte deposit A deposit that preserves an extensive cross-section of the biota that once lived at that locality. Lagerstatte deposits can be Konzentrat-lagerstatte, deposits with very high numbers of fossils, or Preservat-lagerstatte where the fossils are preserved in exceptional detail, including preservation of soft body parts. The Cleveland-Lloyd deposit is a Konzentrat-lagerstatte deposit. The Solnhofen limestone of Bavaria, Germany, the volcanic deposit of Liaoning, China and the Burgess Shale of Canada are all Preservat-lagerstatte deposits.

Lake agassiz. The multi-staged sea-like lake formed by the meltwater of the wisconsinan ice sheet which dominated the southern half of manitoba between approximately 10,000 and 5800 b.c. at various times, this massive body of water occupied parts of saskatchewan, ontario, north dakota, minnesota and even south dakota. During its lifespan, the southern shore shifted generally to the north in accordance with the erratic but inevitable northward retreat of the glacier. In its wake, it left many strandlines and deltas including the 6500 km2 assiniboine delta in the carberry area as evidence of an earlier period of stable water levels. Lakes winnipeg, winnipegosis, manitoba and the many smaller lakes of the pre-cambrian shield of eastern manitoba are but remnants of this once immense lake.

Lakeshore. A preceramic focus or culture defined by vickers (1950). This concept is no longer used and the materials which vickers retrieved are now generally attributed to the mckean complex.

Lalongate Describing the shape of a transversely elongated endoaperture. Example: *Filipendula* (Rosaceae).

Lamarck, Jean An 18th-century naturalist, zoologist, and botanist noted for his study and classification of invertebrates, as well as his evolutionary theories. He traveled extensively throughout Europe and was elected to the Academy of Sciences, where he introduced the principles of heredity and acquired characteristics.

Lamarckian inheritance Historically misleading synonym for inheritance of acquired characteristics.

Lambe, Lawrence M. Lawrence Morris Lambe was a Canadian fossil hunter who named: Chasmosaurus (1914), Edmontosaurus (1917), Eoceratops (1915), Euoplocephalus (1910), Gorgosaurus (1914), Gryposaurus (1914), Panoplosaurus (1919), Stegoceras (1902), and Styracosaurus (1913). He worked for the Canadian Geographical Survey and hunted near Alberta, Canada. Lambeosaurus was named as a tribute to Lambe in 1923.

Lamella A general term for a thin layer (Jackson, 1928).

Laminae Very thin strata.

Laminated Composed of or built up by layers.

Lanceolate leaf-shaped; tapered at one or both ends. In archaeological usage, the term usually refers to long, slender, un-notched chipped stone projectile points.

Land bridge A connection between two land masses, especially continents (e.g., the Bering land bridge linking Alaska and Siberia across the Bering Strait) that allows migration of plants and animals from one land mass to the other. Before the widespread acceptance of continental drift, the existence of former land bridges was often invoked to explain faunal and floral similarities between continents now widely separated. On a smaller scale, the term may be applied to land connections that have now been removed by recent tectonics or sea-level changes (e.g., between northern France and southeastern England).

Landscape archaeology The study of individual features including settlements.

Language A highly flexible and complex system of communication that allows for the exchange of detailed information about both interior and exterior conditions. As a creative and open system, new signals may be added and new ideas transmitted.

Lanston, JR., Wann Wann Langston Jr., is an American vertebrate paleontologist.

Langston has named and described the dinosaurs Acrocanthosaurus (Stovall and Langston, 1950) and Lophorhothon (1960). He received his Ph.D. from the University of California, Berkeley in 1952. Langston was the Director of the Vertebrate Paleontology Laboratory of the Texas Memorial Museum of Science and History at the University of Texas at Austin from 1969 until 1986 (when he retired); before that, he had been the curator at the National Museum of Canada. Dr. Langston was the thesis advisor of Douglas Lawson when Lawson found the giant pterosaur Quetzalcoatlus.

Lapis lazuli Semiprecious blue stone. In ancient times it was mined in Afganistan.

Lapparent, Albert De Albert-Felix de Lapparent (1905 - 1975) was a French Jesuit priest and paleontologist who made nine fossil-hunting trips (mostly alone and later with Philippe Taquet) into the Sahara desert, riding on a camel on his early trips. Lapparent named and described the dinosaurs *Inosaurus tedreftensis* (Lapparent, 1960) and *Lusitanosaurus liassicus* (Lapparent and Zbyszewski, 1957). He also found many new species of known genera, including: *Apatosaurus alenquerensis* (Lapparent and Zbyszewski, 1957), *Astrodon pusillus*(Lapparent and Zbyszewski, 1957), *Brachiosaurus atalaiensis* (Lapparent and Zbyszewski, 1957), *Brachiosaurus nougaredi* (Lapparent, 1960), *Camarasaurus alenquerensis* (Lapparent and Zbyszewski, 1957), *Cetiosaurus mogrebiensis* (Lapparent, 1955), *Elaphrosaurus gautieri* (Lapparent, 1960), *Elaphrosaurus iguidiensis* (Lapparent, 1960), *Megalosaurus pombali* (Lapparent and Zbyszewski, 1957), and *Rebbachisaurus tamesnensis* (Lapparent, 1960). He also discovered the giant crocodile Sarcosuchus (1964). The dinosaur Lapparentosaurus (Bonaparte, 1986) was named to honor Lapparent.

Larter An archaeological site on the Red River north of Winnipeg which has lent its name to the distinctive barbed projectile point style for which it is known as well as to the focus or phase within which these occur. This late Archaic or late Middle Prehistoric Period complex consists of corner-notched projectile points named Larter Tanged, unnotched projectile points, or blanks, a variety of scrapers and knives, drills, gravers, chisels, grooved hammerstones and sinew stones. Larter peoples appear to have followed a seasonal round of activities centered upon the communal hunting of bison, although their diet was supplemented by a variety of other animals as well as fish and shellfish. According to at least one expert (Reeves l983) Larter derives from the earlier McKean Complex.

Larter sub-phase The larter "phase" when viewed simply as the southern manitoba variation of a more widespread and generalized complex (pelican lake) which was distributed over most of the northern grasslands during the late middle prehistoric period (reeves 1983).

Larva (and larval stage) The prereproductive stage of many animals. The term is particularly apt when the immature stage has a different form from the adult. For example, a caterpillar is the larval stage of a butterfly or moth.

Late glacial climate episode The climatic pattern characteristic of the end of the glacial period (roughly ll,000 to 8l00 B.C.).

Late prehistoric period Late Prehistoric Period. the last and most recent of the three "stages" in central North American prehistory. This term is most commonly used for the plains, but is roughly equivalent in terms of its dating to the Woodland Period of the eastern forests (ca. 200 B.C. to the Historic Period). In many ways, the basic lifestyles remained largely unchanged from those of the earlier Middle Prehistoric (or Archaic) Period. Nonetheless, a number of technological and behavioural innovations are present in the archaeological record which allow archaeologists to identify sites of this period with relative ease. These include the manufacturing of pottery, the use of the bow and arrow, the construction of burial mounds and an intensification of the use of bison corrals and (bison) "jumps" as a communal hunting technique.

Lateralization The phenomenon in which the two hemispheres of the brain specialize in regard to different functions.

Latimurate Describing a pollen grain or spore with broad muri.

Latiporate Describing pollen grains with pores in one hemisphere only. Examples: *Juglans*, *Carya* (Juglandaceae).

Latitudinal A general descriptive term, in palynology applied to features which run in lines parallel to the equator.

Antonym: longitudinal.

Laurasia A supercontinent that existed from the Jurassic to Early Tertiary after splitting from Pangea; composed of Laurentia, Baltica, Avalonia, (modern North America, Scandinavia, Greenland, Western and Central Europe); eventually fragmented into Eurasia and North America in the Tertiary with the opening of the North Atlantic Ocea.

Laurel A (now-abandoned) town in northern Minnesota near the Ontario border which has given its name to a distinctive Middle Woodland ceramic ware as well as the complex with which it is associated. Laurel vessels are grit-tempered and manufactured by means of coiling. Vessel shape is conoidal with slightly constricting necks terminating in un-thickened lips. Surface finish is smooth except where decorated. Decorative elements include bosses, dentates, punctates and incisions and these are applied and/or combined in various ways so as to produce pseudo-scallop shell, dragged stamp and various other patterns. The Laurel lithic industry consists of a variety of scrapers, and bifaces, pieces esquillees, netskinkers, hammerstones, anvils, smoking pipes, tools for decorating pottery, mortars, pestles, manos, abraders and pendants. The bone, antler, tooth, claw and shell industries are extremely well represented, and served as media for the production of numerous classes of tools and a variety of personal adornments. Native copper was also utilized for beads, pendants, chisels, fishhooks and knives. Subsistence was based on a wide range of resources including large and small mammals, wildfowl, shellfish, turtles and fish. Plants were also heavily utilized and it is possibly at this time that wild rice first entered the diets of prehistoric Manitobans. One of the most spectacular aspects of the Laurel culture was the construction of burial mounds. The largest of these was originally 36 m in maximum diameter and 14 m high. Within these mounds were placed the deceased together with meager grave goods. The condition of some of the skeletons was suggestive of (ritual) cannibalism. Laurel sites are distributed in a broad arc from east-central Saskatchewan through central Manitoba to northern Minnesota and northwestern Ontario, and eastwards around the shores of Lake Superior to northern Michigan. Radiocarbon dating shows them to date between approximately 200 B.C. and A.D. 1000.

Laurel triangular A projectile point style defined by buchner (1982) which he considered to be diagnostic of the laurel phase. These specimens are isosceles-triangular in outline and range in length from 23 to 33 mm. Bases are slightly convex to rounded and virtually all exhibit basal thinning. They are distinguishable from more recent triangular forms (such as eastern triangular) on the basis of the former's larger size alone.

Laurentia A separate continental plate that existed from the Late Precambrian to Silurian, consisting of the major part of

North America, northwest Ireland, Scotland, Greenland, and pieces of Norway and Russia.

Laurentide ice sheet Late Wisconsin ice sheet that covered all of eastern and central Canada and extended into the north-central and northeastern U.S.A. It may have joined at times with the Greenland and Cordilleran Ice Sheets.

Layer A general term. Applied in palynology to any distinct stratum of the sporoderm (APLF, 1975).

Leaching The removal of water soluble minerals in soil or rock by rainwater. Leaching also accounts for the gradual disappearance of bone from archaeological sites, particularly when the soil is acidic.

Leakey, Maeve A paleoanthropologist at the National Museums of Kenya, Maeve is the discoverer of *Kenyanthropus platyops* and *Australopithecus anamensis*. She is married to Richard Leakey.

Leakey, Mary A British paleoanthropologist described as "a real fossil hunter" and "the real scientist in the family." Her discoveries, some in collaboration with her husband Louis Leakey, included the 1.75-million-year-old skull which first showed the antiquity of hominids in Africa, jaws and teeth of an early *Homo* species, and fossilized footprints of bipedal hominids.

Leakey, Richard The son of renowned anthropologists Louis and Mary Leakey, Richard continued their work on early hominids from 1964 until the 1980s, making a number of significant fossil finds in the Lake Turkana area and serving as Director of the National Museum of Kenya. Later he devoted his energies to conservation and politics.

Lean-to A simple shelter consisting of a covered frame leant against the wall of another structure or against some other vertical face.

Lector priest Translates as "One who bears the ritual book". This priests function was/is to recite from the ritual texts.

Lee, Melanie A molecular geneticist and microbial biologist, who in the 1980s collaborated with Paul Nurse on novel research that demonstrated the commonality of the genetic code between yeasts and humans. Dr Lee later took her molecular skills into the pharmaceutical industry, and was a leader in moving pharmacology away from animal models and towards the use of recombinant DNA technology for screening potential new therapies. She now heads the research division of Celltech, an international biopharmaceutical company, where her team works on drug discovery and development of new therapies, mainly for the treatment of inflammatory and immune diseases.

Legal subdivision system The method of describing parcels of land in terms of "Township, Range, Section, and Quarter Section".

Legitimacy The right to rule on the basis of recognized principles.

Leidy, Joseph Joseph Leidy (1823-1891) was a US anatomist/paleontologist who named the first dinosaurs found in the U.S.A. He excavated the first American dinosaur, Hadrosaurus, in 1858. Leidy named Antrodemus (1870, perhaps Allosaurus), Aublysodon (1868), Deinodon (1856), Diplotomodon (1868), Hadrosaurus (the first nearly-complete dinosaur skeleton and first-known duck-billed dinosaur, 1858), Palaeoscincus (1856), Thespesius (1856), Trachodon (1856), and Troodon (1856). Leidy was also the first scientist to identify many extinct species of camels, horses, sloths, tigers, and rhinoceroses.

Leister A fish spear with three or more points, usually of bone or antler. Each point has a barb on the inner side.

Lek An area of ground divided into territories that are defended by males for the purpose of displaying to potential mates during the breeding season. This form of mating behaviour is known as lekking, and occurs in various bird species (for example the peacock) and also in some mammals. The dominant males occupy the territories at the centre of the lek, where they are most likely to attract and mate with visiting females. The outer territories are occupied by subordinate males, who have less mating success. Over successive breeding seasons, younger subordinate males tend to gradually displace older individuals from the most desirable territories and become dominant themselves. The lek territories do not contain resources of value to the female, such as food or nesting materials, although males of some species may build structures such as bowers that form part of their display.

Lekythos A vessel used by Greeks to hold perfumed oil.

Lemur A small, tree-dwelling primate that belongs to the group called prosimians.

Lemuridae Madagascar prosimian family that includes the femurs.

Lens In archaeology, a biconvex (lenticular) discoloration in a soil profile.

Lenticular Shaped like the cross-section of a lens; biconvex.

Lepidodendron A Pennsylvanian fossil club moss. It was tall (100-feet high), branching, with slender leaves and had distinctive diamond shaped leaf scars along the trunk.

Lept A suffix for thin, indicating that a pollen grain has a leptoma.

Leptoma A thin area at the distal pole of a pollen grain, presumed to function as an aperture. Example: *Pinus* (Pinaceae).

Substitute term for cappula.

Lesser apes The gibbons and siamang of Asia.

Lete A suffix to denote the presence (or absence) of laesura. Examples: alete, monolete, trilete.

Lethal recessive The case in which inheriting two recessive alleles of a gene causes the death of the organism.

Lethals Defects that cause premature death.

Levallois technique The method (common in the Middle Paleolithic) or preparing a tortoise-shaped *core* so that flakes struck from it will be of a predictable shape.

Levant Land area between Mesopotamia and Egypt; it includes modern day Lebanon, Syria, Israel and Jordan

Level bag A bag containing excavated materials from a single level of a single excavation unit.

Level notes Written observations on all significant characteristics of an excavated level.

Level The layer in which archaeologists dig. When they dig down through many layers in one site, it is also a unit of measurement. All sites have different numbers of levels, and even the different units within one site may have different numbers of levels. How do archaeologists decide when they are going into another level? Archaeologists try to judge by cultural clues like floors, but sometimes they will go by changes in soil color or soil type, or even by a specific number of centimeters. For example, an archaeologist might give a different level number every 10 centimeters in a site. Archaeologists want to keep track of levels because this allows them to build a profile (or a cross section) of the units, so that they can look at how the site changed over time.

Leveling mechanism A social or economic practice that serves to lessen differentials in wealth.

Levigate Orthographical variant of laevigate.

Levine, Michael Professor of Genetics and Development in the Molecular and Cell

Biology Department at University of California, Berkeley. Discoverer (with Bill McGinnis) of homeobox sequences in the homeotic genes *Antennapedia* and *Ultrabithorax* while a postdoctoral researcher with Walter Gehring at the University of Basel, Switzerland. His current research involves analysis of gene regulation and patterning in the early *Drosophila* embryo; studies of embryonic development in the tunicate, *Ciona intestinalis*, focused on the specification of the notochord and tail muscles; and a critical test of classical models for the evolutionary origins of the chordate body plan.

Levirate A social custom under which a man has both the right to marry his dead brother's widow and the obligation to provide for her.

Lexicon In linguistics, the total number of meaningful units {such as words and affixes) of a language.

Lexicostatistics The study of linguistic divergence between two languages, based on changes in a list of common vocabulary terms and the sharing of common root words.

Lexigram A symbol that represents a word.

Libation Ritual pouring of a liquid.

Lichenometry The measuremet and study of rates of lichen growth. Lichenometry offers a means of calculating the minimum age of any rock painting or petroglyph which is overgrown with lichen or a petroform if the cobbles from which it is made were overturned during the construction phase. The latter condition is necessary as lichen grows only on the upper, exposed face of boulders and thus inversion will kill the old lichen and allow new patches to begin growing. If the rate of lichen growth in any given region can be determined (usually by long term observation) then a simple calculation will yield the age of the "new" lichen patch and thus the age of the petroform. Similarly, rock paintings and glyphs can theoretically be dated by the amount of lichen on top of them, or by the extent of "new" patches assuming that the rock surface was cleared of "old" lichen before use. Unfortunately lichenometry is not yet sufficiently refined for widespread use by archaeologists.

Life expectency The length of time that a person can, on the average, expect to live.

Life span The theoretical, maximum age.

Lifeway The way of life associated with a culture; the technological and behavioral means of adaptation.

Liftbag A scientific instrument like a giant balloon, which is filled with air from a scuba tank and then used by underwater archaeologists to float heavy artifacts to the surface during underwater archaeological excavations.

Light-table A glass-topped table illuminated from underneath, used in the laboratory photography of archaeological specimens.

Lignite A soft shiny black variety of coal, aboriginally used to manufacture decorative objects.

Limb Synonym of equatorial outline.

Limbus A sharp narrow crease of the saccus or pseudosaccus where the outer and inner exine layers are fused. Example: *Nuskoisporites dulhuntyi*.

Limestone A sedimentary rock made of layers of carbonated sediments (sea life) that thrived in a warm shallow sea. Fossils are often found in this type of rock.

Limonite A substance produced by the oxidation of iron-bearing minerals such as pyrite and magnetite. Limonite is a yellowish brown, soft mineral with no cleavage. It is a widely occurring mineral.

Lindenmeier An archaeological site in north central Colorado which yielded much material of the Folsom Complex. Henry Irwin (1971) defined Lindenmeier as a phase consisting of two sequent subphases, the Folsom and the Midland which followed it. He considered the Lindenmeier Phase to date from 9000 to 8500 B.C.

Lineage A unilineal descent group composed of people who trace their genealogies through specified links to a common ancestor.

Lineal relatives Direct ascendants and descendants.

Linear mound A long, low, linear earthen embankment which may include a right angle. Rounded expansions often occur at the ends of the mounds and at turning points. These are much less common than their oval and circular counterparts and tend to occur most frequently in southwestern Manitoba and adjacent portions of North Dakota. The largest of these was 800 m in length, over 10 m in width and approximately 1 m in height. The function of these structures is something of a mystery. They do not appear to have been used for burying the dead, and although they are occasionally called "entrenchments", they are not usually in easily defensible locations, and would not be particularly effective for that purpose. It is possible that they fulfilled some ceremonial but as yet unknown purpose.

Linear tetrad A uniplanar tetrad in which the four members are arranged in a row. Example: *Typha* spp. (Typhaceae).

Linen Fabric woven from flax fibers.

Lingua franca Any language used as a common tongue by people who do not speak one another's native language.

Linguistic anthropology A subdivision of anthropology that is concerned primarily with unwritten languages (both prehistoric and modern), with variation within languages, and with the social uses of language; traditionally divided into three branches: descriptive linguistics, the systematic study of the way language is constructed and used; historical linguistics, the study of the origin of language in general and of the evolution of the languages people speak today; and sociolinguistics, the study of the relationship between language and social relations.

Linguistics The science which concerns itself with the origin, development, history, relationships and structure of languages. Judicious use of linguistic data can reveal much of direct relevance to archaeology. The degree of similarity between two related languages is a gauge of the amount of contact between the two groups or the amount of time elapsed since they separated. Glottochronology can put an absolute date on the "split", and occasionally archaeology can trace the route of the emmigrant group. Comparison of terms in modern related languages can yield insights into the nature of the common "mother" language from which they all developed. These may include the kind of technological items the speakers of the now-extinct language used and even the geographic region they originally occupied.

Linkage disequilibrium A condition in which the haplotype frequencies in a population deviate from the values they would have if the genes at each locus were combined at random. (When no deviation exists, the population is said to be in linkage equilibrium.).

Linked Of genes, present on the same chromosome.

Linked changes Those changes brought about in a culture when other (interconnected) parts of that same culture undergo change.

Linnaean classification A hierarchical method of naming classificatory groups, invented by the eighteenth century Swedish naturalist Carl von Linné, or Linnaeus. Each individual is assigned to a species, genus, family, order, class, phylum, and kingdom, and some intermediate classificatory levels. Species are referred to by a Linnaean binomial of its genus and species, such as *Magnolia grandjflora.*

Lintel The upper beam of a door or window.

Lira A narrow ridge which forms the murus in a striate pattern.

Synonym of murus and vallum.

Lithic A stone artifact, usually in the form of a stone tool. Archaeologists frequently find lithic artifacts at archaeological sites because humans used to make their tools out of stone before they used metal.

Lithic industry That part of an archaeological artifact assemblage manufactured of stone.

Lithic reduction sequence The entire process of manufacturing stone tools by flaking from the removal of the decortification flakes to the sharpening or retouching of the final product.

Lithic technology The process of manufacturing tools etc. from stone. Most frequently refers to stone flaking.

Lithologic unit Lithostratigraphic unit; a body of rock that is consistently dominated by a certain lithology or similar color, mineralogic composition, and grain size. It may be igneous, sedimentary, or metamorphic and may or may not be consolidated.

Lithosphere The outer skin of the earth, composed of the crust and the uppermost mantle.

Little ice age Period of expanded alpine glaciers and cooler climate during the 16th through 19th centuries. Corresponds with Maunder solar minimum.

Lively, Curtis A professor of biology who studies population biology and the ecology and evolution of host-parasite interactions. His laboratory is involved in detailed studies of the interaction between a parasitic trematode and a freshwater New Zealand snail in which both sexual and asexual females coexist.

Living floor The level within an archaeological site upon which a group of people lived. In the absence of specially prepared ground surfaces, living floors may only be defined on the basis of the depths of artifacts pertaining to that component and the degree of compaction of the floor due to people walking on it. Often, overlying soil will appear to peel off the floor due to the different densities involved. Exposure of entire living floors allows the graphic representation of the various activity areas of a site at a specific point in time.

Living fossils This term, which is by no means unproblematic (indeed it is self-contradictory), is often used for recent species or supraspecific taxa that satisfy the following criteria:

presence of many plesiomorphic character states, compared to the recent sistergroup, and generally only few inconspicuous autapomorphies

Existence of only one or few recent species, compared to numerous fossil species of the same clade or at least a similar grade

Relict type of distribution of the recent representatives, often a disjunct distributional pattern

Recent sistergroup with relatively numerous species

Apparently an intermediate position between two different recent groups or an extinct group and a recent group, caused by symplesiomorphic similarity with one group and a synapomorphic similarity with the other group

Examples for such "living fossils" are within molluscs the genera *Nautilus* (Nautiloidea) and *Neopilina* (Monoplacophora), within actinopterygid fishes the groups Cladistia (e.g. *Polypterus*), Ginglymodi (e.g. *Lepisosteus*) and Halecomorphi (*Amia*), within sarcopterygid fishes the lungfishes (Dipnoi) and the coelacanth (Actinistia: *Latimeria*), within lepidosaurs the tuatara (Rhynchocephalia: *Sphenodon*), within mammals the Monotremata, within arthropods the peripatus (Onychophora), or within dragonflies the genus *Epiophlebia* ("Anisozygoptera"), the family Petaluridae (Anisoptera) and the genus *Hemiphlebia* (Zygoptera). Certain groups of organisms only satisfy some of the mentioned criteria and consequently hardly have been regarded as "living fossils", like for instance all bacteria and all protozoans, or the Acrania (*Branchiostoma* = *Amphioxus*) and hagfishes (Hyperotreta: e.g. *Myxine*). If one would only look at the skeleton of the arms, man would be something like "living fossils" among placental mammals, too, compared to the derived skeletons in bats, whales, ungulates, etc.

Llano complex That complex defined by the association of Clovis projectile points and mammoth remains.

LO analysis A method for analyzing patterns of sexine organization by means of light microscopy.

This method is valuable for elucidating exine patterns. When focused at high level (H), raised sexine elements appear bright (Lux), whereas holes in the tectum are relatively dark (Obscuritas). At lower focus (L) holes become lighter and the sexine elements become darker.

LO pattern A pattern of ornamentation that appears to show "bright islands" at high focus (H) and that become dark at low focus (L), observed when using LO-analysis.

The reverse of OL- pattern.

Loam A fertile topsoil consisting chiefly of sand, clay and silt and partially decomposed organic matter.

Lobate Describing an equatorially aperturate pollen grain with a lobed shape in polar view.

Belongs to a system of shapes combined with the position of the apertures as introduced by Kuyl et al.

Local races Subdivisions of geographical races. One type consists of partially isolated groups, usually remnants of once larger units. The second type includes fairly large subdivisions that contain a degree of variation within them.

Lockport stemmed A poorly defined projectile point style associated with Middle Woodland ceramics at a site in southern Manitoba. These crudely fashioned specimens are approximately 45 mm in length and 20 mm in width and show a tendency toward asymmetry. The stems are short and straight and the shoulders poorly defined.

Locomotor patterns The methods an animal uses to move. These may include such things as swimming, jumping, walking, etc.

Locus The location in the DNA occupied by a particular gene.

Loess A light-colored, un-stratified deposit consisting of fine-grained sands, clays or silts which is for the most part laid down by the wind under cold, dry conditions.

Logan creek Logan Creek. an archaeological site in northeastern Nebraska which has lent its name to a widespread Middle Prehistoric Period complex whose time-depth brackets the warm, dry Altithermal. One of the hallmarks of the complex is the side-notched projectile point which in many areas is the first style to appear after the demise of the Plano lanceolate forms. The former range from approximately 17 to 50 mm in length, have triangular blades, straight or concave bases and are frequently basally ground. Perhaps one of the most diagnostic traits of this complex is the practice of manufacturing side-notched end-scrapers, presumably from fragments of broken projectile points. Other items in the Logan Creek inventory include drills, grinding stones, hammerstones, bone awls, beads, needles, shaft straighteners, and fishhooks, and serrated mussel shells. Sites are widely distributed in the eastern grasslands, particularly near the forest edge, in major river valleys, in outliers of the eastern forests or near perennial sources of water. Presumably this reflects the generalized drought conditions which central North America was experiencing at the time. Faunal remains, found in association with Logan Creek materials tend to be more varied than those of their Paleo-Indian predecessors, although bison seem to have remained the mainstay of these peoples' diets. Radiocarbon dating indicates an age for this culture of 6500 B.C. to 4000 or possibly even 3000 B.C.

Logical class (versus logical individual) A group of things (e.g. organisms) that satisfy certain membership criteria (predicates), that have been defined by man. Logical classes do not really exist in the physical world, they are just inventions of the human mind (concepts). Logical classes usually do not bear proper names (such as "Canada"), but universal names (such as "country"). Paraphyletic and polyphyletic groups of organisms are such logical classes, too.

Logical individual (versus logical class) A particular thing that really exists in the outside world, independent from any human recognition (" the moon is still there even if one does not look at it"). Such logical individuals are for instance the Peter-Paul-Cathedral in Rome, the state Canada, the language English, a particular organism (the individual of everday language), but as well a particular biospecies. Individuals do have a unique origin, a unique fate and a unique end. Furthermore an individual is a cohesive unit, since all its parts are directly or indirectly connected by bonding relations. Individuals can not be defined (like logical classes), but only be discovered, described, and baptized with proper names (nomina propria, contrary to the universal names of logical classes). Since monophyletic groups do have a unique origin, fate and end, but lack the internal cohesion that is required if all parts of an individual shall share the same fate (such as in organisms and biospecies), they should rather be called "historical entities" (*sensu* wiley) than logical individuals.

Logistics The process of transporting, supplying and supporting a field project.

Lolongate Describing the shape of a longitudinally elongated endoaperture. Example: *Rumex* spp. (Polygonaceae).

Long house The long multi-family dwellings of the Iroquois area.

Longi A prefix for long.

Longiaxe Pollen grains with a polar axis longer than their equatorial diameter.

Synonym of prolate (s.l.).

Antonym: breviaxe.

Thompson and Pflug (1953) recognised Longaxones as a group of mid-Cretaceous and younger angiosperm pollen contrasted with Brevaxones.

Longitudinal A general descriptive term, in palynology applied to features which run in lines between the poles.

Antonym: latitudinal.

Looting The removal of artifacts from archaeological context without documentation of their dimensions of variability.

Lophate Describing a pollen grain in which the outer exine is raised in a pattern of ridges (lophae) surrounding depressions (lacunae). Example: *Hieracium* (Compositae).

Lophotrochozoans A clade of animals that share a cluster of HOX genes quite different from those found in the ecdysozoans (and deuterostomes). They share similar sequences in their 18S rRNA genes. The phylogeny of this clade that includes brachiopods, mulluscs, bryozoa, and many other groups is the subject of many competing hypothses.

Lorisidae Prosimian family that includes the lords, potto, angwantibo, and galago.

Loutrophoros Water jar used in weddings and burials.

Lovejoy, Owen A paleoanthropologist and consulting forensic anatomist, Lovejoy is known for his analysis of early hominid fossils. His research includes work on Lucy (*Australopithecus afarensis*).

Low energy budget An adaptive strategy by which a minimum of energy is used to extract sufficient resources from the environment for survival.

Loxocolp(or)ate Describing zonocolp(or)ate pollen with ectocolpi arranged so that they converge in pairs.

Such apertures generally occur in aberrant grains with more than the typical number of apertures.

Lug A protuberance or projecting part on the rim of a pot intended to assist in holding and lifting the vessel.

Lull, R. S. Richard Swann Lull (1867-1957) was a vertebrate paleontologist who headed Yale's Peabody Museum (1922-1936) . Lull said that the best fossils collecting could be done in the basement of the Peabody Museum. He named the following: Anatosaurus Lull and Wright, 1942 (an obsolete name for Edmontosaurus), Anchisauripus (an ichnogenus of theropod dinosaur) in 1904, Diceratops Hatcher vide Lull, 1905, and Proceratops Lull, 1906.

Lumen The space enclosed by the muri.

Lunate Crescent-shaped.

Lycophyte Commonly known as club mosses, lycophytes were among the first seedless plants to appear on Earth. Along with horsetails and ferns, these made the planet more hospitable for the first animals.

Lycopods Scale trees. Vascular plants found from the Devonian to Recent periods. The had simple leaves in spirals. The stem was not jointed. Fossils are common in coal-bearing strata.

Lyell, Charles A 19th-century scientist considered a father of modern geology. Lyell proposed that the geology of Earth is shaped by gradual processes, such as erosion and sedimentation. Lyell's ideas, expressed in his landmark work, *Principles of Geology*, greatly influenced the young Charles Darwin. Darwin and Lyell later became close friends. While Lyell initially opposed the idea of evolution, he came to accept it after Darwin published *On the Origin of Species*.

Lyell's notion of gradual change Also called uniformitarianism, Lyell's notion was that Earth has been shaped by the same forces and processes that operate today, acting continuously over very long periods of time. For example, the ongoing erosion caused by flowing water in a river could, given enough time, carve out the Grand Canyon.

M

Maastrichtian European stage of the Upper Cretaceous, spanning the time between 72 and 66 million years ago.

Macaw Large, brightly colored, tropical American birds closely resembling and related to parrots.

Macedonia Area of ancient Greece north of Mt. Olympus and home of Alexander the Great.

Macroevolution A vague term generally used to refer to evolution on a grand scale, or over long periods of time. There is no precise scientific definition for this term, but it is often used to refer to the emergence or modification of taxa at or above the genus level. The origin or adaptive radiation of a higher taxon, such as vertebrates, could be called a macroevolutionary event.

Macrofossil A fossil that is large enough to be studied without a microscope.

Macromutation Mutation of large phenotypic effect, one that produces a phenotype well outside the range of variation previously existing in the population.

Macrospore General term for the larger spores of heterosporous vascular plants.

Antonyms: microspore (in the general botanical sense); miospore (in the sense of Guennel, 1952).

In dispersed fossil spores where it is not always possible to determine whether the parent plants were heterosporous the term indicates size only (usually >200μm).

Maculate A general descriptive term. In palynology it can be applied to pollen grains or spores with a spotted exine, often due to variations in internal exine structure. Example: *Maculatisporites*.

Maenads Female followers of the god Dionysos.

Magic The supposed art which attempts to influence people or the course of events by supernatural means. The latter may include the use of incantations ("prayers" or songs), graphic representations, or symbols, and the manipulation of objects. Magical beliefs are difficult to infer from the archaeological record, although such items as sucking tubes and medicine bags and other ethnographically known items are occasionally recovered. Some interpret certain examples of rock art, particularly depictions of large game animals, as forms of hunting magic. Painting or engraving the likeness of a prey species on rock may have served to placate the spirit who controlled that species. Alternately, if the animal was represented with arrows in it, the intent may have been to bring about an event by imitating it (imitative magic).

Magnetic dating A method of dating that compares the magnetism in an object with changes in the earth's magnetic field over time. This method is used on baked clay and mud.

Magnetism remnant The magnetism acquired by substances containing magnetic compounds after being heated beyond the Curie Point (670 C) and then being allowed to cool. The heating brings about the loss of whatever magnetism the substance may have contained. As it cools, however, the metallic particles align themselves in accordance with the orientation of the magnetic field of the earth.

Magnetometer A device capable of detecting fluctuations in the strength of the earth's magnetic field at pin point locations. As these variations (or "anomalies") are in part the product of various subsurface features, a magnetometric survey can assist the archaeologist in locating and determining the size and nature of an archaeological site prior to excavation.

Maize (Indian) corn, zea mays. Maize is one of the oldest and certainly one of the most important of the Native American domesticated plants.

Malaria A sometimes-fatal disease transferred to humans by mosquitoes, infecting the bloodstream.

Maleev, E. A. Evgenii Aleksandrovich Maleev (1915-1966) was a Russian paleontologist who named the dinosaur genera Talirurus (1952), Tarbosaurus (1955), and Therizinosaurus (1954); he also named the family Therizinosauridae. The dinosaurs Maleevosaurus (Pickering, 1984) and Maleevus (Tumanova, 1987) were named by to honor Maleev.

Malthus, Thomas A British economist and demographer best known for his treatise on population growth, which states that people will always threaten to outrun the food supply unless reproduction is closely monitored. His theory was in opposition to the utopians of the 18th century.

Mammal like reptile Term once used to describe those animals that were thought to have been reptiles yet thought to have exhibited mammal-like characteristics during life, such as homeothermy. This term has fallen into disfavor as the clade Reptilia has been greatly refined and now excludes those animals formerly called "mammal-like" reptiles. These animals are now referred to the clades Synapsida and Therapsida, with the latter often called "proto-mammals." The Therapsids were the direct ancestors of the true mammals, while Synapsids were the direct ancestors of the Therapsids.

Mammals A group of warm-blooded, hairy vertebrates which live in many environments, including marine, land and in the air.

Mammary glands Only found in mammals, these are specialized glands that can produce milk for feeding young.

Mammoth An extinct elephant having a sloping back and plated teeth that resembled a washboard. Mammoths were grazers and fed on grasses, sedges, and shrubs.

Mandible A part of the bony structure of a jaw. In vertebrates, it is the lower jaw; in birds, the lower bill; in arthropods, one of the paired appendages closest to the mouth.

Maniraptora ("hand robbers") A clade of animals defined as birds and all taxa closer to birds than to *Ornithomimus*. It should be noted that this is not a grouping of dinosaurs that may include birds, it is a grouping of birds that may include certain dinosaurs. If it would be discovered that birds are not dinosaurs but crocodiles, for example, the group would be nested in the crocodilia, retain its name, and be a bird group that no longer includes any dinosaurs but may include certain crocodiles.

Manitoba focus The former name for the blackduck phase.

Manitoba lowlands A physiographic and vegetational zone occupying that portion of the glacial lake agassiz basin which consists of the interlake and the red river valley. This flat, generally poorly-drained land consists of some of the lowest lying terrain in the southern half of the province (approximately 210 to 250 m a.s.l.). Vegetation consists chiefly of black spruce and tamarack with intermittent meadows and swamps, although white spruce, trembling aspen, balsam popular, balsam fir and white birch may occur in some of the better-drained situations.

Manitoba point. A projectile point style defined by pettipas (1972) on the basis of a small number of surface recoveries from the swan river area of west central manitoba. These specimens are lanceolate in overall outline and exhibit highly convex blade margins. Perhaps the most distinctive feature of these points is the set of two or three incipient notches at the basal, lateral edges. Since the time of the original definition, similar points have been recovered as far west as alberta and as far south as wyoming. These are usually surface finds, but are often from the same location as agate basin points. An excavated example of the type from wyoming was radiocarbon dated at 6100 b.c.

Manitoba ware The former name for blackduck ware.

Mano A hand-held stone used as the upper stone (pestle) in milling.

Mantell, Gideon Gideon A. Mantell (1790-1852) was an amateur British fossil hunter, one of the first in the world. This physician named Hylaeosaurus (1833), Iguanodon (1825), Pelorosaurus (1850), and Regnosaurus (1848). Mary Ann Mantell, his wife, is commonly thought to have found the first Iguanodon tooth in 1822; there is no substantiation to this story, however.

Mantle The zone of the earth below the crust and above the core.

Manuport An object, usually a cobble, which has been moved from one place to another by humans. Manuports need exhibit no other evidence of cultural modification to warrant classification as artifacts.

Manzanita Also called bearberry. Manzanita is a low- growing evergreen shrub that is found at elevations of 3500 to 8000 feet. The plant is characterized by its red bark and oval-shaped leaves.

Margo An area of exine around an ectocolpus that is differentiated from the remainder of the sexine, either in ornamentation or by difference in thickness.

Margulis, Lynn A biologist who developed the serial endosymbiosis theory of origin of the eukaryotic cell. Although now accepted as a plausible theory, both she and her theory were ridiculed by mainstream biologists for a number of years.

Marl An english term loosely applied to a wide range of lightly/unconsolidated deposits usually containing fine grained clay and calcium carbonate.

Mars Roman god of war. Called Ares by the Greeks.

Marsh, Othniel Othniel C. Marsh (1831-1899) was a US paleontologist from Yale University who named the dinosaur suborder Theropoda (1881), Sauropoda (1878). He named named roughly 500 new species of fossil animals (they were found by Marsh and his many fossil hunters). Marsh named the following dinosaur genera: Allosaurus (1877), Ammosaurus (1890), Anchisaurus (1885), Apatosaurus (1877), Atlantosaurus (1877), Barosaurus (1890), Camptosaurus (1885), Ceratops (1888), Ceratosaurus (1884), Claosaurus (1890), Coelurus (1879),

Creosaurus (1878), Diplodocus (1878), Diracodon (1881), Dryosaurus (1894), Dryptosaurus (1877), Labrosaurus (1896), Laosaurus (1878), Nanosaurus (1877), Nodosaurus (1889), Ornithomimus (1890), Pleurocoelus (1891), Priconodon (1888), Stegosaurus (1877), Torosaurus (1891), Triceratops (1889), Tripriodon (1889). He named the suborders Ceratopsia (1890), Ceratosauria (1884), Ornithopoda (1881), Stegosauria (1877), and Theropoda. He named the families Allosauridae (1878), Anchisauridae (1885), Camptosauridae (1885), Ceratopsidae (1890), Ceratosauridae, Coeluridae, Diplodocidae (1884), Dryptosauridae, Nodosauridae (1890), Ornithomimidae (1890), Plateosauridae (1895), and Stegosauridae (1880). He also named many individual species of dinosaurs. The dinosaur Othnielia was named in 1977 by P. Galton as a tribute to Marsh, as was *Marshosaurus bicentesmus* (Madsen, 1976).

Marsupial mammals A group (specifically, an order) of mammals whose females give birth to young at a very early stage of development. These newborns complete their development while sucking in a pouch, which is a permanent feature on the female. Examples include kangaroos and opossums.

Martin, Larry Larry D. Martin is a paleontologist, author, professor of ecology and evolutionary biology, and curator of vertebrate paleontology at the University of Kansas. Martin has theorized that birds evolved not from dinosaurs, but from another group of reptiles - this theory has been rejected by the vast majority of paleontologists.

Mary leakey Along with her husband, Louis, and son, Richard, Mary Leakey devoted her life to the recovery and interpretation of the bones and tools of early humans in East Africa.

Mason, Ruth Ruth Mason (-1990) found a huge dinosaur fossils bone bed (a collection of thousands of fossils) on her family's Harding County, South Dakota, USA, ranch when she was 7 years old. Since then, tens of thousands of dinosaur fossils have been found at the "Ruth Mason Quarry," near Faith, SD. The dinosaurs include huge numbers of Edmontosaurus annectens (duck-billed, plant-eating dinosaurs), T. rex teeth, and others.

Massa A specialised structure composed of aborted spores and tapetal material found on certain megaspores and often referred to as a floating apparatus. Applied only to megaspores. Examples: *Azolla* (Azollaceae), *Cytosporites varius*.

Massula A general term for aggregations of pollen grains dispersed as a unit Examples: Mimosaceae, Periplocaceae.

Mastaba The Arabic word meaning; "bench". Used to describe the private tombs of the Early Egyptian Dynastic Period and Old Kingdom. Rectangular in shape with slightly battened sides; the basic form resembled a bench.

Mastodon An extinct elephant having a straight back. They were slightly smaller than the Mammoth. Their teeth exhibit a pattern of cone-shaped cusps ideal for browsing. They ate leaves and branches gathered with their trunk from their forest habitat.

Material culture Those tangible aspects of a culture (such as artifacts, features, architecture, etc.) as opposed to non-material traits (beliefs, attitudes, etc.).

Matrilineal Pertaining to the tracing of descent or the transmission of wealth or authority primarily through females.

Matrilocal Pertaining to the custom of a newlywed couple's taking up residence in the home, village or territory of the bride's family.

Matrix A substance in which rocks, minerals, fossils or other similar materials are embedded. In paleontology, matrix is the rock in which a given fossil is embedded.

Maul A heavy stone implement with blunted or rounded edges used for pounding and/ or crushing. The maul may be grooved to facilitate holding or hafting, and may be classified according to the extent to which the groove travels the circumference of the tool (for example, three-quarter grooved maul, full-grooved maul, etc.).

Maunder minimum Period of solar minimum from A.D. 1645 to 1715, during which sunspots and auroras were much less common and the 11-year sunspot cycle did not exist.

Mausoleum A stately and magnificent tomb, such as the mausoleum of King Mausolos (ruler of Caria, c. 350 A.D.) at Halicarnassus in Asia Minor, and the mausoleum of Augustus (d. 14 A.D.) in Rome.

Maya The Maya originated around 2600 B.C. and rose to prominence around A.D. 250 in present-day Mexico, Guatemala, Belize, Honduras and El Salvador. Inheriting the inventions and ideas of earlier civilizations, the Maya developed astronomy, calendrical systems, hieroglyphic writing, ceremonial architecture, and masonry without metal tools. Maya civilization started to decline around A.D. 900, although some peripheral centers continued to thrive until the Spanish conquest in the early sixteenth century.

Mayan The language group of the Maya peoples, composed of 31 mutually unintelligible languages.

Mayr, Ernst Mayr's work has contributed to the synthesis of Mendelian genetics and Darwinian evolution, and to the development of the biological species concept. Mayr has been universally recognized and acknowledged as one of the leading evolutionary biologists of the 20th century.

McGinnis, William Professor of Biology, University of California at San Diego. Discover (with Mike Levine) of homeoboxes, the sequences of DNA that are characteristic of homeotic genes, which play a central role in specifying body development. His current research uses both genetics and biochemistry to examine such questions as how molecular variations in the Hox genes that specify the head-tail pattern of an organism can generate variety in animal shapes during evolution, and what the molecular changes were that allowed single celled animals to become multicellular.

Mcintosh, Jack Dr. Jack McIntosh is currently the foremost expert on sauropods. McIntosh corrected the naming of many sauropods originally named by O.C. Marsh and E.D. Cope. McIntosh has also done much to popularize the use of the name Apatosaurus (rather than Brontosaurus). His most important contribution to paleontology was identifying the correct skull for Apatosaurus in 1975 (fifty years earlier, Marsh had put a Camarasaurus skull on the Apatosaurus' body).

Mckean. An archaeological site in northeastern wyoming which has given its name to a distinctive late middle prehistoric period projectile point style and the complex with which it is found in association. The mckean lanceolate point is narrow, generally leaf-shaped with a deeply concave, occasionally notched base. Length ranges from 25 to approximately 60 mm. Duncan and hanna projectile points are included within the complex and appear to be somewhat more recent than the more classic lanceolate variety. Other artifacts in the mckean inventory include a variety of scraper styles including those made on tabular fragments of stone, oval blanks or bifaces, crude choppers, polyhedral cores, bone scrapers and beamers and gaming pieces fashioned from coyote teeth. Radiocarbon dates on mckean materials from wyoming may predate 3000 b.c. although those from manitoba range from as early as 1780 b.c. to as recent as 880 b.c. thus it would appear that the complex originates outside the province and that although there is considerable overlap, mckean is generally more recent than oxbow, but older than larter (pelican lake). Bison is the most common faunal element found in association with the complex although it is clear that these people supplemented their diet with a number of other animals, birds and fish. Mckean sites are fairly common over much of the northern plains and in manitoba, excavated sites and surface finds are distributed across most of the southern half of the province.

Medicine society A secret society of Native groups of the central Canadian forests and northern plains. Meetings were held from time to time in specially constructed secret lodges. The ceremonies, conducted by individuals imbued with magical power, centered upon the initiation of new members, healing, and the renewal of the group's contact with the supernatural.

Medicine wheel A circular arrangement of stones with one or more spokes either enclosed by the circle or radiating from it, and occasionally cairns centrally and/or peripherally located. These features, scattered across the northern plains from Wyoming to Manitoba probably served some ceremonial function(s).

Medine A term applied to a faintly lamellated, acetolysis resistant layer considered to be situated between intine and exine.

Mediportal Describes an animal built primarily for slow terrestrial locomotion as a result of its comparatively heavy body weight, but with a few adaptations for speed, such among the rhinoceratidae.

Megafauna Animals weighing more than 100 pounds when alive. The Mammoth, Mastodon, Giant sloth, and Short-faced cave bear would have been megafauna.

Megalith An arrangement or structure of extremely large stones.

Megaspore A general term for large spores of heterosporous vascular plants.

Meiosis A special kind of cell division that occurs during the reproduction of diploid organisms to produce the gametes. The double set of genes and chromosomes of the normal diploid cells is reduced during meiosis to a single haploid set in the gametes. Crossing-over and therefore recombination occur during a phase of meiosis.

Meiospore General term for a spore produced by meiosis.

Meleager Mythical hero who led the hunt for the Calydonian Boar and who died in his youth.

Mellisopalynology The study of pollen in honey or other bee products such as "bee baskets."

Membrana colpi Synonym of colpus membrane.

Membrana pori Synonym of pore membrane.

Meme The word coined by Richard Dawkins for a unit of culture, such as an idea, skill, story, or custom, passed from one person to another by imitation or teaching. Some theorists argue that memes are the cultural equivalent of genes, and reproduce, mutate, are selected, and evolve in a similar way.

Memnon King of Ethiopia killed by Achilles in the Trojan War.

Mendel, Gregor An Austrian monk whose plant breeding experiments, begun in 1856, led to insights into the mechanisms of heredity that are the foundation of genetics today. His work was ignored in his lifetime and only rediscovered in 1900.

Mendelian inheritance The mode of inheritance of all diploid species, and therefore of nearly all multicellular organisms. Inheritance is controlled by genes, which are passed on to the offspring in the same form as they were inherited from the previous generation. At each locus an individual has two genes — one inherited from its father and the other from its mother. The two genes are represented in equal proportions in its gametes.

Mercury Roman name for the Greek god Hermes, messenger of the gods.

Meridional Describing longitudinal features on the surface of a pollen grain or spore which run along lines perpendicular to the equator.

Meridionosulcus A meridional sulcus. Example: *Degeneria vitiensis* (Degeneriaceae).

Mesic Pertaining to or flourishing in conditions of medium dampness.

Mesine A laminated, electron-dense layer considered to lie between the intine and the exine.

Probably the same as the medine, but defined on different methods of microscopy.

Meso A prefix meaning middle.

Mesoamerica The area between northern Mexico and Costa Rica.

Mesoaperture The middle part of a compound aperture in which there is also an ectoaperture and an endoaperture. Example: *Polygonum aviculare* (Polygonaceae).

Mesocephalic Pertaining to a head of medium width in relation to its length; having a cephalic index between 75 and 79.9.

Mesocolpium and mesoporium The area of a pollen grain surface delimited by lines between the apices of adjacent colpi or the margins of adjacent pores.

Mesolithic The period following the Paleolithic during which adaptation involved intensive foraging and experimentation with new food resources.

Mesophyte A plant which thrives under conditions of average moisture.

Mesopotamia Considered to be the birthplace of the first civilizations, Mesopotamia is located in the valley between the Tigris and the Euphrates rivers (present-day Iraq).

Mesozoic The period of time, marked with the great expansion of life in the sea and on the land. The era stretched from 248 to 65 million years ago, and included the evolution and demise of the dinosaurs.

Mesquite A thorny plant that ranges from shrub to tree size. It grows at elevations below 5000 feet from southern Kansas to southwestern California and northern Mexico.

Messenger RNA (mRNA) The molecule transcribed by DNA that then carries genetic information from a gene to the site in the cell where the information determines the order of amino acids in the synthesis of a protein.

Metabolism The chemical processes that occur in a living organism in order to maintain life. There are two kinds of metabolism: constructive metabolism, or anabolism, the synthesis of the proteins, carbohydrates, and fats which form tissue and store energy; and destructive metabolism, or catabolism, the breakdown of complex substances, producing energy and waste matter.

Metallurgy 1. The art and/or science of the extraction of metals from their ores and the working of metals including the use of such techniques as the application of heat.

2. The science concerned with the study of metals.

Metamorphic Pertaining to a rock which has been modified from its original structure by exposure to heat and pressure.

Metamorphic rocks Rocks subjected to metamorphism (High pressure and temperature).

Metamorphosis One or more changes in form during the life cycle of an organism, such as an amphibian or insect, in which the juvenile stages differ from the adult. An example is the transition from a tadpole to an adult frog.

The term "complete metamorphosis" is applied to insects such as butterflies in which the caterpillar stage is distinct from the adult. "Incomplete metamorphosis" describes the life histories of insects such as locusts in which the young go through a series of larval stages, each of which bears similarities to the adult. Metamorphosis in both insects and amphibians is controlled by hormones, and often involves considerable destruction of larval tissues by enzymes.

Metareticulum A reticulum wich is characterized by the consistent presence of one porate aperture in each lumen. Examples: *Froelichia floridana* (Amaran-thaceae), *Vivania rosea* (Vivianiaceae), *Kallstroemia maxima* (Zygophyllaceae).

Metate The lower, stationary stone used in conjunction with the mano for grinding grain.

Metazoans All animals that are multicellular and whose cells are organized into tissues and organs. In the simplest metazoans only an inner and outer layer can be distinguished.

Meyer, Hermann Von Hermann von Meyer was a German paleontologist who named and described Archaeopteryx; (1861), Rhamphorhynchus (1847), Plateosaurus (1837), and Stenopelix (1857).

Micro A prefix for small.

In palynology, generally used to denote features less than 1μm. Examples: microechinate, microverruca.

Microbe A nonscientific and very general term, with no taxonomic significance, sometimes used to refer to microscopic (not visible to the unaided eye) organisms. The term often refers to bacteria or viruses that cause disease or infection.

Microblade A minute parallel-sided blade (sense 4) generally produced by pressure flaking from a specially prepared core.

Microevolution Evolutionary changes on the small scale, such as changes in gene frequencies within a population.

Microforaminifera Calcareous, unicellular marine organisms, with alternating or spirally-arranged progressively larger cells. The chitin ous inner linings are common in pollen preparations of marine or estuarine sediment.

Microfossil A fossil so small that it must be studied with a microscope.

Microlith Small tools which may be any of a variety of shapes, and which have been produced from microblades. Too small to have been used without hafting, some were set edge-to-edge in a groove in a bone or wood shaft and so served as cutting tools, while others would have been functional as barbs

Micrometer A unit of measure. There are one million micrometers in one meter.

Microreticulum A reticulate ornamentation consisting of muri and lumina smaller than 1μm.

Microspore A general term for the smaller spores of heterosporous plants, that is, the spores from which the microgametophyte develops

Antonyms: macrospore, megaspore.

The term is sometimes used in palaeopalynology for small spores generally, whether homospores, micros-pores or even small megaspores. This usage should be avoided, in favour of the terms miospore or small spore, but only if it is not known the plant is homosporous or not. A pollen grain, is the microgametophyte of a seed plant. In developmental studies, a pollen grain is sometimes referred to as microspore until microspore mitosis.

Microsporocyte The mothercell of a microspore or pollen grain .

Midden A heap or stratum of refuse generally located near a habitation site

Middle east The area (sometimes called the Near East) from Afghanistan to Egypt.

Middle kingdom Egyptian chronological term for dynasties 11-13.

Middle prehistoric period The second of the three "stages" in central north american prehistory. The term is most commonly used of the plains sequence, but is roughly equivalent in terms of its dating to the archaic period of the eastern forests. Included within this period in manitoba are the logan creek, oxbow, mckean and pelican lake (or larter) cultures. Some would include part of the besant phase within the middle period also as the bow, the introduction of which is frequently used to mark the onset of the late or woodland period, appears during the besant times.

Midewiwin Also known as the mid, or grand medicine society.

Midland A town in west texas which has given its name to a palaeo-indian projectile point style, to the complex found in association with it, and to the (sub-) phase within which it occurs. Originally, midland points were called "unfluted folsoms" because of similarities in outline shape, manufacturing technology, and a tendency to occur in association with folsom points. It is now clear, however, that many of these are much too thin to have been intended for fluting and most consider midland to be a distinct type. Other artifacts in the complex include a variety of side- and end-scrapers, and knives, spokeshaves, gravers, perforators and denticulates. Midland sites are more or less restricted to the southern plains and are approximately the same age as folsom. Irwin (1971) considers midland to be the second (after folsom) of the two subphases of the lindenmeier phase. He dates the midland subphase at 8750 to 8500 b.c.

Midwestern taxonomic method The system devised by w.c. mckern (1939) for classifying archaeological data. It was he who first explicitly defined the focus, aspect, phase, pattern and base and how these concepts relate to one another. Not all of these terms remain in use, and those which have survived are used in different ways from that proposed by mckern. This scheme is also (incorrectly) known as the midwestern taxonomic system and the mckern taxonomic system.

Migrate To move from one region, climate or environment to another.

Migration 1. The regular movement of animals back and forth between their summer and winter ranges.

2. The movement of a large group of people to a new, permanent residence.

Milankovitch mechanism Mechanism for climatic cycles resulting from changes in latitudinal distribution of solar radiation due to variations in Earth's orbital geometry around the sun.

Miller, Geoffrey Author of *The Mating Mind,* Miller is known for his research on evolutionary psychology and sexual selection. He believes that our minds evolved not only as survival machines, but also as courtship machines — at least in part, to help us attract a mate and pass on genes.

Miller, Ken A cell biologist and professor of biology at Brown University. Miller's academic research focuses on the structure and function of biological membranes. He is the coauthor of widely used high school and college biology textbooks, and he has also written *Finding Darwin's God: A Scientist's Search for Common Ground Between God and Evolution.*

Miller, Veronica A German virologist whose research has focused on HIV-AIDS. Miller was the first researcher to announce that an interruption in drug treatment among AIDS patients may result in reversion of drug-resistant virus to its wild type. This led other researchers and clinicians to explore "structured treatment interruptions" among some patients as an experimental treatment option.

Milnesand A town in eastern new mexico which has given its name to a plano projectile point style. The milnesand point is superficially simlar to the plainview style, but lacks the deeply concave base of the latter, and has more extensive lateral grinding. The generally lanceolate form tapers gently to a straight or slightly concave or convex base which is thinned by the removal of numerous minute flakes. The faces generally exhibit transverse horizontal flaking and length ranges from 4 to 7.5 cm. Although generally viewed as a southern plains style, similar forms have been found as far to the northeast as iowa. At the type site the points were found in association with scrapers and badly decomposed bison bone.

Mimicry A case in which one species looks more or less similar to another species.

Minaret A tall slender tower on or near a mosque, from which a muezzin summons the people to prayer.

Mineral A substance found in the earth that always has the same properties. These properties include color, hardness, shininess, and the way the mineral breaks or splits.

Mineralization The process of the replacement of organic material (as in the case of the fats and gelatin in bone) by minerals. The end product is an inorganic object (which may therefore survive for immense periods of time in the archaeological record) in the shape of the original.

Minerals Inorganic, naturally occurring materials consisting of specific elements. They are usually rocks with a characteristic crystalline structure and other identifiable physical traits. Quartz, talc, and ordinary table salt are examples of minerals.

Minerva Roman goddess of wisdom. Called Athena by the Greeks.

Miocene A epoch of the upper Tertiary period, spanning the time between 23.8 and 5.3 million years ago. It is named after the Greek words "meion" (less) and "ceno" (new).

Miospore A general term for all fossil plant spores smaller than 200μm, regardless of whether they are isospores, microspores, small megaspores, prepollen or pollen grains.

Antonym: macrospore.

Mississippian A period of the Paleozoic era, spanning the time between 360 and 325 million years ago. It is named after the Mississippi River valley, which contains good exposures of rocks of this age.

Mississippian The general name given to the cultural climax of the eastern United States between A.D. 600 and l600. The core traits include small triangular arrowheads, shell-tempered pottery, and flat-topped earthen pyramids. Some of the latter achieved impressive size; the largest at Cahokia in Missouri stands 30 m high, measures 2l5m by 330m and covers l6 acres. Thatched temples were constructed on top of the mounds which together with other encircled a plaza. Beyond these lay a palisaded earthen embankment and beyond that, a moat. These mound clusters likely served as administrative-ceremonial centers for wide areas. Most people lived in "rural" areas, practised maize, bean and squash horticulture, and by their efforts supported the divine rule, the artisans and the governing elite. Sites are located in major valleys of the Mississippi and its tributaries where flooding would regularly refertilize the farmlands. Mississippian craftsmanship in ceramics, stone, copper and shell was superb and in some cases, art styles bespeak influence from or contact with the high cultures of Mexico. Among the most spectacular aspects of the Mississippian are the artifacts and depictions surrounding the Southern Cult which seems to have concerned itself almost exclusively with death and human sacrifice.

Mississippian period 345 to 310 million years ago when Kentucky was covered by a warm, shallow sea in which corals, brachiopods, crinoids, blastoids, bryozoans, and foraminifera (a protozoan) flourished. The sedimentary rocks and most fossil remains of the Mammoth Cave region are from this time period.

Mitochondrial DNA DNA found in the mitochondrion, a small round body found in most cells. Because mitochondria are generally carried in egg cells but not in sperm, mitochondrial DNA is passed to offspring from mothers, but not fathers.

Mitochondrion A kind of organelle in eukaryotic cells. Mitochondria produce enzymes to convert food to energy. They contain DNA coding for some mitochondrial proteins.

Mitosis Cell division. All cell division in multicellular organisms occurs by mitosis except for the special division called meiosis that generates the gametes.

Mixed site A (generally) multi-component site in which the artifacts from the various occupations cannot be distinguished on stratigraphic grounds. Although they may be rich in terms of artifact yield, they may provide little information of significance to the archaeologist due to lack of provenience data.

Mixtec The Native American people populating the region in the modern Mexican state of Oaxaca.

Moai Giant statues of human figures found on the Polynesian island called Easter Island or Rapa Nui.

Mode 1. A way of doing or being.

2. In ceramic analysis, a combination of decorative elements which commonly occur together.

Modern synthesis The synthesis of natural selection and Mendelian inheritance. Also called neo-Darwinism.

Mogollon The prehistoric cultural groups who lived in an area which includes southern New Mexico, southeastern and central Arizona, the El Paso area in Texas, and northwestern Chihuahua Mexico. The Mogollon lifeway began about the same time the Hohokam entered southern Arizona. Although they grew crops, the Mogollon continued to make extensive use of native food resources. Around A.D. 1000, they became influenced by the Anasazi, another southwestern group. The Mogollon culture died out around A.D. 1450, and the people abandoned the Mogollon region, some perhaps merging with the Zuni.

Moh scale A method for describing the hardness of solids on a 10-point scale ranging from the hardness of talc (0) to that of diamond (10).

Mohave basin (sometimes "Mojave") The geographic area encompassing the Mohave Desert in California, north of the Imperial Valley and south of Death Valley.

Mold A cavity in which a substance is shaped. A fossil used to create a replica, or cast.

Molecular clock The theory that molecules evolve at an approximately constant rate. The difference between the form of a molecule in two species is then proportional to the time since the species diverged from a common ancestor, and molecules become of great value in the inference of phylogeny.

Molecular geneticists Scientists who study genes and characters through work with the molecules that make up and interact with DNA.

Molecular systematics A quite unfortunate term that has been introduced for the use of molecular data as characters for phylogenetic analyses. These include methods that are based on overall similarity (Phenetics), like electrophoresis, immuno-distance and DNA-DNA-hybridisation, as well as methods that are based on parsimony (Cladistics), like restriction-site-analysis and sequencing (proteins, RNA, DNA). Actually the term "molecular systematics" is quite misleading, since there exists of course only biological systematics, which is using different methods and different sources of evidence. One could as well coin terms like "osteological systematics" for studies based on skeleton characters, or "behavioural systematics" for studies based on behavioural characters. It has been claimed that molecular data (especially DNA-sequences) are superior to morphological characters, because they are much more numerous, they are easier to define (they are linear rather that multidimensional), and they are representing the basic entity of evolution (the genome). Nevertheless the promises have not been fulfilled yet, and there are only few solutions to old phylogenetic problems that came from molecular evidence, compared to the numerous published "molecular" phylogenies that appear to be quite absurd, e.g. the recent claim that guinea-pigs should not be rodents but the sistergroup of camels!

Mollusk An invertebrate that has a fleshy, muscular body. The phylum Mollusca includes snails, bivalves, squids, and octopuses.

Monad A pollen grain or spore dispersed as an individual unit, rather than in association with others, such as in a dyad, tetrad or polyad.

Monkey trial In 1925, John Scopes was convicted and fined $100 for teaching evolution in his Dayton, Tenn., classroom in the first highly publicized trial concerning the teaching of evolution. The press reported that although they lost the case, Scopes's team had won the argument. The verdict had a chilling effect on teaching evolution in the classroom, however, and not until the 1960s did it reappear in schoolbooks.

Mono A prefix for one.

Monoaperturate Describing a pollen grain or spore with a single aperture.

Examples: monocolpate, monoporate monosulcate

Monodontidae Faimily of cetaceans including the Beluga andthe Narwhal.

Monogamy A reproductive strategy in which one male and one female mate and reproduce exclusively with each other. Contrast with polygyny and polyandry.

Monolete Describing a spore with a single laesura. Example: *Dryopteris* (Dryopteridaceae).

Monolithic Something which is characterized as a single, often rigid, uniform whole.

Monophyla (*sensu* hennig) Are real spatio-temporal entities of nature, that exist totally independent from our possibilities to discover and distinguish them by certain characters. Since they possess some of the properties of logical individuals, but completely lack the internal cohesion that is usually postulated for the latter, they should be regarded as "historical entities" (*sensu* wiley). Groups that include a common ancestor and some (but not all) of its descendants, were often called "monophyletic", too (monophyla *sensu* mayr or remane), but these do not really exist in nature (independent from human recognition), but are mere constructions of the human mind (logical classes), that are arbitrarily defined by certain characters. According to this untenable definition of monophyly, any group of organisms whatever would be "monophyletic".

provided that life originated only once on earth. Example: A group including all roses and all eagles would be "monophyletic", too, since they do posses a common ancestor in the stem-species of all eukaryotic organisms. Of course such a group only makes biological sense if all other eukaryotic organisms, that all descended from the same stem-species, are included in this group, which is then representing the monophyletic taxon Eukaryonta.

Please note: If a monophyletic group is surrounded with a circle in a phylogenetic tree, only one branch is entering the circle, and none (!!!) is leaving the circle.

Monophyletic A group composed of a collection of organisms, including the most recent common ancestor of all those organisms and all the descendants of that most recent common ancestor. A monophyletic taxon is also called a clade.

Monophyletic group A set of species containing a common ancestor and all of its descendants, and not containing any organisms that are not the descendants of that common ancestor.

Monosaccate Describing a pollen grain with a single saccus. Example: *Florinites antiquus.*

Monospecific Describes a grouping, assemblage, collection, etc., that is comprised of only one species.

Monotreme Member of a lineage of egg-laying mammals. The platypus and echidna are the only surviving members of this lineage

Monotremes A group (specifically, an order) of mammals whose females lay eggs. The young hatch and continue to develop in the mother's pouch, which is present only when needed. Two species of spiny anteater and the duck-billed platypus are the only living monotremes.

Monstrous visual signal Large constructions that serve as foci of community integration, such as temples, palaces and sports arenas.

Moore, James The author, with Adrian Desmond, of an authoritative biography of Charles Darwin, Moore has made a 20-year study of Darwin's life. With degrees in science, divinity, and history, he has taught the history of science at Harvard University and at the Open University in the U.K.

Moraine General term refering variously to glacial deposits or landforms. Used in America mainly for glacial landforms of constructional nature, such as end moraine or ground moraine. Used in Europe also to refer to the glacial deposits themselves, as with moraine clay or sand. Also refers to sediment in transport by a glacier, such as medial or lateral moraines. Derived from French word for piles of rubbly deposits in front of glaciers.

Morphologic sequence Refers to a single downstream progression of landforms and stratified drift leading from an ice-contact position to distal river, lake, or sea.

Morphology The physical form or structure of an organism or another object. Species are usually identified from the fossil record based on morphological traits.

Morphon A group of form-species exhibiting continuous variation of morphological characteristics in a single time sequence ("horizontal" variation).

Many authors use the word "complex" in a very similar way.

Morter A (generally) bowl-shaped vessel of some hard material in which objects are crushed, ground and/or mixed with a pestle

Mortlach sub-phase. One of the eight regional subphases which comprises the late middle prehistoric period pelican lake phase. Mortlach (named after a town 80 km west of regina) occupies most of the grasslands of southern alberta and saskatchewan, but makes only a limited incursion into southwestern manitoba. Here, the larter subphase (or phase) is the dominant representative of this larger classification (reeves 1983).

Mortuary cult People who provided funerary offerings for nourishment of the deceased.

Mortuary vessel A vessel specifically manufactured for inclusion with the dead.

Mosaic An art form in which small pieces of colored stone or glass (called tesserae) are cemented onto a surface to create a picture or design. The ancient Greeks often decorated their floors with such pieces of art; later the Romans decorated both their floors and walls with mosaics. The practice was also adopted by Christians in the A.D. 500s and used to decorate churches.

Mosque Arabic for "place of prostration"; a Muslim temple of worship.

Motif A recurring feature or shape within a design or pattern.

Mound builders Any of the prehistoric cultures of eastern North America that erected mounds.

Mound. See burial mound

Mount vesuvius An active volcano located about 6 miles north of the city of Pompeii in what is today Italy. The enormous eruption of Mt. Vesuvius in A.D. 79 destroyed the city of Pompeii and many of the surrounding towns and villas.

Mousterian Old World archeologic culture attributed to Neanderthal Man, 50,000 to 100,000 years BP.

Mudrock, mudstone Siliclastic Sedimentary rocks with an average grain size <0.2mm.

Mueller, Ulrich G. A zoologist and professor whose research aims at understanding microevolutionary forces and macroevolutionary patterns that govern the evolution of organismal interactions, particularly the evolution of mutualisms and the evolution of social conflict and cooperation. Mueller's current research focuses on the coevolution between fungus-growing ants and their fungi and the evolutionary ecology of halictine bees.

Muller A non-stationary grinding stone such as a mano or pestle.

Multibaculate Synonym of pluricolumellate.

Multi-component site An archaeological site which contains the remains of more than one culture. If two or more complexes occur in a stratified site, there is a possibility that the artifacts of each component may be distinguished thus providing an excellent opportunity for the study of cultural change over time. If on the other hand, a number of components co-occur in a mixed site, the information to be gathered is little more than that from a surface collection.

Multiplanar tetrad A tetrad in which the individual members are arranged in more than one plane.

Decussate and tetrahedral tetrads are multiplanar, whereas rhomboidal, linear and T-shaped tetrads are; uniplanar.

Multituberculate Member of a diverse lineage of rodent-like non-therian mammals that were common from the late Jurassic to the early Eocene. Monotremes may be their closest living relatives.

Mummification Artificial process of preserving a corpse through dehydration.

Mummy cave An archaeological site in northwestern wyoming which has lent its name to an early middle prehistoric period complex of the northern plains. On the eastern slopes of the rocky mountains, the mummy cave complex is heralded by the first appearance of side-notched projectile points after the demise of the late plano lanceolate styles. As is the case in the contemporary logan creek complex to the east, the inventory includes side-notched end-scrapers and on this basis, plus other technological similarities, a close cultural relationship has often been suggested. Some sites yield evidence of the communal hunting of bison, while at others (including the type site), bison remains are conspicuously absent. Such discrepancies may be due to differences in the seasons during which these sites were occupied. Radiocarbon dating indicates a timespan from approximately 5800 b.c. to 3400 b.c.

Mummy From the Persian word; "moumiya". A preserved corpse by either natural or artificial means.

Munsell scale A standardized system using colored chips for accurately describing a color in terms of its hue, value and chroma.

Murornate sculpture elements Elevations of the general surface. Examples: cristae, muri.

Murray, Charles An author and policy analyst who has written many controversial and influential books on social policy. He is coauthor with Richard J. Herrnstein of *The Bell Curve: Intelligence and Class Structure in American Life* (1994). He has also written *Losing Ground: American Social Policy 1950-1980* (1984), which argues for the abolishment of the welfare system, *The Underclass Revisited* (1999), and *Income, Inequality and IQ* (1998).

Murus A ridge that is part of the ornamentation and, for example, separates the lumina in a reticulate pollen grain or the striae in striate pollen grain.

Muri in striate patterns are sometimes called valla (Iversen and Troels-Smith, 1950) or lirae.

Mutation A change in genetic material that results from an error in replication of DNA. Mutations can be beneficial, harmful, or neutral.

Mycenae Fortified city and palace which gave its name to a Bronze Age culture.

Mysticete A baleen whale.

Myths and legends Stories passed down through generations, usually about heroic individuals, spectacular events, or powerful gods. Myths represent a culture's beliefs and explain its customs; some are fictional while others may be based on real people and events.

N

N.T.S. Abbreviation for "national topographic series" maps published by the canadian department of energy, mines and resources. The most commonly used in archaeology are the l:50,000 series because of their detail and areal coverage. This scale is equivalent to 2 cm = l km.

Nadene A widespread linguistic phylum which includes the haida, tlingit and athapaskan families. Of these, only the latter is represented in manitoba by speakers of the chipewyan language who occupy the northwestern portion of the province.

Nagel, Ronald A hematologist and professor at Albert Einstein College of Medicine. His research includes molecular, biochemical, and physiological studies of genetic red blood cell defects, including sickle cell.

Nano A prefix for elements smaller than 0.5μm.

Napikwan A cultural tradition of the northern plains which in manitoba consists of the besant and blackduck phases (reeves l970).

Natal group The (residence) group to which one belongs upon birth as opposed to that which one joins by virtue of marriage.

Nation state A political unit consisting of an autonomous state inhabited predominantly by a people sharing a common culture, history, and language.

Natron A naturally occurring salt used as a preservative and drying agent during mummification. It is a mixture of four salts that occur in varying proportions: sodium carbonate, sodium bicarbonate, sodium chloride and sodium sulfate.

Natural environment Climate, terrain, geological resources, and flora and fauna.

Natural selection The differential survival and reproduction of classes of organisms that differ from one another in on or more usually heritable characteristics. Through this process, the forms of organisms in a population that are best adapted to their local environment increase in frequency relative to less well-adapted forms over a number of generations. This difference in survival and reproduction is not due to chance.

Neanderthal A hominid, similar to but distinct from modern humans, that lived in Europe and Western Asia about 150,000 to 30,000 years ago.

Near east The area (sometimes called Middle East) in the vicinity of the eastern end of the Mediterranean Sea.

Necropolis Normally describes large and important burial areas that were in use for long periods. From Greek nekros (corpse) and polis (city).

Negative reticulum A general term used to describe patterns of ornamentation in which sexine areas are separated by narrow, reticulately arranged grooves. Nektonic Used to describe aquatic organisms that swim.

Nelson, Craig A professor of biology and environmental affairs at Indiana University in Bloomington. His research focuses on evolutionary ecology.

Nemes A striped headcloth worn by Pharaohs.

Neo darwinism 1. Darwin's theory of natural selection plus Mendelian inheritance.

2. The larger body of evolutionary thought that was inspired by the unification of natural selection and Mendelism. A synonym of the modern synthesis.

Neo-atlantic A central North American climatic episode dating from A.D. 690 to A.D. 1100 which is believed to have been somewhat cooler and moister than the present.

Neo-boreal A central North American climatic episode, known to some as the "Little Ice Age", which dates approximately A.D. 1550 to 1850. Average temperatures were 1 C cooler throughout this period than the present.

Neoglaciation Series of relatively small glacier readvances during the last few thousand years of the Holocene.

Neo-indian See late prehistoric period.

Neolithic The first era of village farmers in any region.

Nephthys Egyptian goddess and protector of the dead. Sister of Isis, Osiris and Seth.

Neptune Roman god of the sea. Called Poseidon by the Greeks.

Nervous system An organ system, composed of a network of cells called neurons, that allows an animal to monitor its internal and external environment, and to move voluntarily or in response to stimulation.

Net-impressed A surface finish of ceramics produced by impressing the wet clay with a net. Depending upon how this is accomplished, the vessel's surface may be the precise imprint of the mesh or a much more irregular, roughened pattern.

Netsinker A notched or grooved cobble attached to a net in order to sink one edge.

Neural Related to nerves and neurons.

Neutral drift Synonym of genetic drift.

Neutral mutation A mutation with the same fitness as the other allele or alleles at its locus.

Neutral theory (and neutralism) The theory that much evolution at the molecular level occurs by genetic drift.

Neutron activation A technique of nuclear chemistry by which minute amounts of elements in a material are identified.

Névé Zone of accumulation in high portion of glacier where firn is found.

New kingdom Egyptian chronological term for dynasties 17-20.

Newgrange Irish passage tomb of great beauty, built about 3200 BC and re-discovered in 1699.

Newton, Isaac An English physician and mathematician, considered the culminating figure of the scientific revolution of the 17th century. He is best known for his explanation of gravity and for laying the foundation for modern physical optics.

Nexine The inner, non-sculptured part of the exine which lies below the sexine.

Antonym: sexine.

Although the term endexine as originally defined (sensu Erdtman, 1943) was more or less synonymous with nexine it no longer is because endexine is now invariably used in the sense of Fægri (1956). Nexine and sexine are distinguished on purely morphological criteria, whereas ectexine and endexine differ in their staining properties. The two sets of terms therefore have slightly different applications.

Niche The ecological role of a species; the set of resources it consumes and habitats it occupies.

Nicholls, Elizabeth Elizabeth (Betsy) L. Nicholls is a paleontologist and the Curator of Vertebrates at the Royal Tyrrell Museum, Alberta, Canada. Dr. Nicholls studies Triassic period marine reptiles (especially Ichthyosaurs) from North America. Nicholls received her Ph.D. from the University of Calgary, Canada. Nicholls and Jack M. Callaway are the editors of the book, "Ancient Marine Reptiles," 1997. Nicholls has named the ichthyosaur genera Metashastasaurus (Nicholls and Makoto Manabe, 2001) and Parvinatator (Nicholls and Brinkman, 1995).

Nike Winged goddess of victory.

Nile river The largest river in Egypt. Yearly flooding brings water to the land surrounding the river and allows people to grow food along the banks of the river.

Nilometer Staircase descending into the Nile and marked with levels above low water; used for measuring, and in some cases recording, inundation levels. The most famous are on Elephantine island and on Roda island in Cairo.

Nilsson, Dan Erik Professor Nilsson heads the Functional Morphology Division of the Department of Zoology at Lund University in Sweden. His main research interest is the optics and evolution of invertebrate eyes.

Nilsson, Lennart A Swedish photographer who began as a photojournalist, Nilsson soon began exploring new techniques such as the use of endoscopes and electron microscopes to photograph the inner mysteries of the human body. He published a book, *A Child is Born* of his photographs of the beginning of life, and made a number of films, including the mini-series *Odyssey of Life* a coproduction between WGBH/NOVA and SVT Swedish Television.

Nitrogen fixation A chemical process by which nitrogen in the atmosphere is assimilated into organic compounds. Only certain bacteria are able to fix atmospheric nitrogen, which then becomes available to other organisms through the food chains.

Node based clade A clade defined as the most recent common ancestor of two or more taxa and all its descendants. For example, the node-based clade Allosauroidea is defined as the most recent common ancestor of *Allosaurus* and *Sinraptor* and all its descendants

Nodule A small, rounded lump of mineral, usually harder than the surrounding rock.

Nomad A member of a group of people who move according to the seasons from place to place in search of food, water, and grazing land.

Nomadic Having to do with nomads, people who live in no fixed place but move in search of food or grazing land for their animals; of a wandering lifestyle.

Nome From the Greek, nomos; this is an administrative province of Egypt (42 in all). The nome system started in the Early Dynastic Period. During some periods, when there was a highly centralized government the nomes had little political importance.

Nomen ambiguum ("ambiguous name") A name which has been used so long by different authors in different senses that it has become a persistent cause of error and confusion

Nomen conservandum n, pl. nomina conservata ("conserved names") Names classed as available and valid by action of the ICZN exercising its Plenary Powers. Includes rulings to conserve junior synonyms in place of rejected forgotten names (*nomen oblitum*). Such names are entered on the Official Lists

Nomen correctum n, pl. nomina correcta ("corrected names" or "improved names") Available names which are mandatory and allowable emendations of imperfect names or of taxonomic names higher than family (which are not subject to name form and ending regulations). Do not depend on transfer in taxon rank or assignment. (= an emended name).

Nomen dubium n, pl. nomina dubia ("doubtful" or "dubious names") Names which are not certainly applicable to any known taxon or for which the evidence is insufficient to permit recognition of the taxon to which they belong. May possess availability conducive to uncertainty and instability.

Nomen imperfectum n, pl. nomina imperfecta ("imperfect names") Available names which when originally published met all mandatory requirements of the Code (i.e. met all criteria of availability - Articles 10-20) but which contained some defect needing correction, such as names incorporating hyphens, diacritical marks or apostrophes, and higher taxon names using an incorrectly formed stem.

Nomen inquirendum n, pl. nomina inquirenda ("names under enquiry")

Nomen inviolatum n, pl. nomina inviolata ("inviolate names") All available names not subject to any sort of alteration from their originally published form (i.e. they were correct as originally published and need no correction or emendation).

Nomen negatum n, pl. nomina negata ("denied names") Unavailable names which are incorrect original spellings as defined by the Code (Article 32c).

Nomen non rite publicatum Name not properly published.

Nomen novum ("new name") Name which is expressly proposed as a replacement name for a preoccupied name and automatically takes the same type and type locality (= a replacement name or substitute name for a preoccupied name). Commonly applied to names proposed to replace junior homonyms.

Nomen oblitum n, ("forgotten names") Senior synonyms which have remained unused in the literature for more than 50 years. Not to be used unless specifically so directed by the ICZN.

Nomen perfectum n, ("perfect names") Available names which when originally published met all mandatory requirements of the Code and needed no correction of any kind, but which are validly alterable by change of ending.

Nomen substitutum n, ("substitute names") Available names proposed as replacements for invalid names, such as junior homonyms. A substitute name proposed specifically for a preoccupied name is a *nomen novum.*

Nomen translatum n, ("transferred names") Available names which have been altered (usually in endings only), the change depending on transfer from one taxonomic rank to another, or from one taxon to another (e.g. *Spirifer ambiguus* changed to the new combination *Composita ambigua*, Orthidae downgraded to Orthinae, etc.). Authorship and date remains as for original name.

Nomen triviale ("trivial name") An expression used by Linnaeus and others for the specific name (e.g. *rex* in *Tyrannosaurus rex*). Applied by some authors in same sense as vernacular name.

Nomen vanum n, ("vain names") Available names consisting of unjustified but intentional emendations of previously published names. Have status in nomenclature with their own authorship and date.

Nomen vetitum n, ("impermissible names")

Unavailable names published for divisions of the genus group other than genus and subgenus, which are not accepted by the Code (e.g. a subsection of a subgenus).

Non aperturate Synonym of inaperturate.

Non avian dinosaur Describes any member of clade Dinosauria that is not a bird – dinosaur (common usage) *sy.*

Non-literate Pertaining to peoples who lack a written language.

Norman, David David B. Norman is a British paleontologist who has extensively studied ornithischian clades, especially the genus Iguanodon. He noted that Iguanodon's beak structure would let it eat a wide range of plants, not simply soft plants as was thought before. He also realized that Iguanodon's tail was held horizontally, suggesting that it walked on four legs. He has also written many books about dinosaurs, including *The Illustrated Encyclopedia of Dinosaurs* (1985).

Normapolles A group of Cretaceous and Lower Palaeogene pollen, usually triporate, with a complex pore apparatus.

Northern transition zone The northern part of the Boreal Forest which marks the shift from the dense coniferous forest in the south and the open tundra to the north. Black spruce is the most common tree species although white spruce dominates on sand ridges and floodplains. Swamp and muskeg is common throughout. As one proceeds north, trees become less numerous and more stunted and permafrost becomes more widespread.

Notching A V or U-shaped indentation. basal notching. in projectile points, a deep often shallow indentation into the base; narrower and often more pronounced than a basal concavity. ceramic notching. the creation of notches, presumably as a decorative element, into the clay of a vessel prior to firing.

Notochord A flexible skeletal rod running the length of the body in the embryos of the chordates (including the vertebrates). In some simpler types, such as sea-squirts, only the free-swimming larva has a notochord; in others, such as the lancelets and lampreys, the notochord remains the main axial support, and in vertebrates it is incorporated into the backbone as the embryo develops.

Novacek, Michael J. Paleontologist with the American Museum of Natural History. Dr Novacek's research interests include evolution of and relationships among organisms, particularly mammals. Author of *Dinosaurs of the Flaming Cliffs*, an account of AMNH's Gobi Desert Expeditions.

Novas, Fernando E. Fernando E. Novas is an Argentinian paleontologist from the Museum of Natural History in Buenos Aires. He named Frenguellisaurus (1986), conamed Abelisaurus (with J. F.Bonaparte, 1985), Megaraptor (1998), Patagonykus (1996), and conamed Unenlagia (with Puerta, 1997).

NPC classification A morphological system for classifying pollen grains and spores that was based on the number, position and character of their apertures.

Nuclear areas Regions where large, complex societies arose again and again.

Nuclear family A family unit consisting of a married couple and their offspring. A grandparent or a spouse of one of the children may be included.

Nucleotide A unit building block of DNA and RNA. A nucleotide consists of a sugar and phosphate backbone with a base attached.

Nucleus A region of eukaryotic cells, enclosed within a membrane, containing the DNA.

Nudate Synonym of psilate.

Numerical taxonomy In general, any method of taxonomy using numerical measurements. In particular, it often refers to phenetic classification using large numbers of quantitatively measured characters.

Nunatak Eskimo term refering to mountains or lands that protrude through an ice sheet.

Nurse, Paul A pioneer in genetic and molecular studies who revealed the universal machinery for regulating cell division in all eukaryotic organisms, from yeasts to frogs to human beings.

Nutimik concave A projectile point style identified by richard macneish (1958) on the basis of his excavations in southeastern manitoba in the early 1950s. These are described as being long (70 to 80 mm) triangular points with slightly convex blade edges, concave bases, irregular oblique flaking and occasionally a slight amount of lateral grinding. Macneish suspected that the nutimik concave style may have evolved from earlier plano forms and attributed the former to his whiteshell focus which he dated at 3000 to 1500 b.c. the type and focus terms have falled into disuse by archaeologists.

Nutimik focus The later (a.d. 500 to 1000) of the middle or initial woodland cultural-historical periods in macneish's (1958) southeastern manitoba chronology. It was later designated the nutimik phase by mayer-oakes (1967). Hlady (1970) finally advocated the grouping of nutimik with the other middle woodland focus (anderson) due to the general absence of distinguishing artifactual traits and further proposed that this new entity be designated the laurel phase. The suggestion met with universal acceptance and the earlier term is no longer used.

O

Obelisk From the Greek word meaning; "a spit". It is a monumental tapering shaft usually made of pink granite. Capped with a pyramidion at the top. Obelisks are solar symbols similar in meaning to pyramids, they are associated with an ancient stone called benben in Heliopolis. They were set in pairs, at the entrances of temples, and to some Old Kingdom tombs. (e.g. Cleopatra's Needle in New York City) Modern obelisks include such things as the Washington Monument and are still popular as memorial markers in many parts of the world.

Obesity A condition in which a person's weight is 20 percent greater than a sex- and age-specific weight-for-height standard.

Oblate Describing the shape of a pollen grain or spore in which the polar axis is shorter than the equatorial diameter.

This term belongs to the system of shape classes suggested by Erdtman (extended in), based on the measurements of the polar axis (P) and equatorial diameter (E). In this system oblate is defined as a ratio between the polar axis and the equatorial diameter of 0.50-0.75.

Oblate spheroidal Describing the shape of a pollen grain or spore in which the ratio between the polar axis and the equatorial diameter is 0.88-1.00.

Obligate biped Describes an animal whose anatomy requires it to use only its two hindlimbs in walking or similar locomotio A bird is an example of an obligate biped.

Obligate quadruped Describes an animal whose anatomy requires it to use all four limbs in walking or similar locomotio An elephant is an example of an obligate quadruped.

Oblique Diagonal; slanting; lying at an angle.

O'Brien, Stephen J. A geneticist at the National Cancer Institute whose research interests include the evolutionary history of the immunological response in mammals to retroviruses like HIV. With his colleagues, he discovered a mutation that can protect individuals from infection by HIV, the virus that causes AIDS.

Observation Looking at and critically noting the details of a site, an artifact, or cultural behavior.

Obsidian A black, reddish, greenish or brownish volcanic glass which has cooled so quickly from its molten state that no crystals have formed. Due to its excellent flaking properties and the incredibly sharp edge it produces, obsidian was a highly prized lithic resource throughout the world in prehistoric times, and as a consequence was traded widely outside of the areas of its natural occurrence.

Obsidian hydration dating Potentially, a chronometric dating method based on the fact that obsidian, or volcanic glass, progressively develops a thin chemically altered outer layer due to the absorption of water. The thickness of this hydration layer is directly proportional to the amount of time since the rock was formed or since a fresh surface was exposed to the elements. Since the rate of hydration varies between samples from different volcanic sources, this technique could arguably be considered relative rather than chronometric.

Obverse The "head" side of a coin.

Occupation span The length of time a settlement is occupied.

Occupation surface A boundary layer between depositional strata upon which activities were carried out (also called a living floor)

Ocellus Synonym of cata-ulcus.

Ochre A general term for any of the clays or earths containing ferric oxide, silica and alumina. Ranging in colour from yellow through red to brown, ochre was widely used as a prigment for decorative and ceremonial purposes throughout much of prehistory

Ocotillo This plant, also called the coach whip, is characterized by clumps of straight, thorny whip-like stems with no branches. When there is adequate rainfall, the ocotillo leafs out, but loses its leaves when the soil dries. The plant has brilliant red flowers that occur at the tips of its many stems. Ocotillos occur below 5000 feet, from west Texas to southeastern California and northern Mexico.

Oculus The much enlarged part of the pore structure in pollen of the fossil Normapolles group. Example: *Oculopollis*.

The enlarged part is a swelling of the outer wall layer on one or both surfaces of a grain in the region of an exogerminal. An annulus is uniformly thick and completely surrounds the exogerminal region.

Odontocete Toothed whales. Includes Sperm whales, dolphins, porpoises, beaked whales, belugas, etc.

Off-site data Evidence from a range of - information, including scatters of artifacts and features such as plowmarks and field boundaries, that provides important evidence about human exploitation of the environment.

Ogives Alternating bands of dirty and clean ice on surface of valley glacier below an ice fall, annual banding formed by summer and winter movement over ice fall, also known as Forbes bands.

OL pattern A pattern of ornamentation that appears to show "dark islands" at high focus (H) and that become bright at low focus (L).

The reverse of a LO- pattern.

Old copper A late Archaic complex or culture, centered in the western Great Lakes region characterized by well-made copper artifacts. These include socketed, "rat-tail", lanceolate, conical and stemmed projectile points, large crescentic shaped objects, woodworking and hideworking tools, fishing equipment, and occasionally, ornamental objects. Associated artifacts include concave-based side-notched projectile points with square basal edges termed either Raddatz or Osceola, scrapers, drills, "burned" hornstone blades (sense 2), "turkey-tail" blades, bannerstones, triangular cache blades, ground axes, shell beads, bone awls, antler points, notched swan bones, elk antler axes, shell gorgets and antler shaft wrenches. Much of our information on this complex comes from cemeteries and as a consequence, the mortuary pattern is quite well-known. Both primary and secondary (bundle) single and multiple, flexed and extended interments are reported and these may occur in either oval or rectangular pits. Ochre, copper, lithic or faunal grave goods may accompany the deceased and the graves may have been ceremonially burned over. The grave escort or "retainer" phenomenon is also known for this culture. Old Copper peoples appear to have employed a diversified economy eccompassing the collection of nuts, acorns and perhaps wild rice in the fall, fishing in the spring, summer and early fall, the taking of wildfowl in the spring and fall, the trapping of small fur-bearing mammals throughout the year and the hunting of moose and deer. Old Copper artifacts have been found over much of southern Manitoba although they seem to be most common in the forests of the southeastern quarter of the province. The association of finished artifacts, "ingots" of unmodified copper, and waste fragments from tool manufacture suggests that these people imported copper for tool production, rather than receiving finished tools from outside the province. Radiocarbon dates on copper sites argue for a considerable time-depth and a substantial antiquity although most would agree that the period from 3000 B.C. to 1000 B.C. marks the peak of Old Copper usage in its "heartland". Radiocarbon dates from eastern Manitoba fall between 2000 and 1700 B.C. although it seems clear that Old Copper implements were made and used for a much longer period than this.

Old kingdom Egyptian chronological term for dynasties 3-6.

Old world/new world The Old World includes the continents of Europe, Asia, Africa, and Australia and all their associated land. The New World includes the continents of North and South America and their associated land.

Olduvai gorge One of the most important sites for understanding both human evolution and the development of the earliest tools. (around 1.9 million years ago)

Oligocene An epoch of the early Tertiary period, spanning the time between 33.7 and 23.8 million years ago. It is named after the Greek words "oligos" (little, few) and "ceno" (new).

Olivella A genus of small marine snails native to the Pacific coast of North America and elsewhere which were used prehistorically as ornaments and as a medium of exchange.

Olmec A highly elaborate Mesoamerican culture on the Mexican gulf coast which was at its height from 1200 to 600 B.C. The Olmec influenced the rise and development of the other great civilizations of Mesoamerica, such as the Maya, and were probably the first to develop large religious and ceremonial centers with temple mounds, monumental sculptures, massive altars, and sophisticated systems of drains and lagoons. The Olmec were probably also the first Mesoamericans to devise glyph writing and the 360-day calendar.

Olympia Site of Greek sanctuary to Zeus and home of the Olympic Games.

Olympus Mountain in the north of Greece believed to be the home of the gods.

Omniaperturate Describing a pollen grain in which the exine is very thin or absent and the intine is thick, so that no specific apertural region can be distinguished and thus the whole surface can be considered apertural in nature.

Omnivore An organism that eats both plants and animals.

Omnivorous Feeding on both animal and vegetable matter.

Oncus A lens-shaped structure that is not resistant to acetolysis and occurs beneath the apertures of many kinds of pollen grains. Example: *Corylus* (Betulaceae).

Open sites Any site not located in a cave or rock shelter.

Opening of the mouth This Egyptian ceremony was performed at the funeral to restore the senses of the deceased. The ceremony was done by touching an adze to the mouth of a mummy or statue of the deceased, it was believed to restore the senses in preparation for the afterlife.

Operculum A distinctly delimited sexineectexine structure which covers part of an ectoaperture and which is completely isolated from the rest of the sexine.

Opisthokonts A large and believed monophyletic group of eukaryotes, including both the animals and fungi, and some protists that have the common characteristic of flagellate cells to propel themselves.

Opposable thumb An anatomical arrangement in which the fleshy tip of the thumb can touch the fleshy tip of all the fingers.

Optical (cross) section The image seen in optical microscopy when the plane of focus is half way through a palynomorph.

Optical emission spectrometry A technique used in the analysis of artifact composition, based on the principle that electrons, when excited (i.e. heated to a high temperature), release light of a particular wavelength. The presence or absence of various elements is established by examining the appropriate spectral line of their characteristic wavelengths. Generally, this method gives an accuracy of only 25 percent and has been superseded by ICPS (inductively coupled plasma emission spectrometry).

Optically stimulated luminescence (OSL) dating A comparatively new radiometric dating technique similar to thermal luminescence. OSL is based on the fact that minerals in sediment grains are affected by prolonged exposure to light. Specifically, light causes electrons of the mineral atoms to be progressively dislodged. This provides a natural clock. So far, the OSL technique has been used to date silty or sandy water deposited sediments that are 1/2 million years old or younger.

Oral history Verbally transmitted information about past events. Although often providing information about non-written events, such history is subject to the vagaries of human perceptions and mental recall.

Orbicule A general term, applied in palynology for an orbicular granule of sporopolleninsporopollenin.

Synonym of Ubisch body.

Ordeal A painful and possibly life-threatening test inflicted on someone suspected of a wrongdoing.

Ordinate Describing a pollen grain or spore with an arrangement of elements regularly distributed.

Antonym: inordinate.

The elements can be of structural or sculptural origin. Examples: columellae under a tectum forming a reticulum (example: *Vaccaria pyramidata*, Caryophyllaceae); scabrae on the tectum (example: *Alchemilla glabra*, Rosaceae) arranged in a regular pattern.

Ordovician The second earliest period of the Paleozoic era, spanning the time between 505 and 440 million years ago. It is named after a Celtic tribe called the Ordovices.

Organelle Any of a number of distinct small structures found in the cytoplasm (and therefore outside the nucleus) of eukaryotic cells (e.g., mitochondrion and chloroplast).

Organic solidarity The unity of a society formed of dissimilar, specialized groupings, each having a restricted function (Durkheim).

Organism An individual member of a species; a single biological entity.

Orienteering Using a map and compass to navigate your way between checkpoints along an unfamiliar course.

Ornamentation A general term that is useful for describing the organisation of features.

Ornate Describing a reticulate ornamentation consisting of broad, curved muri and lumina that are often anastomosing. Example: *Ceiba aesculifolia* (Bombacaceae).

Ornithischia ("bird hips") Formerly the major grouping of dinosaurs that is comprised of the beaked dinosaurs. Padian and May (1993) formally defined Ornithischia as *Triceratops* and all taxa closer to *Triceratops* than to Saurischia. Dropped by this author in favor of Predentata – ornithischian.

Ornithopoda ("bird feet") Clade comprised of three-toed predentatans –ornithipod.

Orthocolpate Describing a pollen grain with the colpi in the most common position, perpendicular to the equator.

This term is not needed in most instances, where zonocolpate would suffice, but exists to contrast with the term loxocolpate.

Orthodox christians As members of the Eastern Orthodox Church, this branch of christianity is a product of Middle Eastern, Hellenic, and Slavic history and culture. It is seen as holding to traditional teachings and values born in Jerusalem.

Orthodox judaism The branch of Judaism that is committed to keeping its contract with the past. It stresses commitment to Jewish law and observance of all of the Jewish commandments and obligations.

Orthogenesis The erroneous idea that species tend to evolve in a fixed direction because of some inherent force driving them to do so.

Orthograde Vertical posture.

Osborn, Henry Henry F. Osborn (1857-1935) was a US paleontologist who found and named many dinosaurs in Mongolia and the US in the early 1900's. He was the curator of the American Museum of Natural History starting in 1891. He did extensive research on brontotheres (early, rhino-like mammals). Osborn also named and described the following dinosaurs: Albertosaurus (1905), Asiatosaurus (1924), Ornitholestes (1903), Oviraptor (1924), Pentaceratops (1923), Prodeinodon (1924), Psittacosaurus (1923), Saurornithoides (1924), Struthiomimus (1916), Tyran-nosaurus (1905), and Velociraptor (1924).

Osceola. An old copper archaeological site near potosi, wisconsin after which a side-notched projectile point was named. Those from the type site range in length from 3 to l3 cm, have roughly parallel blade edges, exhibit squared-off bases which are slightly concave. Broken points were reworked into smaller points or drills. In most respects, the osceola type is very similar to the raddatz style with which it is generally contemporaneous.

Osiris Egyptian god of the underworld and judge of the dead.

Ossification The process of bone formation.

Osteodontokeratic culture an archaeological culture based upon tools made of bone, teeth, and ivory.

Osteology The scientific study of the development, structure and function of bones.

Ostracoderms Primitive jawless fishes, covered by bony armor, that lived in the Cambrian through Devonian periods.

Ostracon From the Greek word meaning; "potsherd". A chip or shard of limestone or pottery used as a writing tablet. Ostraca are known from all periods.

Ostraderms Primitive jawless fishes, covered by bony armor, that lived in the Cambrian through Devonian periods.

Ostrom, John H. John H. Ostrom is a US paleontologist and author who found (1964) and named (1969) Deinonychus with Grant

E. Meyer et alia. He named Microvenator (1970), Sauropelta (1970), and Tenontosaurus (1970). Ostrom has championed the theory that birds arose from theropod dinosaurs. Ostrom also argued that the dinosaur's must have high body temperatures given their erect posture (1969). Ostrom's books include "The Strange World of Dinosaurs" (Putnam, 1964), "Dinosaurs" (Carolina Biological Supply Co., 1984), "Marsh's Dinosaurs : The Collections from Como Bluff" (with McIntosh and Dodson, Yale Univ Press, 2000), and others. The Cretaceous period bird/dinosaur *Rahonavis ostromi* was named to honor Ostrom.

Outbuilding A term used to refer to all nonresidential structures on a site. These include animal pens, storage buildings, sheds, barns, etc.

Outcrops A term designating the surface exposure of rock layers, which have not been decomposed into soil.

Outgroup In cladistics, a clade considered primitive to a group of organisms in a cladistic analysis of that group.

Outgroup comparison This procedure is more or less a fitting of new characters into well-supported phylogenetic trees (character optimization on a given cladogram). Being based on the principle of parsimony the outgroup comparison represents an indirect method to polarize a new character. The existence of evidence (synapomorphies) for the monophyly of the group in study is essential, as well as the knowledge of some of their closer phylogenetic relatives. If a character has two or more states within the (monophyletic) group in study, the state that is also occurring in the close phylogenetic relatives, which are not members of the monophyletic group in study, is regarded as plesiomorphic state. Example: The monophyly of mammals is well-supported by a lot of derived characters (e.g. hairs, milk glands, heterodont dentition, secondary jaw articulation, etc.). It is also undisputed that all other amniote vertebrates (lizards, snakes, turtles, crocodiles and birds) are more closely related to mammals that all other recent organisms. Within mammals there are two different modes of reproduction: egg laying (ovipary) in monotremes (Prototheria) and vivipary in marsupials (Metatheria) and placentary mammals (Eutheria). Since ovipary is also the mode of reproduction in the other amniotes (with rare exceptions) and even in most other animals, the ovipary of monotremes has to be regarded as a symplesiomorphy, that cannot demonstrate the monophyly of Prototheria, while vivipary is a shared derived character (a potential synapomorphy), that could demonstrate the monophyly of the taxon Theria (Metatheria and Eutheria). This method of phylogenetic outgroup comparison is something fundamentally different from the so-called outgroup-method in computer-cladism, in which an unrooted most parsimonious tree is calculated and a posteriorly rooted by choice of one of the analysed taxa as the outgroup (often a hypothetical ancestor with all character states coded as "0"). The latter procedure should better be termed "outgroup-rooting".

Outline A general descriptive word. Applied in descriptive terms like equatorial outline and outline in polar view.

Outwash Rubble (sand, gravel etc.) deposited by water derived from melting glaciers.

Overthrust Caused when pressure pushes rock strata up until one side folds over onto the second side. This will cause younger rock layers to be located under older layers.

Owen, Richard A 19th-century British comparative anatomist, who coined the word "dinosaur" to describe a breed of large, extinct reptiles. He was the first to propose that dinosaurs were a separate taxonomic group. Owen opposed Darwin's theory of evolution, but ultimately his work helped support evolutionary arguments.

Oxbow A town in southeastern saskatchewan which has lent its name to a middle prehistoric period projectile point style and the complex with which it is found in association. Oxbow points tend to be short and squat with bold side-notching and a basal concavity which lends them an eared appearance. The bases and notches may be ground. Other items in the inventory include "un-notched points" or preforms, a variety of scrapers, knives and bifaces, drills, perforators, choppers, hammerstones, anvils, antler awls, bone beamers and flaking tools and drilled clam shells. Although representations of a variety of floral and faunal species may be found in oxbow components, bison appear to have been central to the economy of these people. Oxbow sites are found across the grassland regions of the canadian prairie provinces and into the forests to the north. Most radiocarbon dates cluster between 3300 and 1000 b.c. although some would argue that it persists much longer. It would seem that the oldest sites are those from the grasslands with dates becoming progressively more recent to the north. This has led some archaeologists to believe that over time, oxbow populations became increasingly dependent upon the resources of the forests for subsistence. The origin of the oxbow complex is controversial and eastern as well as western origins have been suggested.

Oxidation The act of oxidizing or state of being oxidized. Chemically it consists in the increase of positive charges on an atom or the loss of negative charges. Most biological oxidations are accomplished by the removal of a pair of hydrogen atoms (dehydrogenation) from a molecule. Such oxidations must be accompanied by reduction of an acceptor molecule.

Oxidizable carbon ratio (OCR) dating An experimental dating technique based on the fact that organic carbon in soil humus and charcoal progressively convert to oxidizable carbon over time. The ratios of these two forms of carbon vary directly with the age of the material. This technique requires an inexpensive chemical analysis procedure. The effective time range for OCR dating has not yet been established.

Oxygen-isotope ratios Ratio of stable oxygen isotopes in various materials, such as sea water, glacier ice, and fossil shells. Thought to be a reliable indicator of past climatic conditions.

Ozone layer The region of the atmosphere, generally 11-26 km (7-16 miles) above Earth, where ozone forms in high concentrations. The ozone layer absorbs ultraviolet radiation, shielding Earth from its damaging effects.

P

P/E ratio The ratio of the length of the polar axis (P) to the equatorial diameter (E).

Erdtman suggested a widely used system of shape classes defined on the basis of P/E ratios.

Pacific A central North American climatic episode dating from A.D. 1100 to approximately 1550 which is believed to have been characterized by an increased frequency of droughts in the grasslands and adjacent regions.

Paddle-and-anvil A pottery manufacturing method. An object or the potter's hand (the "anvil") is held inside of the roughly formed vessel while the exterior surface is paddled, thus shaping the pot. Additionally, the paddling serves to obliterate seams in instances where the pot was formed by coiling The paddle may be wrapped with a cord or covered with a fabric to prevent it from sticking to the wet clay. This commonly produces a roughened surface.

Paint stone A nodule or fragment of ochre.

Palaeo or paleo A prefix meaning "old".

Palaeoecology The scientific study of the ecology of a given region in ancient times.

Palaeoenvironment The environment of a given region in ancient times.

Palaeo-indian A general term referring to either the earliest inhabitants of north america, or the most ancient of the three stages or periods in north american prehistory. The palaeo-indian period includes the clovis (or llano) complex, the folsom complex and (usually) the plano complexes. The peoples of these cultures are defined as hunters of big game animals, some of which are now extinct.

Palaeopathology The study of the illnesses and injuries of past human populations.

Palaeoserology The study and classification of the blood groups of past human populations.

Paleo tethys ocean A large ocean that originated between eastern Gondwana, Siberia, Kazakhstan, and Baltica in the Ordovician and finally closed in the Jurassic; replaced by the Tethys Ocean as eastern Pangea was assembled.

Paleoanthropologist A type of anthropologist who studies early humans by excavating and looking at fossilized human skeletal remains.

Paleoanthropology The study of the fossil and archaeological record of humans and their primate ancestors. It is also known as human paleontology.

Paleobathymetry The study of ocean depths and topography of the ocean floor in the geologic past.

Paleobiogeography The branch of paleontology that deals with the geographic distribution of plants and animals in past geologic time, especially with regard to ecology, climate, and evolution.

Paleobiology The biological study of fossils.

Paleoceanography The study of oceans in the geologic past, including its physical, chemical, biologic, and geologic aspects.

Paleocene Earliest epoch of the Tertiary period, spanning the time between 65 and 55.5 million years ago. It is named after the Greek words "palaois" (old) and "ceno" (new).

Paleoclimate The climate of a given period of time in the geologic past.

Paleoecology Reconstruction of past environments, especially from a viewpoint of ancient plant and animal life.

Paleoenvironment An ancient environment reconstructed by studying fossils and the rocks in which they were preserved.

Paleo-indian tradition (also called Big Game Hunting Tradition)-a way of life practiced by many of the first human inhabitants of North America, who arrived here between 10,000 and 12,000 years ago. Paleo-Indian means the oldest or first Indians in North America.

Paleolithic The period extending from 2 million to approx 12,000 B.P.

Paleomagnetic dating Dating methods based on the fact that the magnetic north pole wanders around the rotational north pole and has repeatedly reversed position with the magnetic south pole at irregular intervals in the past. There are permanent records of these movements in the form of thermoremanent magnetism found in burned clay and rock as well as geological deposits of volcanic origin. Archaeo-magnetic dating is the term generally used for dating based on the wandering of the magnetic north pole around the rotational north pole over the last 10,000 years or so. Geomagnetic reversal time scale (GRTS) dating and geomagnetic polarity time scale (GPTS) dating are terms applied to the method based on the much longer term reversals of the magnetic poles. The two related paleomagnetic dating methods have been considered by some researchers to be chronometric and calibrated relative by others.

Paleomagnetism Earth's magnetic field of the past as shown by natural remnant magnetism in rocks; paleomagnetic reversals of polarity are a common feature, and reversals can be used as stratigraphic markers.

Paleontologist A scientist who studies fossil remains to learn about life of the ancient past.

Paleontology The scientific study of fossils, their formation and preservation; the formulation of theories and the history and evolution of ancient life that fossils represent. *(Extract from 'The Atlas of the Prehistoric World', Douglas Palmer.)*

PaleopalynologyThe study of fossil palynomorphs, where "fossil" is defined as "any indication of past life" including the term "sub-fossil." Fossil palynomorphs are comprised of sporopollenin, chitin, or pseudochitin little modified from their original compositio Specific applications include stratigraphic palynology, archaeological palynology, and environ-mental palynology. Exclusive of actuo-palynology.

Paleosol Ancient or buried soil, often used as a stratigraphic marker for interglaciation.

Paleospecies A group of similar fossils considered to be members of the same species because the range of their morphological variation does not exceed the range of variation of a living species.

Paleozoic A geologic era that is marked by the culmination of all classes of invertebrates except insects and the appearance of seed-bearing plants, amphibians and reptiles. The Paleozoic era is divided into seven periods designated by inundations of seas. The Mississippian and Pennsylvanian periods of Kentucky occurred during the Paleozoic era.

Palestine An ancient name for the area between Syria, Egypt, the Mediterranean Sea, and the Jordan River; now approximately the same area as that covered by the state of Israel. The state of Palestine was proclaimed in 1988, but in exile, with the land in question under Israeli occupation.

Palisade A strong, high fence or wall made of logs or stakes and intended for defensive purposes.

Palynodebris All palynomorph-sized particles in a sediment excluding those that actually are palynomorphs but including, for example, wood fragments, cuticles and some animal remains.

Palynodeme A group of palynomorph species (form-species) that intergrade and represent the reflection of a known or hypothetical plant species.

As originally used the concept was phylogenetic and referred to characters changing in time ("vertical" change). The term is, however, misused by some authors as if synonymous with morphon and the less formal term "complex".

Palynofacies The assemblage of phytoclasts found in a particular sediment, such as palynomorphs, wood fragments, cuticles, etc.

The term is actually used in two senses, namely the palynolithofacies and palyno-biofacies. Palynogram A diagram summarising the main morphological features of a palynomorph.

Palynologist One who studies plant pollen and spores. Since pollen may be preserved thousands of years it can be used to reconstruct the plant ecology of the past.

Palynology The study of living and fossil pollen grains. Pollen is often well-preserved in the archaeological record and the identification of species and tabulation of their relative frequencies can aid in the reconstruction of past environments

Palynomorph A general term for all entities found in palynological preparations.

In addition to pollen grains and spores, the term encompasses acritarchs, dinoflagellates and scolecodonts, but not other microfossils, such as diatoms, that are dissolved by hydrofluoric acid.

Panporate Synonym of pantoporate.

Pangaea A hypothetical "super" continent. It is believed that Pangaea broke apart about 200 million years ago, during the Mesozoic Era, to form the present day continents.

Pangea The name given to the Supercontinent that existed from the end of the Triassic period. Such was the continent's size, that it stretched from pole to pole, before breaking apart as the Atlantic ocean opened.

Panmixis Random mating throughout a population.

Pannotia A supercontinent that existed in the Late Precambrian and gave rise to the continents of Gondwana, Laurentia, Siberia, and Baltica in the Cambria.

Panthalassic ocean A vast ocean that existed from the Late Precambrian to the Jurassic, circling the globe and connecting to smaller oceans that developed throughout the Phanerozoic; also known as the Panthalassa.

Pantheon Temple built in Rome by the emperor Hadrian to honor all the major gods.

Panto, pan A prefix for global distribution.

Synonym of peri-.

Pantoaperturate Describing a pollen grain with apertures spread over the surface sometimes forming a regular pattern.

Such pollen grains may be, for example, pantocolpate, pantocolporate or pantoporate. The terms based on the prefix peri- as used in the classification of pollen types of Iversen and Troels-Smith (1950) are not recommended.

Papilla A general term, applied in palynology to parallel sided exinous elements with rounded apices, less than 1µm in length.

Papyrus The main Egyptian writing material, and an important export. The earliest papyrus dates to the First Dynasty, the latest to the Islamic Period. Sheets were made by cutting the stem of the plant into strips. These strips were soaked in several baths to remove some of the sugar and starches. These strips were then laid in rows horizontally and vertically. Then it was beaten together, activating the plant's natural starches and forming a glue that bound the sheet together. Separate sheets were glued together to form a roll.

Paracavate An exine in which the intexine is clearly defined but in which its degree of separation from the exoexine is uncertain or indeterminate. Example: *Ancyrospora langii*.

Paradox A seemingly absurd or contradictory, though often true, statement.

Paraisopolar Describing a pollen grain whose polar faces differ only in the attachment of viscin threads to the proximal pole.

Synonym of subisopolar.

Most pollen grains with viscin threads have polar faces of which one is less/more convex than the other.

Parallel flaking A technique used in the production of stone tools that is often found on the earliest projectile points from North Central Texas. Long, consistent chipping scars run parallel on the flat sides of stone tools.

Parallelism An evolutionary pattern that results in the formation of homologous character states. The similar body forms of Old World Monkeys and New World Monkeys are due to their having a common ancestor; their similar body forms are due to parallel evolution rather than convergence.

Parapatric speciation Speciation in which the new species forms from a population contiguous with the ancestral species' geographic range.

Paraphyletic Describes a taxon or clade that does not contain all the descendant species or taxa of their common ancestor.

Paraphyletic group A set of species containing an ancestral species together with some, but not all, of its descendants. The species included in the group are those that have continued to resemble the ancestor; the excluded species have evolved rapidly and no longer resemble their ancestor.

Paraporal lacuna A lacuna of a lophate pollen grain lying in the mesocolpial region adjacent to one side of an equatorial ridge. Example: *Taraxacum officinale* (Compositae).

Paraporal ridge A ridge bounding a pore, extending in a meridional direction. Applied to a lophate pollen grain. Examples: *Tragopogon pratensis*, *Taraxacum officinale* (Composiatae).

Parasite An organism that lives on or in a plant or animal of a different species, taking nutrients without providing any benefit to the host.

Parasyncolpate and parasyncolporate Describing syncolp(or)ate pollen grains in which the apices of the colpi divide into two branches and anastomose towards the poles, delimiting an isolated area known as the apocolpial field. Examples: *Nymphoides peltata* (Menyanthaceae), *Eugenia* (Myrtaceae).

Paratype All of the specimens in the type series of a species or subspecies other than the holotype.

Parish, Amy A biological anthropologist and primatologist whose research focuses on the social behavior of bonobos ("pygmy chimpanzees," or *Pan paniscus*). In addition to comparative work with chimpanzees (*Pan troglodytes*) and endocrinological investigations, Dr. Parish studies reciprocity in chimpanzees, bonobos, and hunter-gatherers.

Parkdale eared A projectile point style defined by r.s. macneish (1958) on the basis of his archaeological investigations in southeastern manitoba. This type is no loner generally used, most archaeologists preferring the term "oxbow" for such specimens.

Parks, William A. William Arthur Parks (1868-1939) was a Canadian paleontologist who described Arrhinoceratops (1925), Dyoplosaurus (1924), Lambeosaurus (1923), and Parasaurolophus (1923). Parksosaurus was named by paleontologist Chas. M. Sternberg in 1937 to honor Parks.

Parsimony The principle of parsimony (also known as "ockham's razor") requires that ad hoc assumptions should be minimized as far as possible in scientific explanations of natural phenomena. This means for Phylogenetic Systematics that from the millions theoretical possible cladograms those should be preferred that minimize the number and/or the weight of necessary assumptions of non-homology (homoplasies). The principle of parsimony is an epistemological principle, and thus should be viewed as a tool, not as a claim that evolution always took the most parsimonious way. This principle is just taking into account that there is no other possibility than parsimony to choose between different alternative hypotheses, that explain singular historical happenings, that can only be reconstructed, but not repeated and tested like scientific experiments. Evolutionary biology in general and Phylogenetic Systematics in particular clearly are historical sciences! Nowadays there exist several software-packages (paup, hennig-86, phylip and MacClade) for the computer-aided calculation of most parsimonious trees (MPT's) from large data sets (numerous taxa and characters), that can be especially large in case of DNA-sequences as characters. The biggest problem of this computer-cladism is the circumstance that characters are either regarded as unweighted (which means in reality "equally weighted"), or that more or less arbitrary discrete weights are assigned to the characters. Since there is no rational way to decide if a character should have the weight 0.3, 1, 17 or 16345, the preference of a most parsimonious tree (in the computer-cladistic sense) which is only some steps shorter than other possible trees appears to be simply nonsense. The subjective choice of characters and the arbitrary delimitation of characters (e.g. lumping or splitting of character complexes) is already representing a (often unreflected) weighting procedure, which makes it quite improbable that all characters have indeed the same weight (viz that we can have the same faith in the correctness of each involved homology-hypothesis), and it makes it impossible to assign discrete weights to the characters, too. Consequently, the over-reductionist view of the principle of parsimony in computer-cladistics, as a mere minimization of the number of homoplasies, has to be dismissed as unwarranted formalism that has no place in a science that is striving for the recognition of natural phenomena.

Parthenogenesis Development from an egg cell that has not been fertilized. The term for a certain form of asexual reproduction that is found in some lizards, insects (notably among aphids), and certain other organisms.

Parthenon The Greek temple dedicated to the goddess Athena that was built on the Acropolis in Athens between 447 and 438 B.C.

Particulate (as property of theory of inheritance) A synonym of atomistic.

Paste The mixture of water, clay and temper from which pottery is made.

Patayan The prehistoric cultural groups that occupied the region west of the Hohokam culture area. The boundaries of the Patayan area are, on the west, the Colorado River Delta north to above Needles, and, on the east, from Gila bend to Prescott. The Patayan practiced a lifeway similar to the Hohokam, although hunting and gathering were more emphasized in the Patayan culture. The Patayan were probably ancestral to the Yuma tribes that occupied the area historically. The culture is first recognized at A.D. 700.

Patella Synonym of patina.

Paternity The identity of the father of an offspring.

Paternoster lakes Chain of small lakes located in successive cirques or rock basins along an alpine glacial valley; may resemble beads on a rosary.

Pathogen A microorganism that causes disease.

Pathological Related to or caused by disease.

Pathology The study of conditions, processes, or results of diseases. Pathology also is used to refer to any abnormal physiological condition.

Patina The discolored skin or rind of a stone resulting from exposure to air, sunlight and soil chemicals.

Patrilineal Pertaining to the tracing of descent or the transmission of wealth or authority primarily through males.

Patrilocal Pertaining to the pattern in which children are raised as part of their father's (as opposed to their mother's) band. Patrilocality is often a result of virilocality among band societies.

Pattern In the Midwestern Taxonomic Method devised by McKern (1939) a pattern referred to a grouping of regional phases with a shared mode of adaptation and a generally similar technology. For example the Woodland cultures of the northeast or the Middle Prehistoric Period bison hunting cultures of the northern plains would constitute two distinct patterns.

Pebble A rounded stone, smaller in size than a cobble.

Pecking A method of shaping stone artifacts by hammering them, thus wearing away the surface. Pecking facets are readily discernible on mauls, axes and Celts unless they have been erased by subsequent polishing.

Pedestrian tactic A method archaeological survey in which surveyors, spaced at regular intervals, systematically walk the area being investigated.

Pedium Synonym of foot layer.

Pedology The study of the formation characteristics and weathering of soil. As various kinds of soil form under different conditions, an analysis of the soil at an archaeological site can yield clues as to the kinds of conditions which prevailed when the site was occupied.

Pelagic Referring to open water marine habitats free of direct influence of the shore or ocean bottom. Pelagic organisms are generally free-swimming (nektonic) or floating (planktonic).

Pelecypods Also called Bivalves. A common fossil in marine rock. Pelecypods have shells with two valves that are the same size and shape. The shells often show growth lines, and radiating ridges and grooves, useful for identification. Oysters, mussels, and clams are living types.

Pelican lake A body of waterlying 75 km west of regina, saskatchewan which has lent its name to a distinctive late middle prehistoric period projectile point style, to the complex of which it is a part, and the phase within which it occurs. Pelican lake points vary in shape somewhat throughout their distributional range but all are characterized by corner-notching. Bases may be concave, straight or convex. The lateral edges may be concave, but are more commonly convex. Associated artifacts include un-notched projectile points, a variety of scrapers and bifaces (some of the latter modified for hafting), drills, perforators and gravers. Pelican lake sites are distributed across the grasslands of the canadian prairie provinces and into the surrounding forests. Sites occur as far south as northeastern colorado. Particularly in the northern areas, pelican lake peoples appear to have employed communal hunting techniques such as bison jumps and bison pounds and likely a seasonal round of movement following the herds. The burial pattern is slowly becoming better known. It is characterized by the painting of the deceased with red ochre and their placement, together with some grave goods beneath a cairn. Radiocarbon dates suggest a total time range of approxmately 1200 b.c. to a.d. 400. The dates from manitoba tend to fall within the earlier half of this range. Reeves (1983) has placed pelican lake within the tunaxa cultural tradition following the mckean and hanna phases and ultimately giving rise to avonlea. Additionally, he divided pelican lake into eight regional subphases. Of these, only two occur in manitoba; the mortlach sub-phase is restricted to southwestern manitoba while the larter sub-phase centres on the red river valley.

Pelican lake focus A now-obsolete term originally applied to a ceramic complex in southwestern manitoba. The pelican lake focus is unrelated to the preceramic pelican lake phase.

Pemmican Dehydrated buffalo meat mixed with fat and occasionally berries.

Pendant A personal ornament which has been perforated or grooved to allow it to hang from a chain or thong.

Penicillin The first antibiotic discovered, penicillin is derived from the mold *Penicillium notatum*. It is active against a wide variety of bacteria, acting by disrupting synthesis of the bacterial cell wall.

Pennsylvanian A period of the Paleozoic era, spanning the time between 325 and 286 million years ago. It is named after the state of Pennsylvania where rocks of this age are widespread.

Pennsylvanian period 310 to 280 million years ago. This was the development of deltas, lowlands, and great swamps. Great trees that formed the coal forests were prominent. Most common were the scale trees (lycopods), seed ferns, horsetails, and chordates. The coal regions of eastern and western Kentucky developed during this period.

Per A prefix for 1. Extremely, and

2. Through.

In palynological terms the prefix is often used for complete (e.g. in pertectate) or very (e.g. in peroblate).

Per reticulate Structural elements fused distally forming an open reticulum.

It applies to the same structure as eureticulate, but is defined on a different basis.

Percussion flaking Removing flakes from a core by use of a hammer. (of stone, bone, antler or wood)

Pererect Synonym of perprolate.

Perforate A general adjective indicating the presence of holes, applied in palynology to holes less than 1μm in diameter and generally situated in the tectum.

Peri Synonym of panto-.

Iversen and Troels-Smith (1950) used the term peri- (as for example, in pericolpate, periporate and pericolporate) in their classification of pollen types, but terms based on panto- are much more widely used.

Periglacial Refers to zone or environment surrounding an ice sheet or cap, often characterized by harsh climate.

Perikles Ruler of Athens in the 5th century BCE during what is called the Athenian Golden Age.

Perine A sporoderm layer that is not always acetolysis resistant and is situated around the exine of many spores. Example: *Pteridium* (Hypolepidaceae).

Synonym of perispore.

The term perine should be used in conjunction with sexine and nexine, whereas perispore should be used with exospore and endospore.

Perinium Synonym of perine.

A latinised form of the term perine.

Period A division of geologic time longer than an epoch and included in an era. A period is usually set off by a significant or striking quality, change or series of events.

Peripatric speciation A synonym of peripheral isolate speciation.

Peripheral isolate speciation A form of allopatric speciation in which the new species is formed from a small population isolated at the edge of the ancestral population's geographic range. Also called peripatric speciation.

Periphery An outlying region; the outermost boundary. For example, Manitoba lies at the periphery of the North American Grasslands.

Perispore Synonym of perine.

Perisporium Synonym of perine.

Peristyle Enclosed courtyard surrounded by columns.

Peritreme Describing a pollen grain with equatorial apertures situated around an outline that is circular in polar view.

The term is not recommended because it is based upon the suffix -treme.

Permafrost The term used to describe an area of ground continuously frozen. Only for very brief periods during the summer will the surface thaw.

Permian The final period of the Paleozoic era which stretched from 290 to 248 million years ago.

Peroblate Describing the shape of a pollen grain or spore in which the ratio between the polar axis and the equatorial diameter is less than 0.50.

Persia Ancient kingdom in the area of present day Iran.

Persian gulf An arm of the Arabian Sea between the Arabian peninsula and Iran. It has been an important trade route since ancient times and gained added strategic significance after the discovery of oil in the region in the 1930s.

Pertectate Synonym of eutectate.

Pertransverse Synonym of perprolate.

Pesticide-resistant insects Insects with the ability to survive and reproduce in the presence of pesticides. These resistant variants increase in frequency over time if pesticides remain present in their environment.

Pestle A club-shaped or cylindrical object used to crush and grind various materials in a mortar.

Petrie, Marion A behavioural ecologist at the University of Newcastle in the UK, Dr Petrie's research interests include the links between sexual selection and speciation, and how males and females assess genetic quality in a mate.

Petrification A process that replaces living materials (wood or bone) with mineral matter. Organic matter is thus turned to stone.

Petrify To convert into stone or a stony substance.

Petroform A geometrical or animal figure formed in outline by the placement of boulders or cobbles. These are particularly common in southeastern Manitoba where they are believed to be the work of Laurel (and perhaps later) peoples.

Petroglyph A figure inscribed onto a rock surface by grinding, pecking or incising. Although petroglyph sites are fairly common in northwestern ontario, none are known in manitoba.

Petrograph A general term for any image produced on stone. Petrograph thus includes both pictographs (rock paintings) and petroglyphs (rock engravings).

Petrography The study of rocks with a microscope(very tedious).

Petrology The study of the origin and provenance of a certain rock.

Ph Abbreviation for "potential of hydrogen". In practical terms, ph measurements provide a scale for the determination of acidity. Pure water has a ph of 7.0; lower readings indicate increasing acidity, while higher ph (to a maximum of 14.0) is associated with minimum acidity.

Phalanx Warriors in close formation.

Phanerozoic The period of time, also known as an eon, between the end of the Precambrian and today, The Phanerozoic begins with the start of the Cambrian period, 544 million years ago. It encompasses the period of abundant, complex life on the Earth.

Pharaoh Wasn't used until 1400 BC but is now used generally for the kings of ancient Egypt.

Pharmacogenomics The application of genomic information to clinical trials to determine which patients, based on their genetic make-up, are most likely to benefit from a certain drug or which are likely to suffer adverse effects.

Pharos lighthouse One of the Seven Wonders of the Ancient World. It was built in Alexandria, Egypt, about 280 B.C., and served as a beacon for ships at sea. It was destroyed in an earthquake in A.D. 796.

Phase An archaeological complex which is sufficiently distinctive so as to be distinguishable from adjacent contemporary complexes, and from those which precede and succeed it. A phase may be viewed as a complex which is bounded in time as well as space.

Phenetic classification A method of classification in which species are grouped together with other species that they most closely resemble phenotypically.

Phenetic species concept A concept of species according to which a species is a set of organisms that are phenotypically similar to one another. Compare with biological species concept, cladistic species concept, ecological species concept, and recognition species concept.

Phenetics A non-phylogenetic approach to biological classification, founded by camin and sokal. It is based on the exclusive criterion of overall similarity, without distinction of plesiomorphic and apomorphic character states. The similarity analysis is generally performed by a computerised statistical analysis (cluster-analysis and nearest-neighbour-joining) of a data matrix with the terminal taxa (OTU's = operational taxonomic units) and a large number of unweighted and unpolarized characters (= Numerical Taxonomy). The false goal of Phenetics was a totally theory neutral and allegedly objective procedure for the generation of biological classifications. Fortunately this anti-biological approach does not play any significant role in current systematic biology anymore, maybe with the exception of microbiology, especially bacteriology. Nevertheless wägele (1996a) has recently demonstrated that mainstream computer-cladistics is in several aspects closer to phenetic methods than to genuine Hennigian Phylogenetic Systematics, since there is no a priori polarisation and homologisation of characters, so that symplesiomorphies and convergences can become group-defining characters, as consequence of an over-reductionist application of the principle of parsimony and a blackbox-like application of the computer analyses.

Phenology Scientific study of dates of plant growth phenomena, such as flowering or ripening of fruit.

Phenotype The physical or functional characteristics of an organism, produced by the interaction of genotype and environment during growth and development.

Phenotypic characters Individual traits that can be observed in an organism (including appearance and behavior) and that result from the interaction between the organism's genetic makeup and its environment.

Phenotypic variance Variance of the phenotype due to genotypic and environmental factors combined. Pheno-typic variance = genetic variance + environmental variance.

Pheromone A chemical substance produced by some organisms and emitted into the environment to communicate with others of the same species. Pheromones play an important role in the social behavior of certain animals, especially insects and some mammals. They are used to mark out territories, to attract mates, to lay trails, and to promote social cohesion and coordination in colonies. Examples are the sex attractants secreted by moths to attract mates and the queen substance produced by queen honeybees, which controls the development and behavior of worker bees. Pheromones are usually volatile organic molecules which are effective at very low concentrations, as little as 1 part per million.

Phidias Athenian sculpture who was responsible for the cult statues of Athena in the Parthenon and the statue of Zeus at Olympia.

Phocoenidae Family of cetaceans commonly known as Porpoises.

Phosphate test A chemical test performed on sediments from archaeological sites. The decomposition of organic material leaves a phosphate residue which is not readily removed by leaching. Thus determination of phosphate levels across a site can reveal which parts of a site were most heavily used. When used in conjunction with archaeological data, it can additional suggest what kinds of activities were pursued in different parts of the site.

Photoreceptor cell A cell, functionally part of the nervous system, that reacts to the presence of light. It usually contains a pigment that undergoes a chemical change when light is absorbed. This chemical change stimulates electrical changes in the photoreceptor that, when passed along and processed by other neurons, form the basis of vision.

Photosynthesis The fundamental biological process by which green plants make organic compounds such as carbohydrates from atmospheric carbon dioxide and water using light energy from the Sun. The process has two main phases: the light-dependent light reaction responsible for the initial capture of energy, and the light-independent dark reaction in which this energy is stored in the chemical bonds of organic molecules. Since virtually all other forms of life are directly or indirectly dependent on green plants for food, photosynthesis is the basis for almost all life on earth.

Phylogenetic relationships (versus tokogenetic relationships) The (generally hierarchical) genealogical relationships between separated biopopulations, biospecies or monophyletic groups, contrary to the reticulate genealogical relationships between the individuals within a biopopulation or biospecies.

Phylogenetic taxonomy The branch of cladistics concerned with the verbal representation (rather than the reconstruction or estimation) of phylogenetic relationships. It was developed by Kevin de Queiroz and Jacques Gauthier in a series of papers in the early 1990s.

Phylogeny The evolutionary development and history of a species or higher taxonomic grouping of organisms. The phylogeny of an organism is sometimes called a "family tree." Cladistics is used to construct hypothetical phylogenies, called clades - phylogenetic.

Phylum A major taxonomic division. In linguistics a phylum is composed of a number of families, which in turn comprise various languages, each of which may have numerous regional dialects. In biology, phyla are even more all-encompassing. In descending order of inclusiveness the scheme is as follows: kingdom, phylum, class, order, family, genus, species, variety (or sub-species).

Physical anthropology The branch of anthropology which concerns itself with the origins, evolution, structure and variation of human populations.

Physiography 1. Physical geography; the study of the earth's crust.

2. The earth's surface; terrain.

Physiological pathway The sequence of biochemical interactions among proteins that regulates the function of living cells.

Phytoclast A general term for plant-derived, more or less resistant-walled, particle occurring in a sediment.

Phytoliths The siliceous secondary cell walls of certain terrestrial plants. Phytoliths are destroyed by hydrofluoric acid so they are never present in routine palynological preparations, and phytoliths processing involves treatment with strong oxidizing agents, which destroy polle Studied phytoliths generally range from ca. 20 - 50 µm.

Phytoplankton Microscopic aquatic organisms that, like plants, use photosynthesis to capture and harness solar energy.

Pick Archaeologists use small picks to remove delicate items from archaeological units. For example, animal bones or human bones are very fragile and should be removed by picking around them and then brushing them off with a light paintbrush before removing. Old dental picks are very useful tools to archaeologists.

Pickford, Martin A paleontologist at the College de France in Paris. In 2000, Pickford and Brigitte Senut discovered *Orrorin tugensis*, a proto-hominid dated at 6 million years old.

Pictograph A painting on rock. Unlike petroglyphs, rock paintings are numerous in manitoba, particularly in the eastern and northern parts of the province.

Piece esquillee. Literally "stepped piece"; a stone artifact, often wedge-shaped, with evidence of bipolar battering and reduction. These may have served as wedges for the splitting of bone or antler, or they may simply be exhausted cores. In the later instance, the crushing at one end may be the result of the necessity of resting the core on an anvil due to its diminutive size.

Pigment Coloring agents derived from plants and minerals.

Pigweed A common, coarse weed that grows in disturbed soil. It usually appears after summer rains at elevations below 5500 feet.

Pile dwellings Prehistoric lake or riverside settlements where a substantial part of the buildings were built on poles, in or near the water. The remains of such settlements have been found underwater. It's often difficult to say whether the submersion is a result of raising sea level or the buildings actually were built standing in the water. Thus some pile dwellings may originally have been built on shore, later submerged, and today we find the remaining house poles underwater.

Pilum A sexine element, usually standing directly on the nexine, consisting of a rod-like part (columella) and a swollen apical part (caput).

Pinker, Steven A psychologist and professor with a special interest in language, linguistic behavior, and cognitive science. Pinker's publications include the popular science books *The Language Instinct* and *How the Mind Works.*

Pinon pine A pine tree that grows at elevations of 5000 to 7500 feet. The piñon bears large edible nuts.

Pit An excavation recovery unit.

Pithouse A wooden dwelling in which the lower portion is a pit.

Pitted A general term for small depressions .

Placental mammals A group (specifically, an order) of mammals in which the young develop inside the mother, attached to her and nourished by a specialized structure called the placenta. In placental mammals, the young are born in an advanced stage of development. Compare with marsupial and monotreme.

Placoderms A peculiar group of primitive armored jawed fish, found almost exclusively in rocks from the Devonian Period.

Plains Extensive tracts of level or nearly level land. In north america, the plains extend from the southern canadian prairie provinces to texas, and from the foothills of the rocky mountains in the west to the mississippi river.

Plains side-notched One of the small side-notched projectile point styles of the northern plains during the late prehistoric period. These specimens, which were probably used as arrowheads, range in length from approximately 10 to 34 mm and bear small, deep, narrow rectangular side-notches above frequently squared-off bases. Bases are generally straight or occasionally concave and at least as wide as the blade (sense 3). Basal grinding is often present (kehoe 1973).

Plains traingular A late prehistoric period/ protohistoric projectile point style defined by macneish (1954) on the basis of his investigations in southwestern manitoba. These range in length from 14 to 32 mm and in width from 11 to 21 mm. Maximum length may be equal to or greater than the maximum breadth. The blade edges and the bases are straight or convex. This and similar styles are fairly common over much of north america in this general time period.

Plains village pattern. A late prehistoric period cultural pattern comprised of a number of complexes along the eastern margins of the north american grasslands. Common to all were permanent settlements often protected by ditches and palisades. These horticulturalists grew maize, beans and squash with the aid of the bison scapula hoe in creek and river bottomlands, and supplemented their diets with hunting and fishing and the collection of plants. The artifactual inventory of these people included pottery and a greater variety of tools of bone, horn, shell and stone than that of their predecessors. The proliferation of underground storage pits is suggestive of a food surplus stemming from a highly efficient exploitation of their environment. The earthlodge villages of the middle missouri are part of this pattern and have been attributed to the ancestors of the mandan, hidatsa and arikara peoples.

Plainview A town in north-central texas which has given its name to a palaeo-indian projectile point style, the complex of artifacts with which these are associated and the phase within which these occur. The lanceolate points have parallel or convex edges and concave bases and are irregularly flaked although parallel flaking occasionally characterizes the tips. Basal and lateral edges are almost inevitably ground and length ranges from 45 to 77 cm. Associated artifacts include a variety of side- and end-scrapers, perforators and blades (sense 4). The position of plainview in the early prehistoric period chronology is a matter of some debate. Some place it between clovis and folsom (at roughly 9000 b.c.) while others would consider it to be an early member of the plano complexes which follow folsom.

Plan of nature The philosophical theory that nature is organized according to a plan. It has been influential in classification, and is a kind of idealism.

Planaperturate Describing a pollen grain with an angular outline, in which the apertures are situated in the middle of the sides when seen in polar view, rather than at the angles. Example: *Tilia* (Tiliaceae).

Antonym: angulaperturate.

The term is useful for describing the position of apertures, but should be avoided as a description of equatorial outline.

Plane table A surveying instrument consisting of a drafting board mounted horizontally upon a tripod. Generally used with a sighting instrument such as an alidade, the plane table allows direct plotting of objects onto a map.

Plankton Minute or microscopic animals (zooplankton) and plants (phytoplankton) that float and drift in water, usually near the surface. In the top meter or two of water, both in the sea and in freshwater, small plants can photosynthesize, and abundant microscopic life can be observed. Many organisms that are sessile (attached to a surface) as adults disperse by means of a planktonic larval stage.

Planktonic Used to describe aquatic organisms that float.

Plano A general term which refers to the late early prehistoric (or palaeo-indian) complexes of the north american plains. Plano thus embraces the numerous non-fluted projectile point styles such as agate basin, alberta, angostura, eden, hell gap, manitoba, midland, milnesand and scottsbluff as well as their associated complexes. Most would consider plainview to be a part of this series also. Plano peoples were nomadic big game hunters who made use of the bison jump, but they also exploited a range of other animals of modern species. Perhaps one of the most important differences between the subsistence patterns of plano peoples and the earlier llano (clovis) and folsom peoples is the occasional recovery of milling stones which may suggest a heightened reliance upon plant foods.

Plasmid A genetic element that exists (or can exist) independently of the main DNA in the cell. In bacteria, plasmids can exist as small loops of DNA and be passed between cells independently.

Plate tectonics The theory that the surface of the earth is made of a number of plates, which have moved throughout geological time resulting in the present-day positions of the continents. Plate tectonics explains the locations of mountain building as well as earthquakes and volcanoes. The rigid plates consist of continental and oceanic crust together with the upper mantle, which "float" on the semi-molten layer of the mantle beneath them, and move relative to each other across the earth. Six major plates (Eurasian, American, African, Pacific, Indian, and Antarctic) are recognized, together with a number of smaller ones. The plate margins coincide with zones of seismic and volcanic activity.

Platea Areas of the inner wall layer separated by a triradiate channel extending between the endogerminals of a Normapolles pollen grain.

Example: *Pompeckjoidaepollenites*.

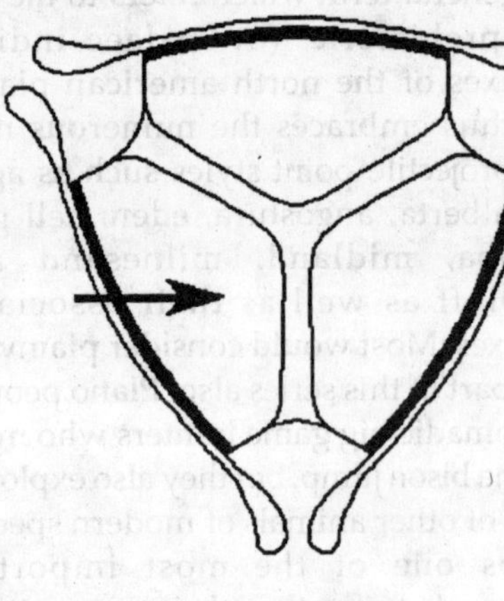

Fig. Platea

Platea luminosa Synonym of groove.

The term was introduced for use in striate pollen and to contrast with lumina, which was.restricted to reticulate pollen.

Pleurotreme Synonym of planaperturate.

Plates tectonics The term used to describe the process which results in the movement of the earths crust. Over time the semi-molten mantle within the earth creates convection currents, which cause heated rock to rise and cooling rock to descend. At the earths surface this process results in the movement of the continental plates. The resulting movement means that the UK and US are moving apart at roughly the same rate that an average persons finger nails grow.

Platform burial See buriai, scaffold.

Plato Ancient Greek philosopher of the 5th and 4th century B.C., who wrote many works and taught about the ideal way of life for the ancient Greek citizen.

Pleistocene A geologic period, usually thought of as the Ice Age, which began about 1.6 million years ago and ended with the melting of the large continental glaciers creating the modern climatic pattern about 11,500 years ago.

Pleistocene epoch Subdivision of the Quaternary Period. Characterized by almost modern fossils (70% fossil mollusks still living organisms), major glaciations of the mid-latitudes, and with associated climatic and environmental changes. Ended roughly 10,000 years BP with the end of the last major glaciation.

Plesiomorph Evolutionarily primitive character state. The long neck of the giraffe is apomorphic; the short neck of its ancestor is plesiomorphic; a character present throughout a taxon because it is present in an outgroup of the taxon – plesiomorphy – plesiomorphic.

Plesiomorphic (versus apomorphic) A "primitive" character state, that is taken over from an ancestor without change, is termed the plesiomorphic state. This property is relative (just like apomorphic), since it is depending on the compared character state and the hierarchical level in focus. A plesiomorphic state is only plesiomorphic compared to a derived state, but it can be apomorphic compared to an even more plesiomorphic state. Example: The fore leg of non-flying tetrapods is plesiomorphic compared to the wings of pterosaurs, birds and bats, but it is apomorphic compared to the pectoral fins of "fishes". Shared plesiomorphic similarities can be homologous (symplesiomorphies) or non-homologous (reversals). klausnitzer & richter (1981) suggested the (synonymous) terms "plesiotypic" and "symplesiotypic", since the suffix "-morphic" is strictly speaking only correct for morphological characters. The term "plesiomorphic" or even "primitive" (the latter term should be avoided anyway) should only be used for character states, but never for organisms or taxa (taxa that originated from very old splitting events and have retained a lot of plesiomorphic states should rather be called "ancient groups"), since all taxa always possess a mixture of plesiomorphic and derived character states (compare living fossils).

Plica A general term for a fold, applied in palynology to ridge-like folds of the exine in *Ephedra* (Ephedraceae) and *Lusatipollis*.

Pliny the elder Roman writer who witnessed the eruption of Mt. Vesuvius in A.D. 79. He died when he returned to Pompeii during the eruption in order to try to save some of the Roman navy, which was stationed in the harbor.

Pliny the younger The nephew of Pliny the Elder who wrote about the eruption of Mt. Vesuvius and the destruction of the city of Pompeii in A.D. 79.

Pliocene Final epoch of the Tertiary period, spanning the time between 5.3 and 1.8 million years ago. It is named after the Greek words "pleion" (more) and "ceno" (new).

Plow zone The mixed surface layer of soil resulting from cultivation. Although no longer *in situ*, cultural material from a plow zone can yield some insights into the original shape and extent of an archaeological site.p plumb bob. A weight which when suspe-nded from a (plumb) line is used to establish the verticality of walls or to set up a piece of surveying equipment directly over a datum point.

Plumb line The line from which a plumb bob is suspended.

Plummet A weight which may be used as a plumb bob, netsinker or for any similar purpose.

Pluricolumellate With the columellae arranged in several rows beneath each murus.

Pluvial/interpluvial Period in arid regions much wetter than at present when large freshwater lakes existed. Interpluvials are drier periods (like the present) between pluvials. Pluvials are thought to correspond approximately to major Pleistocene glaciations.

Podzol A greyish, generally acidic soil which is heavily leached in its upper layers. Such soils are characteristic of coniferous forests.

Poisson distribution A frequency distribution for number of events per unit time, when the number of events is determined randomly and the probability of each event is low.

Polar area Synonym of apocolpium.

In spores often used for the area around a pole.

Polar area index (PAI) Synonym of apocolpium index.

Polar axis The straight line between the distal and proximal poles of a pollen grain or spore.

Polar distance The vertical distance from the equator to the pole (pd).

Polar field Synonym of apocolpium.

Polar lacuna A lacuna situated at the pole of a lophate pollen grain.

Polar view A view of a pollen grain or spore in which the polar axis is directed towards the observer.

Antonym: equatorial view.

Polarity The condition of having distinct poles

The polarity of palynomorphs may be determined from their orientation in tetrads, or by inference from the distribution of apertures, or other features.

Polarity of characters The polarity of a character specifies which of its character states has to be regarded as plesiomorphic and which as apomorphic. The polarity can be directly deduced from the stratigraphic or ontogenetic precedence of states, or indirectly deduced with the method of outgroup comparison. Naturally all these criteria and methods might fail, since the fossil record is fragmentary and potentially misleading, and ontogenesis is complicated and can be misleading, too (e.g. in case of heterochronies or caenogenetic larval adaptations). Since the most popular method of outgroup comparison is based on a pre-existing knowledge of the phylogenetic relationships, it is not suited as a final criterion of character polarity (problem of infinite regression, also well-known as "Münchhausen-Trilemma"). From an epistemological point of view these final criteria are the stratigraphic evidence (the oldest fossils a protozoans and fossil man are rather young) and the ontogenetic evidence (all multicellular organisms develop from a single celled stage), as well as the theoretical conclusion that a scientific explanation of life necessarily implies an evolution from simple to more complicated organisms (as general tendency, not in every particular case!), since a reverse view would imply a creational act by some kind of god.

Pole Either of the two extremities of the polar axis.

Pollen The male sex spores of flowering plants. Pollen particles are usually nearly invisible, except in large concentrations, due to their extremely small size. They can survive for hundreds of thousands of years in silty, anaerobic soils, such as mud sediments at the bottom of a lake. When excavated from early human sites, they can be used to reconstruct what the climate was like at the time of the occupations as well as relatively date them.

Pollen analysis The study of assemblages of dispersed palynomorphs such as those isolated from samples of peat.

Pollen cement Synonym of pollenkitt.

Pollen class An artificial grouping of pollen grains that share a distinctive character, or suite of characters. Such classes are useful in identification keys and may be subdivided into more restrictive categories, pollen types and pollen groups. Example: tricolpate class.

Pollen coat Synonym of pollenkitt.

Pollen group A pollen morphological category, subsidiary to a pollen type, including a number of pollen grains that show intergrading characters but no distinguishing characters.

Pollen mother cell Synonym of microsporocyte.

Pollen type A pollen morphological category, subsidiary to a pollen class, and including pollen grains which can be distinguished either by one distinct character or by a unique combination of characters.

Poilenkitt A sticky material, produced by the tapetum, that may hold pollen grains together during dispersal.

Pollination ecology The branch of agriculture or ecology concerned with the distribution of pollen by wind or animals and its efficacy in fertilization and seed set.

Pollinium A general term for aggregations of many pollen grains, which form dispersal units Examples: Asclepiadaceae, Orchidaceae.

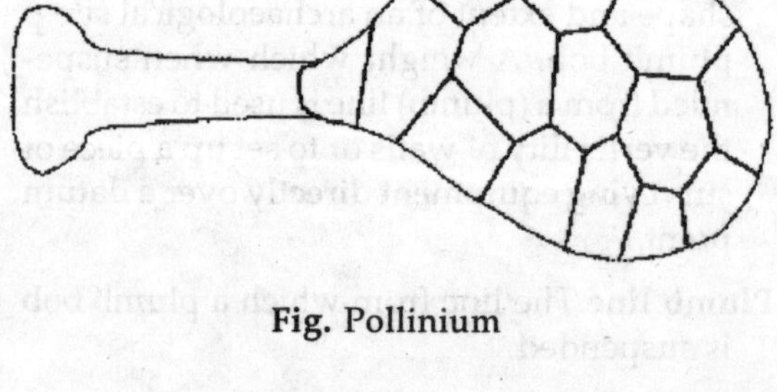

Fig. Pollinium

Polumbra A darkened triangular or subcircular area centred on the proximal pole. Example: *Retusotriletes distinctus*.

The feature appears to be most commonly observed in specimens that have lost a perisporal outer exoexinal layer.

Polyad A dispersal unit comprising more than four pollen grains. Example: *Acacia* (Mimosaceae).

Polyandry A reproductive system in which one female mates with many males. Seahorses and jacanas are examples of polyandrous species, which are less common than monogamous or polygynous species.

Polyannulus A structure in which the sexine of the outer aperture has multiple layers each with its own thickening. Example: *Atlantopollis*.

This term refers especially to pollen grains of the fossil Normapolles group.

Polychotomosulcate Describing a pollen grain with a many branched sulcus.

Polygenetic Originating in various ways or from various sources.

Polygyny Reproductive strategy in which one male mates with several females. Lions, peacocks, and gorillas all have polygynous mating systems. Compare with polyandry and monogamy.

Polymorphism A condition in which a population possesses more than one allele at a locus. Sometimes it is defined as the condition of having more than one allele with a frequency of more than five percent in the population.

Polynya Open seawater between pack-ice and the land or edge of a glacier, Arctic or Antarctic regions.

Polyphyletic Describes a taxon or clade that does not include the common ancestor of the taxon's or clade's species. Polypheletic groups are often of different species of animals that do not share a recent common ancestor. An example of this is the Pachydermata, an early grouping no longer used, that united rhinos, hippos and elephants. All of these animals evolved their similar forms from different ancestors, rather than from the same shared ancestor, thus making the Pachydermata polypheletic.

Polyphyletic group A set of species descended from more than one common ancestor. The ultimate common ancestor of all species in the group is not a member of the polyphyletic group, usually because the common ancestor lacks the characteristics of the group.

Polyplicate Describing a pollen grain with more than three meridional ridges (plicae) separated by deep grooves. Example: *Ephedra* (Ephedraceae).

Polyploid An individual containing more than two sets of genes and chromosomes.

Polyporate With many pores.

Synonym of pantoporate and zonoporate.

Pompeii The famous city in southern Italy destroyed by the eruption of Mt. Vesuvius in A.D. 79. The circumstances of destruction allowed the city to be remarkably well-preserved. Archaeologists have been excavating the site for about 250 years to uncover and learn about ancient Roman culture.

Ponderosa pine Also called the yellow pine. This large tree grows at elevations of 5500 to 8500 feet.

Pontil mark. A "scar" left on the base of a bottle by the rod (the pontil) that was used to hold it while the upper portion was being shaped.

Pontoperculum A type of operculum that is not completely isolated from the remainder of the sexine but linked to it at the ends of the aperture. Example: *Sanguisorba officinalis* (Rosaceae), *Passieflora tetandra* (Passiefloraceae).

Popol vuh A book written in the Maya language of Quiche shortly after the Spanish Conquest in Spanish. However, it covers many stories of Maya mythology.

Population A group of organisms, usually a group of sexual organisms that interbreed and share a gene pool, and are normally relatively isolated from other groups of the same species.

Population genetics The study of processes influencing gene frequencies.

Population parameter A characteristic of a population.

Populational viewpoint The idea that if two fossils have major similarities they should be categorized as being members of the same species. From this perspective, minor anatomical differences within the same population are expected since the members of living species have individual variation. People who advocate this viewpoint are also referred to in the biological sciences and "lumpers".

Poral lacuna A lacuna of a lophate pollen grain surrounding an endoaperture, which communicates with adjacent abporal lacunae via interlacunar gaps. Example: *Cichorium intybus* (Compositae).

Pore A general term, applied in palynology to a circular or elliptic aperture with a length/breadth ratio less than 2.

Pore canal The space between the ectopore and the endopore.

Pore membrane The aperture membrane of a pore.

Poro-colpate Describing a pollen grain with an arrangement of apertures in which colpi alternate with pores round the equator. Example: *Pardoglossum* (Boraginaceae).

Pororate Describing a pollen grain with compound apertures in which both the ectoaperture and the endoaperture are pores and the two are not congruent. Example: *Myrica gale* (Myricaceae).

Pollen grains that have congruent ectopores and endopores are generally simply referred to as porate.

Porphyry Any igneous rock characterized by large cyrstals embedded in a finer matrix.

Porus Synonym of pore.

Poseidon Greek god of the sea. Called Neptune by the Romans.

Post glacial Of or pertaining to a period after the end of glaciation, particularly the period following the pleistocene (ie. The holocene)

Post hole The pit in which an upright stake or pole is placed. In the archaeological record, these are recognizable as holes filled with rotted wood, or in cases where the post has later been removed, as circular patches of differentially-coloured soil.

Post mould A circular patch of earth or clay which was packed or tamped in the hole around a post when it was erected.

Postulate A basic principle.

Postzygotic isolation A form of reproductive isolation in which a zygote is successfully formed but then either fails to develop or develops into a sterile adult. Donkeys and horses are postzygotically isolated from one another; a male donkey and a female horse can mate to produce a mule, but the mule is sterile.

Potassium/argon dating A chronometric technique for dating in which the proportion of radioactive potassium (K^{40}) to that of its decay product, argon is calculated.

Pothunter A person who illegally excavates for and/or collects artifacts solely for their intrinsic value without regard for their provenience or scientific worth.

Potsherd Any fragment of a ceramic vessel.

Pottery Containers made out of a combination of clay and sand that can be hardened in the heat of an oven, called a kiln. Pottery is one of the most common ways that archaeologists date occupation levels of sites since pottery chronology is well-established at many sites. Archaeologists can date a piece of pottery from a site by comparing it to other pieces from the same site that have already been dated. They then use the piece of pottery to date the entire level of the site.

Prairie In north america, that portion of the northern and eastern plains occupied by tall grasses. Trees are absent or sparse except along watercourses.

Prairie side-notched. One of the small side-notched projectile point styles of the northern plains during the late prehistoric period. These specimens range in length from ll to 41 mm and bear large, wide, rounded but shallow side-notches above squared or rounded bases. Bases are generally straight but may be slightly concave or convex. Generally, the basal width is less than the maximum width of the blade (sense 3). Basal grinding is sometimes present (kehoe 1973).

Praxiteles Greek sculpture best known as the creator of the Aphrodite of Knidos.

Pre A prefix for before.

As for example in prepollen, pretectum.

Preadaptation The possession of a biological or cultural trait which confers an advantage to certain individuals under changed environmental circumstances.

Pre-boreal A central north american climatic episode dating from 8080 to 7350 b.c. which although cooler than present marks the end of the wisconsin glacial period and the beginning of the warming trend which culminates at approximately 5000 b.c. (wendland 1978).

Precambrian All geologic time before the beginning of the Paleozoic era. This includes about 90% of all geologic time and spans the time from the beginning of the earth, about 4.5 billion years ago, to 544 million years ago. Its name means "before Cambrian."

Precambrian era The oldest and longest of the eras. It began when Earth was first formed 4.6 billion years ago. Few fossils of this era have been found.

Preceramic Of or pertaining to that period of time prior to the introduction or invention of ceramics in a given region.

Precision In sampling, the degree of correspondence among statistics obtained in repeated trials of the same sampling technique

Precocial Describes organisms that are born capable of surviving with little or no parental care. Sea turtles are an example of a precocial organism.

Pre-columbian Of or pertaining to that period of time prior to columbus' first voyage. For the new world this correlates with prehistory.

Predation The behavior of capturing and feeding on another organism.

Predator An organism that captures and feeds on other organisms.

Predentata ("predentary-boned (dinosaurs)") Major grouping of dinosaurs that is comprised of the beaked dinosaurs. Formerly called the Ornithischia, the name was dropped by this author in favor of Predentata as this new name is based on a character that predentatans share with no other dinosaurs, a predentary bone. Padian and May (1993) formally defined Ornithischia, now Predentata, as *Triceratops* and and all taxa closer to *Triceratops* than to Saurischia –predentatan.

Pre-dorset. A middle prehistoric period archaeological complex or culture the remains of which have been found in the eastern canadian arctic and greenland. Such typical arctic small tool tradition artifacts as burins, side- and end-blades and microblades) are found in this complex as are a variety of scraper forms, drills and knives. With some stylistic and technological modifications, these classes persist into the more recent dorset complex. Additionally, some sites contain large tools such as gouges and adzes which bear witness to the importance of the woodworking industry. Judging from the distribution of sites and the classes of artifacts found within them, it seems possible that these people employed a seasonal round of activity which focused upon the taking of sea mammals, caribou hunting and fishing alternating with inland caribou hunting. While on the coast, pre-dorset populations lived in circular or square skin-covered houses which in the archaeological record are documented by the arrangements of boulders used to hold down the edges of the tents. These immigrants from the western arctic occupied northern manitoba from approximately 1500 to 500 b.c. some archaeologists identify this culture as the ancestors of the modern inuit.

Predynastic Period of Egyptian history that predates writing and unification of Egypt into one state. The prehistoric period before 3100 B.C.

Prehistoric The time before recorded history.

Prehistoric hunter Gatherers-Humans who lived prior to written history and depended upon the hunting of wild animals and the gathering of natural plant foods for their livelihood

prehistoric sites Locations where people who were alive before modern written records existed once lived, hunted, camped, or were buried. Painted or carved rock outcrops are considered sites as well.

Prehistory The period before written records in a given area, or the study of that era.

Preparator Scientists who strip and clean the matrix or encased rock from fossils in order to prepare them for reconstruction.

Prepollen The microspores of certain extinct seed plants characterised by proximal apertures and presumed proximal germination, rather than the distal, equatorial or other typical apertures of seed plant pollen grains.

Preserve To keep safe and protect from injury, harm, or destruction; to keep alive, intact, or free from decay; to save from decomposition.

Pressure flaking Producing flakes by applying pressure to a core with a tool (often an antler)

Prey An organism that is hunted and eaten by another organism.

Prezygotic isolation A form of reproductive isolation in which the two species never reach the stage of successful mating, and thus no zygote is formed. Examples would be species that have different breeding seasons or courtship displays, and which therefore never recognize one another as potential mates.

Prickly pear Several species of cacti with flat stems and oval, flat, leaf-like pads. Prickly pears grow in semiarid and arid western North America. The fruits of the cactus are often referred to as "tunas."

Primary burial See burial, primary.

Primary context Referring to the position in a site where an object was originally deposited. Finding an artifact or fossil in primary context allows a researcher to apply the principle of association in dating and interpreting it. This can not be done if the object is in a secondary context, which is to say that it was moved to a new location after the original deposition.

Primary forest efficiency. A subsistence strategy which involves the selective harvesting of nuts, berries and other vegetables and possibly fish and shellfish as these become seasonally available and supplementing these with the continuous year-round hunting of land and water animals (caldwell 1958). Such a means of exploiting the natural environment is particularly efficient in terms of the amount of energy expended and the nutritional value and diversity of the food collected. The diversity of the resource base also allows people who pursue such a strategy greater security in the event that one particular resource becomes depleted. Primary forest efficiency often results in a seasonal round of activity which is more restricted in its geographical extent than that of a culture which relies more heavily on a single, migratory big game species such as bison. It also allows people to become more intimately familiar with the resources available within their own territory. Primary forest efficiency is often used as a major defining characteristic of the archaic period in eastern north america. Thus the regional diversity and the proliferation of complexes in this region at this time is ultimately explicable in terms of the increasing regionalism which resulted from restricted wandering after the close of the palaeo-indian period.

Primary refuse Trash discarded at the location of use.

Primate A mammal belonging to the order Primates (about 195 species), which includes prosimians, monkeys, apes, and humans. Primates probably evolved from insectivorous climbing creatures like tree shrews and have many adaptations for climbing, including five fingers and five toes with opposable first digits (except in the hind feet of humans). They have well-developed hearing and sight, with forward-facing eyes allowing binocular vision, and large brains. The young are usually produced singly and undergo a long period of growth and development to the adult form. Most primates are arboreal, but the great apes and humans are largely terrestrial.

Primates/primatologist Any of an order of mammals that includes humans, apes, monkeys, lemurs, and related forms. One who studies primates.

Prime cause A factor, such as population growth, used by itself to explain changes in social complexity.

Primexine A developmental precursor of at least part of the exine (the sexine/ectexine), formed during the tetrad stage that is composed largely of polysaccharides and therefore lacks resistance to acetolysis.

Primitive Describes a character state in the common ancestor of a clade. A five-digit forefoot is the primitive forefoot state in amniotes.

Principle of association The assumption that if two objects are found in their primary context in the same strata of a site, they very likely date to the same time period.

Principle of superposition The assumption that if there are layers in a sedimentary deposit, those laid down first will be on the bottom and those laid down last will be on the top. This is the basis of using stratigraphy as a relative dating method.

Pro A prefix indicating a developmental precursor of a feature. Example: probaculum.

Probability sampling Sampling in which sample units are selected at random, as by flipping a coin or consulting a table of random numbers.

Probe array Used in microarray technology - a silicon or glass chip to which small pieces of DNA, or probes, have been attached, with each probe representing a sequence of bases unique to a known gene or gene fragment. Also known as a microarray or DNA chip.

Proboscidea The order to which modern elephants as well as the now-extinct mammoth and mastodon belong.

Procurement Obtaining or extracting resources from the environment.

Profile A profile is a picture of the layers of a unit. It's sort of like if you were to make a five-layer cake with each cake layer a different color. If you looked at it from the side you would see all the different levels. This is very useful to archaeologists because they can see changes in soil color or composition (for example, sand and then black soil and then rock). A profile of a unit helps archaeologists understand the levels that were excavated, as well as changes in human activity over time.

Projectate Describing a pollen grain in which the apertures are borne on the ends of strongly projecting arms. Example: *Aquilapollenites*.

Projectile point Artifacts such as arrowheads and spearheads, used mostly for hunting animals.

Prokaryote A cell without a distinct nucleus. Bacteria and some other simple organisms are prokaryotic. Compare with eukaryote. All prokaryotes form a paraphyletic grouping.

Prolate Describing the shape of a pollen grain or spore in which the polar axis is larger than the equatorial diameter.

This term belongs to the system of shape classes suggested by Erdtman (1943, and extended in 1952), based on the measurements of the polar axis (P) and equatorial diameter (E). In this system prolate is defined as a ratio between the polar axis and the equatorial diameter of 1.33-2.00.

Prolate spheroidal Describing the shape of a pollen grain or spore in which the ratio between the polar axis and the equatorial diameter is 1.00-1.14.

Prosimian One of the group of primates that includes lemurs and lorises; the other two primate groups are tarsoids and anthropoids.

Protaspis The earliest stage recognized in larval trilobite (Trilobita) development. The larva is small, often spiny, and grows through successive moult stages. Initially it is a small disc but size and segmentation increase with each successive moult.

Proteins Any of a large number of complex organic molecules that are composed of one or more chains of amino acids. Proteins can serve a wide variety of functions through their ability to bind to other molecules. Proteins may be enzymes, hormones, antibodies, structural components, or gas-transporting molecules.

Proterozoic The final era of the Precambrian, spanning the time between 2.5 billion and 544 million years ago. Fossils of both primitive single celled and more advanced multicellular organisms begin to appear in abundance in rocks from this era. Its name means "early life."

Protist An organism that belongs to the kingdom Protista, which includes forms with both plant and animal affinities, i.e., protozoans, bacteria, and some algae, fungi, and viruses.

Protosaccus A saccus which is completely filled with an alveolar structure. This feature is used in the description of pollen in the Perm - Trias. Example: *Luekisporites virkkiae*.

For extant saccate pollen grains don't show this character it is considered to be primitive.

Protozoa A group of unicellular, usually microscopic, organisms now classified in various phyla of the kingdom Protoctista. They were formerly regarded either as a phylum of simple animals or as members of the kingdom Protista. Most feed on decomposing dead organic matter, but some are parasites, including the agents causing malaria (*Plasmodium*) and sleeping sickness

(*Trypanosoma*), and a few contain chlorophyll and carry out photosynthesis, like plants.

Provenience The three-dimensional location of an artifact or feature within an archaeological site, measured by two horizontal dimensions, and a vertical elevation.

Proximal 1. The end of a bone nearest the midline of the body when the individual assumes a normal standing posture.

2. The end of an artifact nearest the user or observer.

3. The end of a lithic artifact which was originally a part of the striking platform of the core.

Proximal face That part of a palynomorph which faces towards the centre of the tetrad, between equator and proximal pole.

Antonym: distal face.

Proximal pole The centre of the proximal face.

Antonym: distal pole.

Proximocavate An exine in which the exoexine is detached, or partly detached, from the intexine only on the proximal face.

Pseudoaperture A thinning of the exine which, although superficially resembling an aperture, is not associated with a thickening of the intine and is presumed not to function as an exitus. Pseudocolpus A colpus-like pseudoaperture. Example: *Myosotis* (Boraginaceae).

Pseudogene A sequence of nucleotides in the DNA that resembles a gene but is nonfunctional for some reason.

Pseudopore A pore-like pseudoaperture.

The term has also been used for the leptoma of certain coniferous pollen.

Pseudosaccus An extensive, saccus-like separation in the wall of a spore resembling a saccus, but lacking the characteristic alveolate infrastructure. Example: *Grandispora spinosa.*

Pseudo-scallop shell A pottery decoration made by pressing the edge of a notched stick, bone or stone into the wet clay before the pot is fired. The impression resembles the edge of a scallop shell.

Psilate Describing a pollen or spore with a smooth surface.

Psilolophate Describing a lophate pollen grain which lacks spines.

Psoralea ssp The prairie turnip, as it is commonly referred to, is actually several species of plants which grow underground tubers. Only one species was actually recorded as having been used by Plains Indians, and only recently has any archaeological evidence for the use of this plant been recovered.

Pterosauria ("winged lizards") Clade of animals, partly distinguished by a greatly elongated fourth finger that supported a membranous wing, that first evolved in the latter third of the Triassic and survived until the end of the Cretaceous. Pterosaurs were not dinosaurs, but were closely related to both dinosaurs and crocodiles -pterosaur.

Pterosaurs The name given to a group of giant flying reptiles which lived between the Late Triassic and Cretaceous periods.

Ptolemy Greek general, who after the death of Alexander the Great became ruler of Egypt.

Ptychotreme With apertures situated in invaginations of the outline, when seen in polar view.

The term is not recommended because it is based upon the suffix -treme.

Public archaeology A movement to increase public awareness and education about archaeology which advanced the legislative attempts to provide funding and protection for archaeological sites.

Pueblo A type of village (common to the Southwest United States) composed of clusters of rectangular chambers of stone or adobe blocks roofed with logs.

Pumice A type of rock formed by volcanic eruptions. Pumice is light in weight because it is full of holes—as lava surfaces and cools, water vapor is expelled at high temperatures, creating these holes.

Punctate A decoration on pottery produced by pressing the end of a stick or similar object into the wet clay before the vessel is fired. The raised or mounded feature formed on the opposite side of the vessel wall is a boss.

Punctates Impressions in the surface of ceramic vessels made by implements or by fingernails as a form of decoration.

Punctum A rounded or elongate tectal perforation, less than 1μm in length or diameter.

Pupa The third stage of development in the life cycle of some insects, including flies, butterflies (in which it is the chrysalis), ants, bees, and beetles. During the pupal stage locomotion and feeding cease and metamorphosis from the larva to the adult form takes place. The adult emerges by cutting or digesting the pupal case after a few days or several months.

Purine A kind of base in the DNA; adenine (A) and guanine (G) are purines.

Purposive sampling Sampling in which sample unite are selected on the basis of specific criteria.

Pylon From the Greek word meaning "gate" It is a monumental entrance wall of a temple. Pylons are the largest and least essential parts of a temple that is usually built last. Some temples have more than one set, the temple at Karnak has 10 Pylons.

Pyramid texts Spells and magical inscriptions found inside Egyptian pyramids from dynasty V to Dynasty VI.

Pyramidion Pyramid-shaped tip of obelisk, or capstone of a pyramid.

Pyramids of giza The pyramids located in Giza, Egypt, built by the pharaohs Khufu, Khafre, and Menkaure.

Pyrimidine A kind of base. In DNA, cytosine (C) and thymine (T) are pyrimidines. In RNA, cytosine (C) and uracil (U) are pyrimidines.

Pyxis Storage container.

Q

Qanat An irrigation system formed by tunneling into a hillside to find a water-bearing formation, and then channeling the water out of the hill to the surface for use in irrigation. Much of the water used in the plateau regions of Iran, for example, is brought into use in this way.

Quadrant generally refers to one-quarter of an excavation unit or level, e.g. "the northwest quadrant of excavation unit N. 2-4, E. 4-6".

Quadrat A rectangular sampling unit.

Quadruped An animal with four feet.

Quadrupedal The collective name given to land dwelling creatures which walk on all four legs.

Quadrupedalism Locomotion on four limbs.

Quantitative character A character showing continuous variation in a population.

Quarry. An archaeological site from which copper, stone or clay were acquired by either surface collection or mining.

Quarry site A site where lithic raw materials have been mined.

Quarter rudder A.k.a. side rudder. Ships of antiquity had two side rudders. Nordic ships had one, placed on the starboard (right) side. During the middle ages all European ships slowly changed to using one stern rudder.

Quartz. A very hard (7 on the moh scale), common mineral composed of silicon dioxide in a hexagonal crystalline form. Present in most rocks (particularly granite and sandstone), this colourless, glasslike stone exhibits conchoidal fracturing properties thus lending itself to flaking. Vein quartz. Quartz which occurs naturally as a vein with a matrix of some other mineral.

Quartzite A stone which was formed in water deposited sediments and consists of sand grains which have been cemented together. It can be chipped, but is difficult to work.

Quasisaccus Synonym of protosaccus.

Quasitectate A spore exine in which the outer and inner, more or less homogeneous, layers aer seperated by a clearly defined mesexinous layer of discontinous columellate elements, simulating those that characterise many angiosperm pollen.

Quaternary palynology The branch of environmental palynology dealing with palynomorphs of Quaternary age, ca. the last 2.5 million years; typically using radiometrically- or paleomagnetically-dated sedimentary sequences.

Quaternary period Last period of the geologic time scale, commonly subdivided into Pleistocene and Holocene Epochs. Began 1.6 million years BP and characterized by frequent and rapid environmental changes. The evolution of the genus *Homo* (man) is probably the single most important development of this period, thus the term Anthropogene is used by some for this period.

Quaternary The second period of the Cenozoic era, spanning the time between 1.8 million years ago and the present. It contains two epochs: the Pleistocene and the Holocene. It is named after the Latin word "quatern" (four at a time).

Quern A stone used for grinding corn or grain.

Quetzal A plumed jungle bird, highly prized for its tail feathers in Maya costumes.

Quiver A container for arrows or darts.

Quran The collection of Islamic scriptures. It is made up of 114 sections that are believed to have been revealed verbatim to Muhammad over a period of time through the angel Gabriel.

R

R selection Natural selection typically occurring in situations that favor a high reproductive rate (r). This tends to result in a species which survives by producing many young with minimal individual resource investment by the parents, who typically abandon their eggs or newborns —r-selected.

Race A subgroup of human population that shares a greater number of physical traits with one another than they do with those of other subgroups.

Racemization The process by which amino acids change from the L-molecule form to their mirror image D-molecule form. This occurs naturally following the death of cells.

Raddatz An archaeological site in central wisconsin which has given its name to an archaic side-notched projectile point style. The raddatz side-notched type ranges in length from 38 to 64 mm and characteristically bears u-shaped side-notches close to the base which are set at right angles to the long axis of the blade (sense 2). The lateral edges of the bases are squared-off while the bases themselves are usually straight and bear evidence of grinding. Only a small minority are either slightly concave or convex. The blade (sense 3) edges were originally nearly parallel to one another although they may converge more rapidly to the tip as a result of the reworking of broken specimens. In many respects, raddatz points are similar to osceola although the latter tend to be larger and more finely made. The type site was multi-component, but the raddatz points occurred in levels 5 through 12. These levels also contained stemmed projectile points, symmetrical biface knives, scrapers, a drill, a grinding stone, hammerstones, bone awls, a chisel fashioned from a beaver tooth, turtle shell bowl fragments, shell spoon fragments, antler flaking tools, an antler anvil and a barbed antler projectile point. Radiocarbon dates from the type site indicated that this assemblage dated between approximately 4500 and 1800 b.c. in manitoba, projectile points similar to those termed raddatz have been found in the southeastern part of the province.

Radial A general term describing features radiating from a centre. Applied in palynology to the region of a spore beyond the ends of the laesurae.

Radial symmetry The arrangement of parts in an organ or organism such that cutting through the centre of the structure in any direction produces two halves that are mirror images of each other. The stems and roots of plants usually show radial symmetry, while all animals belonging to the Cnidaria (e.g. jellyfish) and Echinodermata (e.g. starfish) are radially symmetrical - and typically sessile - in their adult form. The term actinomorphy is used to describe radial symmetry in flowers.

Radially symmetric Describing a pollen grain or spore with two or more vertical planes of symmetry, but, if only two such planes are present, then their axes are of equal length.

Radioactive dating A dating method that measures the amount of radioactive decay that has taken place in the rocks being studied.

Radioactive decay the regular process by which radioactive isotopes break down into their decay products with a half-life which is specific to the isotope in question.

Radioactivity The emission of energy due to changes in the nucleus of an atom. Such spontaneously released radiation is a characteristic of certain elements and at some levels can be harmful.

Radiocarbon dating (C 14) The absolute dating technique which more than any other revolutionized archaeology. Cosmic radiation from the sun constantly bombards the earth's upper atmosphere producing a radioactive form (isotope) of carbon (c-l4) from non-radioactive nitrogen (n-l4). C-l4 mixes with the earth's atmosphere and oceans and behaves in precisely the same manner as does the non-radioactive (stable) form c-l2. Because all living organisms absorb carbon through water or by breathing, the ratio of c-l4 to c-l2 within them is the same as that in the atmosphere, and this ratio has remained more-or-less constant over time. When an organism dies, however, it ceases to be part of the carbon cycle and thus no new c-l4 is absorbed. Like all radioactive substances, c-l4 is unstable, which is to say that over time it will break down (or decay) into a more stable form — in this case c-l2. The rate of this decay is known and is also constant. Consequently, by measuring the ratio of c-l4 to c-l2 in a piece of wood, bone or other organic matter from an archaeological site, one can determine the length of time since the sample last absorbed new c-l4. Recently, dendrochronology has been used to check the accuracy of radiocarbon dating, with actual material from the tree rings being used as samples.

Radioimmunoassay A method of protein analysis whereby it is possible to identify protein molecules surviving in fossils which are thousands and even millions of years old.

Radiometric dating Referring to techniques for chronometric dating based on known half-lives of particular isotopes or the rate of other cumulative changes in atoms resulting from radioactivity. Examples include electron spin resonance, fission track, potassium-argon, radiocarbon, and thermoluminescence dating.

Radius A general term, used in palynology for a laesura in trilete spores.

Raised beaches These are remnants of former coastlines, usually the result of processes such as isostatic uplift or tectonic movements.

Rak, Yoel An Israeli paleoanthropologist and anatomist whose research interests include facial morphology of fossil hominids. Rak was part of the team that found a 2.3-million-year-old skull fragment from the genus *Homo* at Hadar, Ethiopia.

Ramada An open-air shade built of upright posts that are covered with a flat roof. The Pimas and Papagos also made use of the ramada as a focal point of family activity.

Ramon tree A nut-bearing tree common in the lowland rain forests of the Yucatan Peninsula.

Rancheria A form of village arrangement in which individual dwellings are widely separated. This can be contrasted with the Pueblo sale of architecture in which dwellings within a village are attached to one another in an apartment-like complex.

Random drift Synonym of genetic drift.

Random mating A mating pattern in which the probability of mating with another individual of a particular genotype (or phenotype) equals the frequency of that genotype (or phenotype) in the population.

Ranked societies Societies in which there is unequal access to prestige and status e.g. chiefdoms and states.

Rational economic decisions The weighing of available alternatives and calculation of which will provide the most benefit at the least cost.

Re Egyptian creator god. Sun god of the sky, earth and the underworld.

Re-atum A manifestation of Re as the setting sun.

Reaves Bronze Age stone boundary walls, for instance on Dartmoor, England, which may designate the territorial extent of individual communities.

Rebellion An attempt within a society to disrupt the status quo and redistribute the power and resources.

Recanted Withdrew a statement or opinion; disavowed a former assertion.

Recapitulation A partly or wholly erroneous hypothesis stating that an individual, during its development, passes through a series of stages corresponding to its successive evolutionary ancestors. According to the recapitulation hypothesis, an individual thus develops by "climbing up its family tree."

Receptors Proteins that can bind to other specific molecules. Usually on the surface of a cell, receptors often bind to antibodies or hormones.

Recessive An allele (A) is recessive if the phenotype of the heterozygote (Aa) is the same as the homozygote (aa) for the alternative allele (a) and different from the homozygote for the recessive (AA). The allele (a) controls the heterozygote's phenotype and is called dominant. An allele may be partly, rather than fully, recessive; in that case, the heterozygous phenotype is nearer to, rather than identical with, the homozygote for the dominant allele.

Reciprocity A mode of exchange in which transactions take place between individuals who are symmetrically placed, i.e. they are exchanging as equals, neither being in a dominant position.

Reclamation Activities that move objects from archaeological context back into systemic context.

Recognition species concept A concept of species according to which a species is a set of organisms that recognize one another as potential mates; they have a shared mate recognition system. Compare with biological species concept, cladistic species concept, ecological species concept, and phenetic species concept.

Recombination An event, occurring by the crossing-over of chromosomes during meiosis, in which DNA is exchanged between a pair of chromosomes of a pair. Thus, two genes that were previously unlinked, being on different chromosomes, can become linked because of recombination, and linked genes may become unlinked.

Recording unit A level, feature, or occupation surface used as a reference point to designate the recovery location (provenience) of artifacts at a site.

Recovery theory The principles on which are based the choice of techniques to be used in a given archaeological survey and excavation.

Recovery units The two-dimensional (in surface collection) and three-dimensional (in excavation) spaces in which archaeological materials are recovered and recorded at a site.

Rectimurate Describing a pollen grain or spore with more or less straight muri.

Recurrent fauna A group of animal species that is found in a particular environment. As the environment recurs, so does the fauna.

Refitting Sometimes referred to as conjoining, this entails attempting to put stone tools and flakes back together again, and provides important information on the processes involved in the knapper's craft.

Refugium A geographical area which remains unaltered or changes much more slightly than surrounding regions in response to a change in climate. A refugium therefore constitutes a haven for plants and animals which were once more widely distributed .

Refutationist view Approach which holds that science consists of theories about the empirical world, that its goal is to develop better theories, which is achieved by finding mistakes in existing theories, so that it is crucial that theories be falsifiable (vulnerable to error and open to testing). The approach, developed by Karl Popper, emphasizes the important of testability as a component of scientific theories.

Regelation Process of high-pressure melting and refreezing of water at temperatures below 0°C due to pressure changes at base of glacier.

Regional dimension The artifacts that occur in close association with each other.

Regional survey Locating and describing the remains of settlements in a region.

Regional system The behavioral component made up of communities related to each other by trade, political alliances, and other forms of social interaction.

Regolith Deeply weathered bedrock produced by prolonged and intense decay of bedrock usually under tropical conditions. Remnants of preglacial regolith found in glaciated regions of Canada and Fennoscandinavia.

Reinforcement An increase in reproductive isolation between incipient species by natural selection. Natural selection can directly favor only an increase in prezygotic isolation; reinforcement therefore amounts to selection for assortative mating between the incipiently speciating forms.

Relationship (versus similarity) It must be distinguished between a mere phenetic relationship, which is based on overall similarity, and a true genealogical relationship, which is based on common ancestry. Furthermore one has to distinguish between horizontal indirect relationship, which is based on shared ancestors (sistergroup-relations) and vertical direct relationship, which is based on parental relationships (ancestor-descendant-relations). A confusion of these terms has often led to misunderstandings. Phylogenetic Systematics deals almost exclusively with indirect relationships, since direct ancestors are hardly discoverable out of theoretical and practical reasons

The so-called "genetic relationship" (*sensu* mayr) means, that two species are regarded the more related the more similar are their genotypes (mayr: "we do not classify phenotypes but genotypes"). Genetic relationship in this sense is nothing but a phenetic relationship, based on the overall similarity of the genome. This phenetic attitude leads to curious consequences, e.g. one-egged twins would be regarded as closer related with each other than each of them with its parents or children.

In Phylogenetic Systematics the term (phylogenetic) relationship is defined as follows: A taxon *A* is more closely related with a taxon *B* than with a taxon *C*, only if *A* and *B* are the descendants of a common stem-species, which is not a stem-species of C. Phylogenetic relationships are therefore defined by the "recency of common descent", thus in exactly the same way as the genealogical relationships (tokogenetic relationships) between individual organisms (the common sense notion of relationship in everyday life). The logical consequence of the above given definition is, that statements about phylogenetic relationships only make sense if they refer at least to three species or monophyletic groups ("three taxa statements"). Statements like "taxon A is closely related to taxon B", which are still common place in biology textbooks, are rather meaningless, if it is not specified compared to which other taxon these two taxa shall be more closely related.

Relative date A date that gives the time of an event only with reference to another event that is not worldwide in scale. It only indicates that one event occurred earlier or later than another. For instance, the observation that strata 2 is younger than strata 1 beneath it in a geological deposit does not provide information about how many years ago strata 2 was laid down. It only indicates its age relative to strata 1. In addition to the use of stratigraphy, relative dating methods include biostratigraphy and fluorine analysis dating.

Relative dating The process of ordering fossils, rocks, and geologic events from oldest to youngest. Because of the way sedimentary rocks form, lower layers in most series are older than higher layers, making it possible to determine which fossils found in those layers are oldest and which are youngest. By itself, relative dating cannot assign any absolute age to rocks or fossils.

Relative time scale Placing rocks and events in the chronological order in which they occurred.

Relativism The concept that a cultural system can be viewed only in terms of the principles, background, frame of reference, and history that characterize it.

Relief A form of sculpture where the figures stand out from, but are still attached to the background.

Religion A framework of beliefs relating to supernatural or superhuman beings or forces that transcend the everyday material world.

Remote sensing Observation and recording at a distance, especially by specialized serial photography of sites and natural environment.

Replica A copy or reproduction of the original.

Reproductive character displacement The increased reproductive isolation between two closely related species when they live in the same geographic region (sympatry) as compared with when they live in separate geographic regions. A kind of character displacement in which the character concerned influences reproductive isolation, not ecological competition.

Reproductive isolation Two populations or individuals of opposite sex are considered reproductively isolated from one another if they cannot together produce fertile offspring.

Reptilia All descendants of the most recent common ancestor of turtles, lepidosaurs (tuataras and squamates), and archosaurs (crocodilians and birds). In the past, animal groups were defined by physical traits such as hair, scales or whether they were "warm-" or "cold-blooded." Animals are now grouped by ancestry thus making birds reptiles, or more formally members of the clade Reptilia, despite the fact that they carry few of the traits that defined the Class Reptilia of the old Linnean system of taxonomy. Because birds in particular do not share the characteristics that many people today think all reptiles should have, many scientists favor dropping the name

"Reptilia" and adjective "reptile" completely.

Rescue archaeology A term applied to the emergency salvage of sites in immediate danger of destruction by major land modification projects such as reservoir construction.

Research design The plan for an archaeological investigation.

Resharpening flakes Usually small flakes removed from the edges of chipped-stone cutting or scraping tools to rejuvenate the effectiveness of the edge.

Residential mobility The degree of movement of household social units from dwelling to dwelling, particularly of hunting-gathering communities

Resistivity A pre-excavation method for determining the size, shape and occasionally the nature of an archaeological site. It works on the principle that different deposits offer differential resistance to an alternating current passing through them, largely in response to the amount of water contained within them. Thus a damp, buried ditch will provide less resistance while a wall or foundation ill offer more. Mapping electrical resistance across a site can greatly aid the archaeologist in deciding where to dig.

Resistivity survey A pre-excavation method for determining the size, shape and occasionally the nature of an archaeological site. It works on the principle that different deposits offer differential resistance to an alternating current passing through them, largely in response to the amount of water contained within them. Thus a damp, buried ditch will provide less resistance while a wall or foundation ill offer more. Mapping electrical resistance across a site can greatly aid the archaeologist in deciding where to dig.

Restoration The process of cleaning and studying an artifact and attempting to return it to its original form (before it was buried).

Reticule A frame with a grid used as an aid in mapping. Those used in Manitoba consist of a square wooden frame with a network of strings set at regular intervals. These may be placed over a feature thus allowing more accurate sketch maps to be made. Locally, reticules have found their greatest utility in the recording of petroforms.

Reticuloid With bacula arranged in a more or less reticulate pattern.

Synonym of retipilate.

Reticulum A network-like pattern consisting of lumina or other spaces wider than 1μm bordered by elements narrower than the lumina.

Retina The back wall of the eye onto which images are projected. From the retina, the information is sent to the brain via the optic nerve.

Retipilate Describing a reticulum formed by rows of pila instead of muri. Example: *Callitriche* (Callitrichaceae).

Retoucher A small tool (such as that found on the body of Otzi the Iceman) made especially for pressure flaking flint and chert tools. In the iceman's case, his retoucher was at first mistaken for the stub of a pencil.

Retusoid Describing a spore with prominent contact areas and curvaturae. Example: *Retusotriletes*.

Reverse "Tail" side of a coin.

Reverse stratification Earlier materials overlying later materials.

Revolution An attempt to overthrow the existing form of political organization, the principles of economic production and distribution, and the allocation of social status.

Reworking Modification of an object so that it may be used again. The reworking of stone artifacts after breakage was particularly common in times and places where suitable lithic resources were scarce. A projectile point which had been broken at the time could easily be resharpened and some were probably reworked several times in this fashion, becoming shorter relative to their widths each time. Drills and side-notched end-scrapers with bases similar to those of the projectile points with which they are associated are often considered to be examples of reworking.

Rhizopod A protozoan of the class Rhizopoda that has pseudopodia.

Rhomboidal tetrad A uniplanar tetrad in which the proximal faces of two individual members are in direct contact and the remaining two are separated, giving a rhomboidal outline to the tetrad. Example: *Epipactis palustris* (Orchidaceae).

Rhyolite A fine-grained igneous rock of the same composition as granite (quartz, feldspar and mica), but capable of being worked by flaking into tools. Characteristically, rhyolite has a glossy appearance and small inclusions of quartz and feldspar within the finer matrix.

Ribosomal RNA (rRNA) The kind of RNA that constitutes the ribosomes and provides the site for translation.

Ribosome The site of protein synthesis (or translation) in the cell, mainly consisting of ribosomal RNA.

Rich, Thomas H. Dr. Thomas H. Rich is an Australian paleontologist and author. Rich has unearthed many important finds in Victoria, Australia, Alaska, USA, and Patagonia, South America. He co-named (with his wife P. Vickers-Rich) the dinosaurs Leaellynasaura (1989) and Timimus(1994) (both named for his children), Atlascopcosaurus (1989), Qantassaurus (1999), Tehuelchesaurus (1999, also with Gimenez, Candúneo, Puerta, and Vacca), the important early Australian mammal Ausktribosphenos, and others.

Ridge, oceanic A major submarine mountain range.

Rift system The oceanic ridges formed where tectonic plates are separating and a new crust is being created; also, their on-land counterparts such as the East African Rift.

Rift zone A zone of volcanic features associated with underlying dikes. The location of the rift is marked by cracks, faults, and vents.

Rimsherd A fragment of pottery which originally formed part of the rim of a vessel.

Rimula The sub-equatorial aperture that encircles pollen grains of the Classopolles group. Example: *Corollina* (*Classopollis*).

Potonié (1934) used rimula to refer to short colpi, but the term is no longer used in this sense.

Ring of fire The regions of mountain-building earthquakes and volcanoes which surround the Pacific Ocean.

Ring species A situation in which two reproductively isolated populations living in the same region are connected by a geographic ring of populations that can interbreed.

Riparian Pertaining to the bank of a river.

Rite of solidarity any ceremony performed for the sake of enhancing the level of social integration among a group of people.

Rites of intensification Rituals intended either to bolster a natural process necessary to survival or to reaffirm the society's commitment to a particular set of values and beliefs.

Rites of passage Rituals that mark a person's transition from one set of socially identified circumstances to another.

Ritual Behavior that has become highly formalized and stereotyped.

Riverine archaeology Maritime archaeology, specializing in the study of artifacts of early river traffic.

Rock alignment Any artificial arrangement of rocks or boulders into rows or other patterns.

Rock art A general term for figures or designs painting or engraved on rock or formed through the placement of boulders. Rock art thus includes petroforms, petroflyphs, petrographs and pictographs.

Rock cut tomb Method of excavating tombs that is fairly common world wide. The most famous are in Egypt and were begun during the Middle Kingdom. The burials in the Valley of the Kings are perhaps the best known Rock-cut tombs.

Rock lake focus An early ceramic focus defined by vickers (1949) on the basis of his investigations in southern mantoba. The majority of this material is now considered to be part of the laurel complex.

Rock painting A more common term for pictograph.

Rock shelter A small cave or overhang of rock which afforded some degree of protection from the elements either as a permanent camp or temporary location of activity.

Rocker stamping A pottery design produced by rocking a straight or notched edge from one end to the other, while moving the tool sideways across the surface of the wet clay. The result is a zigzag pattern of curved lines.

Rodinia A supercontinent that existed during the Late Precambrian before the supercontinent Pannotia; the oldest supercontinent for which we have a good record; Russian for "homeland.".

Role A set of behavioral expectations appropriate to an individual's social position.

Roman colosseum Built in Rome, Italy between A.D. 70 and 80, the four-story structure was the site of bloody fights between animals, gladiators, or animals versus gladiators for more than 300 years. After A.D. 404, the gladiator battles ended, but scuffles between animals in front of about 50,000 spectators remained a popular sport until the 6th century A.D.

Roman copies of greek originals Replicas of famous Greek statues made for the Roman market.

Roman republic Period of Roman history from the expulsion of the Etruscan kings till the establishment of the empire by Augustus Caesar.

Roof comb The part of a Mesoamerican temple that extends over the usable area. Frequently carved with glyphs and figures.

Rosetta stone A basalt slab discovered at Rosetta (Rashid), at the western mouth of the Nile, during Napoleon's occupation of Egypt. It is now in the British Museum. The text is an honorific decree of Ptolemy V (196 B.C.) that was written in ancient Greek, hieroglyphics, and demotic, which is a later form of ancient Egyptian. The stone enabled the French explorer and linguist Champollion to decipher ancient Egyptian hieroglyphics by comparing the unknown text of hieroglyphics to the known and translated text in Greek and demotic.

Rotational north pole The point on the northern extremity of the earth where the axis of rotation is located. Compared to the magnetic north pole, the rotational one is relatively stable.

Rov Remote Operated Vehicles are remote controlled underwater robots used instead of divers for difficult operations, e.g. on great depth.

Rudist An extinct bivalve mollusk from the Jurassic and Cretaceous that had two different sized and shaped shells; they usually were attached to the substrate and were either solitary or in reeflike masses.

Ruga Synonym of colpus.

Rugulate Describing a type of ornamentation consisting of elongated sexine elements more than 1μm long, arranged in an irregular pattern that is intermediate between striate and reticulate. Examples: *Sedum* (Crassulaceae), *Ulmus* (Ulmaceae), *Camptotriletes corrugatus.*

Rule of deviation During an allopatric speciation by geographic separation of a stem-population, the latter is usually not divided into two populations of similar size (dichopatric speciation), but only a small peripheral population is separated from the remaining population (peripatric speciation). Because of a phenomenon that is generally known as genetic drift the small founder population shows a stronger morphological divergence from the original state than the large rest population that stays more or less unchanged. This "founder effect" was called by willi hennig the "rule of deviation". The morphological hardly changed rest population cannot be regarded as "surviving stem-species", because of theoretical reasons , and therefore has to be regarded as non-divergent descendant species.

Russell, Dale A. Dr. Dale Alan Russell is a vertebrate paleontologist and author. He is a Research Professor at North Carolina State University and Senior Curator of Paleontology at the North Carolina Museum of Natural Sciences. Russel has named the following dinosaurs: Alxasaurus (Russell and Dong, 1994), Archaeornithomimus (1972), Atlasaurus (Monbaron, Russell and Taquet, 1999), Cristatusaurus (Taquet and Russell, 1998), Daspletosaurus (1970), Dromiceiomimus (1972), Dyslocosaurus (McIntosh, Coombs and Russell, 1992), Lurdusaurus (Taquet and Russell, 1999), Sigilmassasaurus (Russell, 1996), and Sinornithoides (Russell and Dong, 1994). Russell has written many papers and books, including, "Odyssey in Time: The Dinosaurs of North America" 1989 and "Systematics and Morphology of American Mosasaurs" (1967). In 1971, Russell and the physicist Wallace Tucker published a paper (in the journal Nature) called "Supernovae and the extinction of the dinosaurs," which theorized that a supernova caused the K-T Extinction. This was the first theory to put the blame on an extra-terrestrial phenomeno

S

S ype tetrad A tetrad in which only one members is fully developed. Example: *Leucopogon fasciculatus* (Epacridaceae).

Saag, Michael Dr. Saag is director of the AIDS Outpatient Clinic and Associate Professor of Medicine at the University of Alabama, Birmingham. He is also associate director for clinical care and therapeutics at the UAB AIDS Center. Dr. Saag's research activities focus on both clinical and basic aspects of the human immunodeficiency virus. He serves on several state and national advisory panels, including the NIH/NIAID AIDS Clinical Trials Group Executive Committee.

Saccoid A saccus-like expansion of the exine with a complex infrastructure build up by a three dimensional network of sexine elements, extending to and fused with the nexine.

Saccus A sac formed by an expansion of the exine of a pollen grain and at least partly filled with an alveolate infrastructure.

Saff tomb An Arabic word that means "row", it describes the rock-cut tombs of the early 11th Dynasty that consisted of a row of openings on the hillside. (Egypt)

Sagittal crest A ridge of bone projecting up from the top midline of the skull, running from front to back. It serves as a muscle attachment area for the muscles that extend up the side of the head from the jaw. The presence of a sagittal crest indicates extremely strong jaw muscles.

Saguaro One of the largest cacti, the saguaro has one central trunk with one or more upward curving branches. This tree-size cactus played an important role in the economy of the Hohokam and the Pima and Papago cultures.

Salmon river side-notched. One of the projectile point styles of the bitterroot phase of idaho.

Salvage archaeology Archaeology conducted primarily because a site or area is in imminent danger of destruction by natural forces or by construction or development. The British equivalent to this term — rescue archaeology — is self-explanatory.

Sample Any part of subset of a population.

Sample size The number of sample units chosen for a study.

Sample statistics The characteristics of a sample.

Sample unit A member of a population examined.

Sampling Selecting from within regions those areas to be surveyed and, within sites, those areas to be excavated.

Sampling bias The tendency of a sample to exclude some members of the sampling universe and overrepresent others.

Sampling error In population genetics, the transmission of a nonrepresentative sample of the gene pool over space or time due to chance.

Sampling universe The largest entity to be described, of which the sample is a part.

Sanction Any means used to enforce compliance with the rules and norms of a society.

Sand A loose material composed of minute (0.2 to 2 mm in diameter) particles of weathered rock, especially quartz.

Sandstone A general name for any of the common sedimentary rocks composed of sand grains cemented together with clay, iron oxide or calcium carbonate.

Sandy lake ware A cord-impressed or smoothed-over late woodland ceramic ware of southeastern manitoba, northern minnesota and northwestern ontario. These globular vessels are small when intended for use in mortuary contexts, while utilitarian vessels may approach 15l in volume. Decoration consists of notching at the lip or punctates where it is present at all (anfinson 1979). In many respects, sandy lake ware is similar to the cemetery point corded type defined by macneish in southeastern manitoba.

Santonian European stage of the Upper Cretaceous, spanning the time between 88 and 84 million years ago.

Sapir-worf hypothesis The notion that a person's language shapes her or his perception and view of the world.

Sarcophagus From the Greek word meaning; "flesh eater". It was the name given to the stone container within which the dead were placed.

Sardis Capital of the Lydian kingdom; located in western Anatolia.

Sassanian A Persian dynasty that ruled from 224 to 651 C.E. The Sassanians sought to obliterate Greek cultural influences and revive Persian traditions. At one point the Empire controlled territory from Indus to the Nile, and from Yemen to the Caucasus.

Satyr Half human half beast follower of the god Dionysos.

Sauropods A group of giant four legged dinosaurs which are characterised by a small head, long neck, immense torso and long tail. The Sauropods were not predators, instead they used their height to feed on the vegetation out of reach to smaller dinosaurs.

Scabrate Describing elements of ornamentation, of any shape, smaller than 1μm in all directions. Examples: *Quercus* (Fagaceae), *Artemisia* (Compositae).

Ornamentation elements larger than 1μm are described according to their shape, for example, baculum, clava, gemma, verruca.

Scandic A central North American climatic Episode dating from A.D. 270 to A.D. 690 which represents the warming trend which culminated in the Neo-Atlantic Climatic Episode.

Scanning electron microscope (SEM) A microscope in which a finely focused beam of electrons is scanned across a specimen, and the electron intensity variations are used to construct an image of the specimen. This type of microscope is ideal for magnifications from 200 to 35,000.

Scansorial Describes an animal with unspecialized limbs allowing it to grasp and climb as well as walk and run.

Scaphopoda An order of molluscs aslo known as tusk shells.

Scapula hoe A hoe fashioned from a scapula (shoulder blade).

Scarab The dung-rolling beetle was, to the ancient Egyptians, a symbol of regeneration and spontaneous creation, as it seemed to emerge from nowhere; in fact it came from eggs previously laid in the sand. Seals and amulets in scarab form were very common and were thought to possess magic powers.

Scarce resources A central concept of Western economics which assumes that people have more wants than they have resources to satisfy them.

Scarp An escarpment, cliff or other steep slope, such as the slope between fluvial terraces.

Scavenging The removal of materials from archaeological context by the residents of a settlement.

Scent marking Marking territory by urinating or defecating or by rubbing scent glands against trees or other objects.

Schist Any of a group of metamorphic rocks which may be split into plates or needle-shaped grains. Constituent elements may include mica, chlorite, talc or iron.

Schneider, Chris A biologist and professor at Boston University whose research focuses on the evolution of vertebrate diversity in tropical systems and the scientific basis for conservation of tropical diversity. He uses a variety of molecular genetic methods, such as DNA sequencing, to study speciation, systematics, and biogeography of terrestrial vertebrates, with an emphasis on reptiles and amphibians.

Schultz, Ted R. An ant systematist at the Smithsonian Institution, Dr. Schultz studies the evolution of the symbiosis between fungus-growing ants and the fungi they cultivate.

Science a method of learning about the world by applying the principles of the scientific method, which includes making empirical observations, proposing hypotheses to explain those observations, and testing those hypotheses in valid and reliable ways; also refers to the organized body of knowledge that results from scientific study.

Scientific method A method of obtaining information by systematic observation, hypothesis testing, and experimentation.

Scientific theory A statement that postulates ordered relationships among natural phenomena.

Scientism The belief that there is one and only one method of science and that it alone confers legitimacy upon the conduct of research.

Sclerine A term encompassing both exine and perine that can be used whether a perine is present or not; sporoderm excluding the intine.

Scott, Eugenie C. A human biologist specializing in medical anthropology and skeletal biology. As executive director of the National Center for Science Education, Scott is an advocate of church/state separation in schools, and speaks widely about science, evolution, and natural selection.

Scott, Matthew P. A professor and researcher whose work in developmental biology explores how homeotic genes orchestrate differentiation and multicellular organization.

Scottsbluff A town in western nebraska which has given its name to a nearby archaeological site and to the distinctive stemmed plano projectile points which were recovered from it. The latter may be of two styles designated types i and ii. The scottsbluff type i projectile points are parallel-sided or slightly triangular in outline and have poorly-defined shoulders above wide stems. Flaking is either transverse parallel or more irregular and length ranges from 5 to 10 cm. In many respects, these are similar to the eden points with which they are sometimes associated save that the scottsbluffs tend to be wider relative to their length. Scottsbluff type ii points are similar to the type i specimens, but have more triangular blades and more pronounced shoulders. The associated artifacts and fauna and the distribution of the points are those of the cody complex.

Scraper An artifact used to remove the fat from the underside of a hide or to smooth wood. Unless otherwise specified, these are generally understood to be made of stone. combination (side- end-) scraper. one which is sharpened on at least one side and one end. dome (-shaped) scraper. a scraper which is roughly circular as seen from above, dome-shaped in cross-section and sharpened around most or all of its circumference. end-scraper. one which is worked at one or both ends. The "end" may be defined as the shorter of the edges in a rectangular specimen or the end which initially formed part of the striking platform or the edge opposite it. flake scraper. technically any scraper fashioned from a flake, but in general usage a scraper made on a thin flake; a raclette. keeled scraper. a scraper with an intact dorsal ridge. scraper plane. a large, often oval scraper with a flat ventral face and dome-shaped dorsal face which is often sharpened on one side. side-scraper. a scraper which is sharpened on one or both sides only. The "side" may be defined as one of the longer edges, or one of the edges adjacent to the striking platform. tabular scraper. a scraper fashioned from a flake with flat, often parallel dorsal and ventral faces. thumb(nail) scraper. a small, often domed scraper shaped like a thumbnail.

Screen A wire mesh mounted on a frame which is used to sift the soil from an archaeological excavation. The screen may be used manually or shaken by means of a small motor and serves to catch the specimens which are too small for the excavator to collect practically. The size of the mesh depends at least partially on the fineness of the deposits on the site. In Manitoba, sizes from 1/16" to 1/4" are the most common. Experiments have shown that different mesh sizes produce dramatically different results in the relative frequencies of objects of different sizes.

Scribes Ancient Egyptians trained to read and write.

Scrobiculus Synonym of punctum.

Sculptine A term encompassing both exine and perine (sclerine) but excludes the nexine and so provides a neutral term for the sculptured layer when there is doubt whether the pattern belongs to the exine or the perine.

Sculpture The external texture or pollen grains and spores . For example spines or reticulations.

Sculpturing The surface relief, or topography, of a pollen grain or spore.

Praglowski (1975) provided a circums-cription of this term which would restrict its application to tectate pollen grains.

Sculpural density The estimated number of sculptural elements in an area of 100 μm2 of the surface of the exine.

Sea peoples A group of people who migrated to eastern shore of the Mediterranean Sea around 1400 BC.

Seafloor spreading The mechanism by which new seafloor crust is created at oceanic ridges and slowly spreads away as plates are separating.

Seamount A submarine volcano.

Seasonal isolation A form of reproductive isolation in which the breeding seasons of two closely related populations do not exactly correspond.

Seasonal round. 1. The sequence of places or different ecological zones which a group of people exploit throughout the course of the year.

2. The cycle of activities pursued throughout the year.

Secondary burial A human interment which was moved and re-buried

Secondary center of ossification An area of ossification, usually near the end of a long bone.

Secondary datum A local base measuring point at a known distance from the main horizontal or vertical datum points.

Secondary deposit a body of natural or cultural sediments which have been disturbed and re-transported since their original deposition.

Secondary refuse Trash deposited at other than the location of use.

Secondary retouch finishing or resharpening flaking done after the basic shape of a lithic tool has been completed.

Secondary sexual characteristics physical features other than the genitalia that distinguish males from females after puberty.

Secondary state A society that achieved a state level of organization through contact with another state.

Section A one-square mile unit in the legal subdivision system.

Secular trend The tendency over the last hundred or so years for each succeeding generation to mature earlier and become, on the average, larger.

Secularization The act of transferring control or ownership from religious to civil authorities.

Sedentary A term applied to human groups leading a settled, non-migratory lifestyle.

Sedentary pastoralism Animal husbandry that does not involve mobility.

Sedentism The practice of establishing a permanent, year-round settlement.

Sediment Solid unconsolidated rock and mineral fragments that come from the weathering of rocks and are transported by water, air, or ice and form layers on the Earth's surface. Sediments can also result from chemical precipitation or secretion by organisms.

Sedimentary Pertaining to stone which has been formed from loose deposits which have been subsequently compacted by the weight of overlying deposits or water.

Sedimentary beds Beds, or layers, of sediments; also called strata.

Sedimentary facies An accumulation of deposits that exhibits specific characteristics and grades laterally into other sedimentary accumulations that were formed at the same time but exhibit different characteristics.

Sedimentary rocks A type of rock formed of mechanical, chemical, or organic sediments transported from their source and deposited as sediment in layers on lake or river bottoms, river sandbars, beaches, and oceans. The sediments can be formed of fragments, formed by precipitation or solution, or formed from inorganic remains (shells or skeletons) of organisms.

Sedimentation The accumulation of geological or organic material deposited by air, water, or ice.

Sedimentology A subset of geomorphology concerned with the investigation of the structure and texture of sediments i.e. the global term for material deposited on the earth's surface.

Seed ferns An extinct plant that developed seeds on its leaves and grew to a height of nine feet. Flowering plants may have evolved from seed ferns. Most fern-like fossils are not true ferns, but seed ferns.

Seed megaspore A large, functional megaspore associated with three small, presumably aborted spores present in some fossil lycopsids. Example: *Cystosporites*.

Seeley, Harry G. Harry Govier Seeley (1839-1909) was a British paleontologist. In 1887, he divided the dinosaurs by hip structure, into the orders Saurischia and Ornithischia. He named Agrosaurus (1891), Anoplosaurus (1878), Aristosuchus (1887), Craterosaurus (1874), Macrurosaurus (1869), Orthomerus (1883), Priodontognathus (1875), Rhadionsaurus (1881), and Thecospondylus (1882).

Segmentary lineage A descent group in which minimal lineages are encompassed as segments of minor lineages, minor lineages as segments of major lineages, and so on.

Segmentary societies Relatively small and autonomous groups, usually of agriculturalists. who regulate their own affairs; in some cases, they may join together with other comparable segmentary societies to form a larger ethnic unit.

Seismic reflection profiler An acoustic underwater survey device that uses the principle of echo-sounding to locate submerged landforms; in water depths of 100 m, this method can achieve penetration of more than 10 m into the sea-floor.

Selection Synonym of natural selection.

Selectionism The theory that some class of evolutionary events, such as molecular or phenotypic changes, have mainly been caused by natural selection.

Selective agent Any factor that brings about differences in fertility and mortality.

Selective attention Unconscious focusing on and response to stimuli that are perceived to be important, to the exclusion of other stimuli.

Selective pressures Environmental forces such as scarcity of food or extreme temperatures that result in the survival of only certain organisms with characteristics that provide resistance.

Self organization The product of a theory derived from thermodynamics which demonstrates that order can arise spontaneously when systems are pushed far from an equilibrium state.

Self-reducing tacheometer A major surveying instrument (transit or alidade) which allows the direct read-out of true vertical and horizontal distances within the eye-piece without the use of trigonometric formulae or tables.

Selkirk chert A white to cream-colored medium to fine-grained limestone chert which occurs naturally along the banks of the Red River. Most nodules are dense, gray, and often mottled in appearance while those of poorer quality are chalk-like. Due to its conchoidal fracturing properties, Selkirk Chert was a popular material for the manufacture of stone tools throughout most of the province's prehistory (Leonoff 1970).

Selkirk focus The original term for the selkirk or mistikwas phase.

Selkirk side-notched. The projectile point style associated with the selkirk culture. These range in length from 1.9 to 3.0 cm and have convex bases and wide side-notches near the base.

Selkirk ware. The more common term for winnipeg fabric-impressed ware – the pottery style associated with the selkirk culture.

Semantic domains Groups of related categories of meaning in a language.

Semantics The study of the larger system of meaning created by words.

Semaphoront The object of a phylogenetic-systematic study is according to willi hennig not the individual organism as such, but always only an organism at a certain, relatively short, period of time or even only a point of time in its individual development (ontogenesis). This study object has been termed semaphoront by hennig, which means " character bearer". The complete set of characters of a semaphoront has been termed its holomorph. A minor problem of the semaphoront-concept is the simple circumstance that it was mainly developed for morphological charaxcters, while some classes of characters do only exist in the time dimension (e.g. physiological processes, behaviours, and ontogenetic sequences, etc.) (R. Schoch. pers. comm.), and other classes of characters even are totally independent from any semaphoront-stage (e.g. sequences of DNA, RNA, or proteins).

Semi erect Synonym of subprolate.

Semitectum A partially discontinuous tectum in which the tectal perforations are equal to or wider than the muri and usually larger than 1μm in diameter.

Antonym: eutectum.

Semites People who spoke a Semitic language which originated in Arabia and Mesopotamia. Some of the peoples include the: Arabs, Hebrews, Canaanites, and Phoenicians.

Semitransverse Synonym of suboblate.

Senescence Old age.

Senut, Brigitte An anatomist at France's National Museum of Natural History. In 2000, Senut and Martin Pickford discovered *Orrorin tugensis*, a proto-hominid dated at 6 million years old.

Separate creation The theory that species have separate origins and never change after their origin. Most versions of the theory of separate creation are religiously inspired and suggest that the origin of species occurs by supernatural action.

Sequence The order of the chemical bases - adenine (A), thymine (T), cytosine (C), and guanine (G)-in a gene. The base sequence determines the order of amino acids making up the protein encoded by the gene.

Sereno, Paul C. Paul C. Sereno (1958 -) is a US paleontologist from the University of Chicago who has worked in South America, Asia and Africa. He discovered the first complete skull of Herrerasaurus, excavated a giant Carcharodontosaurus (1996), found and named Afrovenator (with others, 1994), named the oldest-known dinosaur, Eoraptor (with others, 1993), Suchomimus, found the second oldest fossils bird, Sinornis ("Chinese bird"), in 1991, Jobaria, and Nigersaurus. Sereno named: Deltadromeus (1996) and Marasuchus (with Arcucci, 1994). He has rearranged the dinosaur family tree, reorganizing the ornithischians and naming the clade Cerapoda (1986), formed from the ornithopods and marginocephalians.

Serial monogamy An exclusive union followed by divorce and remarriage, perhaps many times.

Seriation A relative dating method based on the fact that artifact types change through time in frequency as a result of new technologies, styles, and available construction materials. The frequency of artifact types in a strata can be compared to known frequency changes previously recorded for an ancient culture. In this way, the strata can be dated relative to other strata or sites. When a seriation sequence has been cross-calibrated with reliable chronometric dating methods, it can be considered a calibrated relative technique.

Serpent mound A snake-shaped site in south-central Ohio built by Native Americans of the Fort Ancient culture between A.D. 900 and 1600.

Serpentine A metamorphic mineral altered from limestone or basic igneous rocks such as olivine and amphibole. One form of serpentine-chrysolite-is a common source of asbestos.

Serrated Having a notched, toothed or saw-like edge.

Seth Egyptian god associated with evil. Brother and murderer of Osiris.

Settlement A cluster of activity areas used by a community and forming a camp, village, town, or city.

Settlement pattern 1. The spatial relationship of artifact classes within a single site. Such studies serve to indicate the activity areas at a site and may allow inferences of social and political organization to be made. For example, such elements as the presence of more exotic trade goods, or a greater diversity of faunal remains associated with a larger house structure might indicate that its occupants were more affluent and perhaps more influential than their "poorer" neighbors.

2. The spatial relationship of a number of sites of the same culture and the environmental setting within which each occur. The investigation of such patterns can provide useful information on the relationship between subsistence, technology and ecology.

Settlement system The entire set of settlements used by a community; for example, all the base camps and hunting camps used by a band of hunter-gatherers.

Sex chromosome The chromosome or chromosomes that influence sex determination. In mammals, including humans, the X and Y chromosomes are the sex chromosomes (females are XX, males XY). Compare with autosome.

Sexual stratification The ranking of people in a society according to sex.

Sexually dimorphic When males and females of a species have considerably different appearances, which may include size, coloration, or other features, such as special plumage.

Shaft smoother A coarse-grained object (usually stone) with a groove for an arrow or spear shaft. The shafts are smoothed by abrasion after the fashion of sanding. Shaft smoothers may come in pairs which fit together around the shaft.

Shaft straightener A naturally or artificially perforated tool of bone or stone. The wooden shaft, while still in a "green" state, is inserted into the hole and the implement is then used as a lever to bend the shaft straight.

Shale A dark, fine-grained sedimentary rock formed of clay with cleavage planes parallel to the bedding.

Shaman The Siberian term for medicine man. By extension, the term is now used for an individual in any society who derives power directly from the supernatural and uses it for healing or for interpreting unusual phenomena. Only rarely do shamans use their power to harmful ends.

Shape classes Categories of pollen and spore shape based on the relations between polar axis (P) and equatorial diameter (E).

Share cropping Working land owned by others for a share of the yield.

Shared ancestral feature Compared with a shared derived feature, a homology that did not appear as recently and is therefore shared by a larger group of species.

Shared derived feature A recently appearing homology that is shared by a relatively small group of closely related taxa.

Sharia Meaning "path" in Arabic, Sharia law tells devout Muslims what to eat, how to pray, and how to behave in public. Muslims believe the Sharia are revelations from God (Allah). Under strict Sharia law, people may not consume alcohol, borrow money at interest, or allow women in their family to go out in public without a veil.

Sharing clusters Among chimpanzees, temporary groups that form after hunting to eat the meat.

Shawabty Also called "ushabty." A statuette placed in an Egyptian tomb to labor for the deceased in the afterlife.

Shell midden An archaeological accumulation of shells such as those found on the Hebredean islands.

Sherd (or shard) A fragment of any brittle substance, especially pottery.

Shield archaic An archaeological tradition associated with the Pre-Cambrian Shield-Boreal Forest of the eastern half of Canada throughout the Archiac Period. According to the chief proponent of this construct, J.V. Wright (1972), Shield Archaic populations developed from an earlier northern Plano (Agate Basin, Keewatin Lanceolate) base in the Keewatin District. The recession of glacial ice allowed plants and animals to invade the Shield which in turn provided the impetus for early Shield Archaic peoples to expand throughout the area. Ongoing climatic change, however, together with the expansion of Pre-Dorset peoples which it favored, forced the earlier residents from Keewatin at approximately 1000 B.C. Once adopted to life in the Boreal Forest, Shield Archaic peoples' lifeways remained essentially unchanged for thousands of years with little if any influence from cultures in adjacent environmental zones. Continuities in lifeways, artifact styles and the geographic area occupied, lead Wright to suggest that Shield Archiac people spoke a language ancestral to modern Algonkian. Artifacts recovered from Shield Archaic sites include a variety of notched and lanceolate projectile points, various scraper and biface forms, uniface blades (sense 2), wedges, flake knives, hammerstones, preforms, ground slate objects, drills, abraders, manos, metates, paintstones, spokeshaves, anvils, ground clets, points and knives, backed blades (sense 2), flaked adzes, and copper fishhooks, barbs and gaffs. For Wright (1972), the relative frequencies of the major artifact classes, projectile points (16.3%), biface blades (25.8%) and scrapers (41.6%) are diagnostic of Shield Archaic sites. The Shield Archaic is poorly dated. Wright suggests a date on the order of 5000 B.C. for the beginnings of the tradition and accepts a radiocarbon date of A.D. 727 from a site on the north shore of the Gulf of St. Lawrence as representative of a late survival of the Shield Archaic. Unfortunately, bone preservation is extremely poor throughout the Shield and thus radiocarbon dates are few and the reconstruction of subsistence patterns is a highly inferential matter.

Shifting cultivation (swidden, slash and burn) a form of plant cultivation in which seeds are planted in the fertile soil prepared by cutting and burning the natural growth; relatively short periods of cultivation on the land are followed by longer periods of fallow.

Ship burial A ship burial or boat grave is a burial in which a ship or boat is used either as a container for the dead and for grave goods, or as part of the actual grave goods. If the ship is really small, it's called a boat grave.

Shipworm This little animal eats wood. It only resides in salt water. Thus wooden wrecks in the Baltic Sea and freshwater lakes are spared.

Short faced bear A flesh-eater that lived 1,000,000 to 10,000 years ago. The most powerful land predator of its day, it combined size, strength, and speed with stout fangs and vise-like jaws to hunt large prey such as bison and giant ground sloths.

Shouldered Having a lateral extension or protrusion. This term may be used to describe knives, projectile points or ceramic pots.

Shovel screening A rapid excavation procedure in which the site matrix is shoveled directly through a screen (usually 1/4" mesh).

Shovel test A small scale, generally informal test excavation to ascertain the nature of the deposits, to determine the presence or absence of an archaeological site, or to delimit the boundaries of a known site.

Shubin, Neil A paleontologist who is known for his work on early tetrapods (any creature with four limbs). He presented a hypothesis of general patterns of the development of tetrapod limbs which changed the way scientists think about this field. The study of limbs is crucial to evolutionary science; one example of why this is important is that human development would have been impossible without limbs.

Siberia A separate continental plate that existed from the Latest Precambrian to the Carboniferous, composed of a large part of central Russia, namely Siberia.

Sickle cell anemia A disease in which poorly formed red blood cells cannot bind correctly to oxygen, resulting in low iron, blood clotting, and joint pain.

Side-blade A sharp, bifacial tool set into the side of a harpoon.

Silica A chemical combination of silicon and oxygen.

Silk road A widespread network of caravan trails crossing Europe and Asia from the Miditerranean coast to China. By the middle of the second century B.C.E., the Silk Route began to function as a regular diplomatic and commercial artery, carrying silk from China, spices and gemstones from India, silver goods from Iran, Byzantine cloth, and many other goods.

Silt A fine, loose sediment deposited by water and composed of particles between 0.02 and 0.002 mm in diameter.

Siltstone Coarse grained mudrock, intermediate between sandstone and claystone.

Silurian A period of the Paleozoic, spanning the time between 440 and 410 million years ago. It is named after a Celtic tribe called the Silures.

Simonsen An archaeological site in northwestern iowa which produced artifacts of the early side-notched point tradition and more specifically of the logan creek complex. In fact, some designate these assemblages as the simonsen (rather than logan creek) complex. The excavators believed that the site was used as a bison (occidentalis) drive. Radiocarbon dating showed the site to have been occupied at 6480 b.c.

Simplibaculate Synonym of simplicolumellate.

Simplicolumellate With a single row of columellae under each murus. Example: *Viburnum opulus* (Caprifoliaceae).

Simpson, George Gaylord One of the most influential paleontologists of the 20th century and a leading developer of the modern synthesis. He wrote hundreds of technical papers in addition to many widely read popular books and textbooks, and was a leading expert on Mesozoic, Paleocene, and South American mammals.

Simulation The formulation and computer implementation of dynamic models i.e. models concerned with change through time. Simulation is a useful heuristic device, and can be of considerable help in the development of explanation.

Sinew stone A soft, abrasive stone against which a length of sinew is drawn back and forth. This thins the sinew and makes it more uniform. Over time, deep grooves are worn into the stone from repeated use.

Sinew Tendon; the tough tissue which attaches a muscle to a bone.

Single component site An archaeological site which contains the remains of only one culture.

Single nucleotide polymorphism (SNP) A site on the DNA at which the base sequence differs among individuals. (The abbreviation is pronounced "snip.").

Sinkhole A surface depression created by underground collapse.

Sinu aperturate Describing a pollen grain in which the equatorial apertures are situated in the middle of concave sides when seen in polar view.

The term belongs to the system of ambs defined by Erdtman.

Siouan A language family which, among others includes Crow, Hidatsa, Mandan, Assiniboine and Dakota.

Sirenia Mammal herbivors adapted to aquatic coastal environments such that arms are modified into paddles, have no legs, are fat, and have highly modified skulls with greatly reduced dentition. Their nipples are under their arms (e.g. the extant manatees).

Sistergroups Two monophyla or two species, that together form a monophyletic group, thus originated from the same speciation event (mostly dichotomic splits of a stem-species), have been termed sistergroups by hennig. AX (1984) suggested the (synonymous) term "adelpho-taxa" which fortunately never became very popular. The discovery of sistergroup relationships is one of the foremost goals of Phylogenetic Systematics. Since the fossil record will never be complete, the term sistergroup usually only makes sense if it is restricted to recent taxa. The fossil relatives are referred to as stem-group representatives. Example: Crocodiles are the sistergroup of birds, while dinosaurs (paraphyletic!), which are closer related to birds than crocodiles, are belonging to the stem-group of birds.

Sister's hill. A single component hell gap site in north central wyoming which produced radiocarbon dates of 7700 and 7650 b.c. the sister's hill phase as the term is used by irwin (1971) includes the agate basin and hell gap complexes as sequent subphases in the interval between 8500 and 7500 b.c.

Sistrum The sistrum was a sacred noise-making instrument used in the cult of Hathor. The sistrum consisted of a wooden or metal frame fitted with loose strips of metal and disks which jingled when moved. This noise was thought to attract the attention of the gods. There are two types of sistrum, an iba, was shaped in a simple loop, like a closed horse-shoe with loose cross bars of metal above a Hathor head and a long handle. The seseshet had the shape of a naos temple above a Hathor head, with ornamental loops on the sides. The rattle was inside the box of the naos. They were usually carried by women of high rank.

Site A distinct spatial clustering of artifacts, features, structures, and organic and environmental remains. as the residue of human activity.

Site catchment analysis A type of off-site analysis which concentrates on the total area from which a site's contents have been derived; at its simplest, a site's catchment can be thought of as a full inventory of artifactual and non-artifactual remains and their sources.

Site survey The process of searching for and describing archaeological sites in a given area.

Slag The material residue of smelting processes from metalworking. Analysis is often necessary to distinguish slags derived from copper smelting from those produced in iron production. Crucible slags (from the casting process) may be distinguished from smelting slags by their high concentration of copper.

Slash and burn agriculture A method of farming, also called *swidden agriculture,* by which fields are cleared, trees and brush are burned, and the soil, fertilized by the ash, is then planted.

Slate A fine-grained metamorphic rock formed of mudstone or shale and generally dark in colour. Slate may be sculpted or ground to produce a sharp edge.

Slavery A practice that permits some people within a society to own other persons and to claim the right to their labor.

Slip A creamy or watery mixture of clay and water which is applied to the surface of a pot before it is fired. This serves to smooth the vessel surface and renders it more waterproof by clogging the pores of the earthenware. A slip may also change the color of the pot to virtually any desired shade.

Slope distance in mapping the inclined distance (as opposed to true horizontal or vertical distance) between 2 points.

Slump The collapsing of a cliff or the walls of an excavation.

Small, Meredith A professor of anthropology. Her research interests include primate behavior and ecology; mating strategies; reproduction; and the evolution of human behavior. Small's publications include Female Choices: Sexual Behaviour of Female Primates, What's Love Got to Do With It?, and The Evolution of Human Mating.

Smith, John Maynard An eminent evolutionary biologist and author of many books on evolution, both for scientists and the general public. A professor emeritus at the University of Sussex, his research interests include evolution of human mitochondrial DNA sequences and investigation of evidence for extensive recombination.

Smith, Tom An ornithologist and conservation biologist, Smith is executive director of the Center for Tropical Research at San Francisco State University. His work combines basic research in ecology and evolutionary science with applied research in conservation biology. Among other issues, Smith is interested in the role of ecological gradients in speciation and maintaining species diversity.

SMOW Stands for "standard mean ocean water", used as a chemical and isotopic reference standard, for example in oxygen-isotope analysis.

Smudge A small fire which produces smoke and drives away insects.

Snow knife A large flat-bladed knife often fashioned from bone which is used by the Eskimo to cut the blocks of snow for igloos.

Social class A group of people having similar social standing.

Social darwinism A doctrine that applies the principles of selection to the structure of society, asserting that social structure is determined by how well people are suited to living conditions.

Social differentiation The number and variety of social roles and classes in a community.

Social environment Human adaptive systems - communities, regional systems, or empires - surrounding a society.

Social organization The way individuals and social units interact to form a society.

Social pressure A means of social control in which people who venture over the boundaries of society's rules are brought back into line.

Social role The activities carried out by an individual in a social unit.

Social unit A group of people organized to carry out particular activities.

Society A human adaptive system.

Sociobiology The study of the biological control of social behavior.

Sociocultural anthropology A branch of anthropology that deals with variations in patterns of social interaction and differences in cultural behavior.

Socio-function The role of an artifact in a society's social organization.

Sociolinguistics A branch of anthropological linguistics that studies how language and culture are related and how language is used in different social contexts.

Soil The loose layer at the earth's surface which is composed of weathered rock particles, water, humus and air and which is capable of supporting rooted plants.

Soil resistivity A method of subsurface detection which measures changes in conductivity by passing electrical current through ground soils. This is generally a consequence of moisture content, and in this way, buried features can be detected by differential retention of groundwater.

Soil texture The relative proportion of clay, silt and sand sized particles in a soil.

Soil-sieves Small, precision metal screens, used for determining the proportions of different sized particles in a soil sediment sample.

Soil-sounding radar A method of subsurface detection in which short radio pulses are sent through the soil, such that the echoes reflect back significant changes in soil conditions.

Sole Synonym of foot layer.

Solifluction The slow downslope movement of surface sediments in a saturated condition. Prevalent in permafrost areas due to the seasonal thawing of the surface of the permafrost zone. Can cause complete mixture of site stratigraphy and archaeological components.

Solution A substance dissolved in another substance. Sugar dissolved in water is "in solution".

Somatic A term that refers to the body.

Sonoran desert a region stretching from Parker, Wickenburg, and the Tonto Basin in Arizona on the north; to the Rio Culiacan in Sinaloa, Mexico on the south; east to the Sierra Madre Occidental in Mexico, and the Huachuca, Santa Catalina, Rincon, and Pinaleño Mountains in Arizona, and west to the Gulf of California, the Colorado River delta, and the Salton Sea in California. The desert includes parts of the state of Sinaloa, most of Sonora, and the northeast corner of Baja California in Mexico; and the southern half of Arizona and the southeastern corner of California in the United States. The area is characterized by vegetation ranging from creosote bush and bursage at lower elevations to palo verde, mesquite, and saguaro at higher elevations.

Southern cult Southern Cult (also Buzzard Cult, Death Cult) The name attached to the ceremonial artifacts and art style shared by Mississippian sites. Wood, copper, clay, stone and shell were used to fashion figurines, headdresses, earspools, celts, plaques, axes, masks, effigy rattles, maces, and effigy pipes. Depicted through these media are birds of prey, vultures, skulls, spiders, dancers with snakes, winged warriors holding human heads, the hand-eye symbol, the sun symbol and the weeping eye motif. Included too are depictions of priest-like figures holding daggers to the throats of their victims and various other scenes suggestive of human sacrifice. Fairly clearly of Mexican inspiration, the ceremonial artifacts are more like one another over a broader area than are the more mundane tool and weapon styles. For this reason, they are seen as representative of rituals and beliefs which are shared by a large number of peoples irrespective of the differences in the other aspects of their lives.

Spacer region A sequence of nucleotides in the DNA between coding genes.

Spall A flake which has been produced naturally (such as by exposure to heat) or by human design.

Spear A weapon consisting of a long shaft and a sharp point which may be thrown (as a javeline) or thrusted.

Special wall Pollen mother-cell wall

Specialization The limited range of activities in which a single individual is likely to be engaged.

Specialized pastoralism The adaptive strategy of exclusive reliance on animal husbandry.

Specialized species A species closely fit to a specific environment and able to tolerate little change in that environment.

Specialized trait A structure used basically for one function.

Speciation Changes in related organisms to the point where they are different enough to be considered separate species. This occurs when populations of one species are separated and adapt to their new environment or conditions (physiological, geographic, or behavioral).

Species (biological) A group of organisms that can reproduce and produce viable offspring.

Species An important classificatory category, which can be variously defined by the biological species concept, cladistic species concept, ecological species concept, phenetic species concept, and recognition species concept. The biological species concept, according to which a species is a set of interbreeding organisms, is the most widely used definition, at least by biologists who study vertebrates. A particular species is referred to by a Linnaean binomial, such as Homo sapiens for human beings.

Specific gravity The ratio of the density of a material to the density of water.

Speech community A socially distinct group that develops a dialect; a variety of language that diverges from the national language in vocabulary, pronunciation, and grammar.

Spheroidal Describing the shape of a pollen grain or spore in which the polar axis and the equatorial diameter are approximately equal.

This term belongs to the system of shape classes suggested by Erdtman (1943), and extended in 1952), based on the measurements of the polar axis (P) and equatorial diameter (E). In this system spheroidal is defined as a P/E ratio of 0.88-1.14.

Sphinx A figure with the body of a lion and the head of a man, hawk or a ram.

Spina Synonym of Spine.

Spine A general word, applied in palynology to long and tapering pointed elements, exceeding 1μm.

In Erdtman's (1952) definition spines were defined as more than 3μm long (in contrast to spinules which were shorter than this) but since the maximum size of features that may be prefixed with micro- is 1μm the present definition provides consistency with other terms.

Spinule Small spines, less than 3μm in length.

The size distinction made by Erdtman between spinules and spines is not consistent with other size criteria used in palynology.

Spiraperturate Describing a pollen grain with one or more spiral apertures. Example: *Eriocaulon aquaticum* (Eriocaulaceae).

In the pollen classes of Iversen and Troels-Smith (1950), spiraperturate pollen was included in the syncolpate class.

Spirit helper A personal guardian or protective spirit who was often acquired by questing and privation and who usually communicated through dreams or visitations.

Spirit possession The supposed control of a person's behavior by a supernatural spirit that has entered the body.

Spokshave A scraper with a pronounced concave working edge used for scraping arrow or spear shafts or bows, etc.

Sponge A member of the phylum Porifera, marine and freshwater invertebrates that live permanently attached to rocks or other surfaces. The body of a sponge is hollow and consists basically of an aggregation of cells between which there is little nervous coordination, although they do have specialized sets of cells that perform different functions. One set of cells causes water to flow in through openings in the body wall and out through openings at the top; food particles are filtered from the water by these cells. Other cells construct a stiffening skeletal framework of spicules of chalk, silica, or fibrous protein to support the body.

Spontaneous generation An old and incorrect idea that complex life forms could be spontaneously created from nonliving material.

Spore A general term for the usually microscopic, unicellular, asexual or sexual reproductive units of cryptogams and fungi

Sporoderm The entire wall of a pollen grain or spore.

Sporomorph A general term for spore-like palynomorphs.

Sporopollenin The carotenoid-like polymer comprising the outer wall of pollen grains and certain spores , and apparently dinoflagellates and acritarchs. Highly resistant to forms of deterioration other than oxidatio

Square tetrad Synonym of tetragonal tetrad.

Stability The ability of an ecosystem to return to equilibrium after disturbances.

Stabilizing selection Natural selection where species survive by maintaining existing characters. Individuals that exhibit extreme variations of any critical characters are eliminated, and the population tends to become narrow in the expression of adapted traits. Stabilizing selection does not result in evolutionary changes; rather, the existing state of adaptation is maintained.

Stade/interstade Glacial advance of brief duration within a glacial stage. An interstade is a brief retreat or minor deglaciation within a glacial stage.

Stadial rod A graduated staff used in conjunction with a transit or theodolite in surveying. The distance from the instrument to the rod may be calculated in the same units with which the rod is calibrated (meters or feet).

Stage A distinct "level" or period of development of an organism or a culture.

Stamnos Storage jar with two handles.

Standard deviation Statistical measurement of the amount of variation in a series of determinations; the probability of the real number's falling within plus or minus one standard deviation is 67 percent.

Standing wave technique Acoustic method, similar to bosing, used in subsurface detection.

State Term used to describe a social formation defined by distinct territorial boundries, and characterized by strong central government in which the operation of political power is sanctioned by legitimate force. In cultural evolutionist models, it ranks second only to the empire as the most complex societal development stage.

Statigraphy The study of rock strata.

Statistical analysis The application of probability theory to quantified descriptive data.

Statistics 1. The science which concerns itself with the collection, classification and use of quantitative data and with the application of probability theory to estimate aspects of a population from a sample.

2. The numerical data themselves.

Status Position in a pattern of reciprocal behavior.

Steatite A soft gray-green or brown talc which was worked into smoking pipes and bowls by grinding and polishing; also known as "soapstone".

Stegosaurs The name given to a group of dinosaurs that lived during the Jurassic and Cretaceous periods. These dinosaurs are characterised by their bony plates which ran across their back and along their tail.

Stela An upright stone monument in the form of a slab or pillar, often carved, and having a rectangular cross-section

Stem based clade A clade defined as one taxon and all taxa sharing a more recent common ancestor with that taxon than with another. For example, the stem-based clade Carnosauria is defined as *Allosaurus* and all taxa sharing a more recent common ancestor with *Allosaurus* than with "birds." Despite appearances, stem-based clades, like node-based clades, include the actual common ancestor of the taxa in the clade; the common ancestor of the taxa in the clade is considered to be ancestral to itself. For example, the actual species from which the entire Carnosauria evolved, which we will call the ur-carnosaur, shares a more recent common ancestor with *Allosaurus* than it does with "birds," thus making it a carnosaur. That common ancestor of *Allosaurs* and the ur-carnosaur is the ur-carnosaur itself.

Stem-group / stem-group representatives A group of all the fossil species, that are closer related to a recent monophyletic group than to its recent sistergroup, but are older and more basal than the last common stem-species of all recent representatives, has been termed the "stem-group" of a recent monophylum. That means that none of the species of the stem-group is closer related to any recent subgroup of the monophylum. All organisms that belong to the stem-group are referred to as "stem-group representatives" of the recent monophylum. Consequently, no recent species can be a stem-group representative, and every fossil species can at best belong to the stem-group of one recent monophylum. To recognize stem-group representatives as such, they have to possess at least one of the autapomorphies of the recent monophylum, but they usually will not possess its complete set of autapomorphies (compare heterobathmy). A stem-group is usually a paraphyletic group, since it is including all fossil relatives that existed between the split of a recent monophylum from its recent sistergroup and its division into the recent subgroups. These two events can be quite remotely separated in time; e.g. the origin of the monophylum Mammalia (= Synapsida) is very old (Upper Carboniferous), while its subdivision into Monotremata and Theria (Marsupialia plus Placentalia) is relatively young (Upper Cretaceous). Because of their paraphyletic status stem-groups are not formally named as taxa, but the stem-group concept is still very useful, since it allows a fuzzy phylogenetic classification of fragmentary fossils, of which the precise position in the phylogenetic tree might never been determined with sufficient certainty. Stem-group representatives can be either direct ancestors of the recent monophylum (species of the stem-line = stem-species), or they can belong to extinct side-branches of the stem-line. If stem-group fossils lack any autapomorphies that would identify them as side-branch-species, it is principally impossible to decide if they indeed represent stem-species, that of course never had autapomorphies, or if they are side-branch-species of which the autapomorphies were simply not preserved (absence as artefact). A group of fossil organisms, that was incorrectly united because of numerous symplesiomorphies, but includes representatives of the stem-groups of different recent monophyla, has been termed "false stem-group" by hennig (e.g. the Thecodontia, that include stem-group representatives of all archosaurs, as well as of crocodiles and of birds). The monophyletic group of the recent monophylum plus its complete stem-group has been termed the "pan-monophylum" by lauterbach, while the recent monophylum (without its stem-group) has been termed the crown-group by jeffries.

Stenopalynous Describing plant taxa characterized by only a slight variation in their palynomorphs.

Antonym: eurypalynous.

Step trenching Excavation method employed on very deep sites, such as Near Eastern tell sites, in which the excavation proceeds downwards in a series of gradually narrowing steps.

Stephano Synonym of zono-.

Fægri and Iversen (1950) used the prefix (as for example, in stephanocolpate, stephanocolporate, stephanoporate) in their classification of pollen types.

Stepped cline A cline with a sudden change in gene or character frequency.

Sterile In archaeology, lacking in any evidence of human activity.

Sternberg, Charles H. A fossil hunter who found many dinosaurs for E.D. Cope, mostly in Alberta, Canada from 1912-1917. He worked with his sons Charles M., George, and Levi.

Sternberg, Charles M. Charles M. Sternberg (son of Charles H. Sternberg, who collected fossils for E. D. Cope, working mostly in Alberta, Canada from 1912-1917) was a US fossil hunter who named the following dinosaurs: Brachylophosaurus (1953), Edmontonia (1928), Macrophalangia (1932), Montanoceratops (1951), the Pachycephalosaurid family (1945), Pachyrhinosaurus (1950), Parksosaurus (1937), and Stenonychosaurus (1932).

Sternberg, George George Sternberg (son of Charles H. Sternberg), found a "mummified" duck-billed dinosaur in Wyoming in 1908.

Stockade An enclosure or a strong, high barrier made of stakes or timbers.

Stone boiling. A method of cooking. Hot stones or clay balls are dropped into a basket, or other container or into a pit which contains water and the food to be cooked. The stones are replaced as they cooi. This technique was used in instances in which the container could not be directly exposed to an open flame.

Stone/bronze/iron ages General classification of archeologic cultures based on tool materials, devised in Denmark.

Stonehenge Near the town of Avebury in Wiltshire, England. Stonehenge is perhaps the finest of the British megalithic monuments. It stands in the center of Salisbury Plain, surrounded by a complex of cemeteries and ritual sites. Stonehenge is believed to date to about 2780 B.C., near the end of the Neolithic period. The function of the monument is thought to have been spiritual, but the arrangement of the megaliths also suggest possible astronomical uses, such as a calendar.

Stoneware A kind of pottery made of a clay high in vitreous minerals and fired at such a high temperature (l200 c) that it fuses into an extremely dense, non-porous substance. Stoneware is very hard and impervious to acids and scratching.

Storage pit Circular excavations usually less than 3 m in diameter assumed to have aboriginally functioned as storage "cellars".

Strandline Shoreline; an active beach or one which marks the edge of a now-extinct body of water.

Strata The name given to the formation of layers of rock by the deposition of sediment. Over many millions of years these layers are compressed to form tightly pack bands, evident in many areas of the world.

Stratification Layers of deposits that provide archaeologists with one of the major tools or clues for interpreting archaeological sites (stratigraphy). Over time, debris and soil accumulate in layers. Color, texture, and contents may change with each layer. Archaeologists try to explain how each layer was added – if it occurred naturally, deliberately (garbage), or from the collapse of structures – and they record it in detailed drawings so others can follow.

Stratified random sampling A sampling method in which the area or site to be tested is first divided into a number of smaller sections (strata) and then each of these is sampled randomly. Excavation is such a time-consuming business that archaeologists rarely have the time or funds to excavate an entire site. The problem becomes completely insurmountable when the region to be investigated comprises a township, valley or county. For that reason, archaeologists must concern themselves with methods of taking a sample, the object of which is to achieve a body of data which is representative of all that the site or region contains. To avoid human bias, an archaeologist might take a sample randomly, but too often, truly randomized units cluster together leaving some areas untested while "oversampling" others. Stratified random sampling ensures that the entire length and breadth of a site or all of the ecological zones within an area are tested while at the same time retaining an acceptable degree of randomness for most purposes.

Stratified society Society in which extensive subpopulations are accorded differential treatment.

Stratigraphic The study of rock strata, especially of their distribution, deposition, and age.

Stratigraphic palynology The use of palynomorphs, their identification, distribution, and abundance to correlate among sedimentary sequences of any age, or to provide chronological control for these sedimentary sequences. Exclusive of environmental palynology,, but the study of sedimentary sequences often includes both stratigraphic and environmental palynology.

Stratigraphy Scientific classification of rock or sediment strata according to attributes of the strata:

Lithostratigraphy = Organization of strata into units based on their lithological character.

Biostratigraphy = Organization of strata into units based on their fossil content.

Chronostratigraphy = Organization of strata into units based on their age (time).

Morphostratigraphy = Organization of strata into units based on their surface morphology (landforms).

Stratotype A specific sequence of sediment or rock, in which a stratigraphic unit or boundary is defined. The stratotype should exhibit the typical characteristics for the stratigraphic unit or boundary. The stratotype serves as a global standard.

Stratum A level or layer, particularly when part of a series of layers.

Striae Grooves between elongated sculpturing elements.

Striate A general descriptive term applied in palynology to elongated, generally parallel elements separated by grooves.

The positive elements of striate ornamentation may also be referred to as muri.

Striation (striae) Small (mm sized) scatches on bedrock and stones made by glacial scouring, often accompanied by grooves, polish, and fracture marks. Indicators of ice movement direction.

Striato reticulate Describing a pattern in which parallel or subparallel muri are cross-linked to form a reticulum in the grooves. The connections between the muri lie on a single level or different levels. Example: *Gentiana pneumonathe* (Gentianaceae).

Strike-a-light An object of stone, iron or steel which is struck against a stone to produce a spark for igniting gunpowder or tinder.

Striking platform That portion of a core (sense l) which is struck in order to remove a flake.

String grid A square meter (approximately three square feet) frame divided into small ten-centimeter squares (less than four inches).

Stromatolites Morphologically circumscribed accretionary growth structures with primary lamination that is, or may be, biogenic.

Stromatoporoid Stromatoporoids, once thought to be related to the corals, are now recognized as being calcareous sponges. Sponges similar to fossil stromatoporoids are found in the oceans today. Like modern sponges, stromatoporoid created currents to pump water in and out of their body, where they filtered out tiny food particles. Fossil stromatoporoids can be massive, chocolate-drop in shape, tabular, encrusting, cylindrical, or even arm-shaped ("ramose"). There are two main groups of fossil stromatoporoids that lived in different eras, the Paleozoic and the Mesozoic. After their appearance in the Ordovician, the Paleozoic stromatoporoids were dominant reef builders for over 100 million years. The second group of stromatoporoids, from the Mesozoic, may represent a distinct group with a similar growth form. They were also important contributors to reef formation, especially during the Cretaceous.

Stromer von reichenbach, Ernst Ernst Stromer von Reichenbach (1870-1952) was a paleontologist/geologist from Munich, Germany. He found Egyptian dinosaurs between 1911 and 1914, in the Bahariya Oasis 180 miles southwest of Cairo. He found and named: Aegyptosaurus (1932), Bahariasaurus (1934), Carcharodontosaurus (1931), and Spinosaurus (1915). He also named the family of Spinosaurids (1915). The original specimens of these 94-million-year-old dinosaurs were destroyed in the Bayerische Staatssammlung museum when the allies bombed Munich in 1944 (towards the end of WWII).

Structural functionalism The theory that the central function of the various aspects of a society is to maintain the social structure—the society's pattern of social relations and institutions.

Structuralist approaches Interpretations which stress that human actions ate guided by beliefs and symbolic concepts, and that underlying these ate structures of thought which find expression in various forms. The proper object of study is therefore to uncover the structures of thought and to study their influence in shaping the ideas in the minds of the human actors who created the archaeological record.

Structure The internal construction of the pollen or spore wall.

Sturgeon triangular. A projectile point style identified by richard macneish (1958) on the basis of his excavations in southeastern manttoba in the early 1950s. These points are 36 to 70 mm in length, and have straight or slightly convex bases and parallel or slightly convex sides. Many of these lanceolate specimens are crudely fashioned; in fact, macneish suspected that some of them were not projectile points at all, but rather quarry blanks. They were found in association with artifacts of the whiteshell and larter foci. The type name and one of the foci with which it was associated (whiteshell) are no longer used by manitoba archaeologists.

Style According to the art historian, Ernst Gombrich, style is "any distinctive and therefore recognizable way in which an act is performed and made." Archaeologists and anthropologists have defined "stylistic areas" as regional units representing shared ways of producing and decorating artifacts.

Styx River that separated the world of the living from the world of the dead in Greek and Roman myth.

Sub A prefix for under, or less than.

Sub layer Synonym of stratum.

Subarctic 1. Of or pertaining to regions south of the Arctic Circle; these regions themselves.

2. Of or pertaining to the coniferous (Boreal) forest zone which lies south of the tundra and north of the grasslands and deciduous forests; this zone itself.

Sub-atlantic A central North American climatic episode dating from 810 B.C. to A.D. 270 which was cooler and moister than the present.

Sub-boreal A central North American climatic episode dating from 3110 to 810 B.C. which marks the transition from the hot, dry Atlantic to the cool, moist Sub-Atlantic episode.

Subduction zone A zone where rocks of an oceanic plate are forced to plunge below much thicker continental crust, along margins between adjoining plates. As the plate descends it melts and is released into the magma below the earth's crust. Such a zone is marked by volcanoes and earthquakes.

Subisopolar Describing a pollen grain or spore in which the proximal and distal faces are slightly different. Examples: one face is convex and the other is less convex (*Banksia*, Proteaceae); on one face viscin threads are connected (*Oenothera*, Onagraceae).

Suboblate Describing the shape of a pollen grain or spore in which the ratio between the polar axis and the equatorial diameter is 0.75-0.88.

Subprolate Describing the shape of a pollen grain or spore in which the ratio between the polar axis and the equatorial diameter is 1.14-1.33.

Subsistence Livelihood; the means by which an individual or group maintains life.

Subsistence pattern The basic means by which a human group extracted and utilized energy from its environment.

Subsistence. Livelihood; the means by which an individual or group maintains life.

Subspheroidal Describing the shape of a pollen grain or spore in which the ratio between the polar axis and the equatorial diameter is 0.75-1.33.

This shape class includes suboblate, oblate spheroidal, prolate spheroidal and subprolate.

Substantivism A school of economic anthropology that seeks to understand economic processes as the maintenance of an entire cultural order.

Substitution The evolutionary replacement of one allele by another in a population.

Subtransverse Synonym of oblate spheroidal.

Subtropical Bordering on the tropics or nearly tropical.

Subturma A group of form-genera of fossil spores or pollen in the turma-system of Potonié.

Successiform Referring to a phyletic series of pollen types with increasing numbers of apertures, ranging from tricolpate to pantocolpate and pantoporate.

Succession The things that follow each other in sequence. For an ecosystem, it is the unidirectional change created as competing organisms and (especially) the plants respond to and modify the environment. For example, an open field would have a development from weeds to grass to forest community, in that order.

Sucking tube A hollow cylinder, often made from a cut section of longbone, through which a shaman or other healer magically withdraws an intrusive object from a patient. The belief that illness is a result of some foreign object within a person is fairly widespread. Sometimes the shaman will hide a small object within his mouth beforehand so that after his treatment he may produce it as proof of the cure.

Sues, Hans dieter Hans-Dieter Sues is a paleontologist at the University of Toronto and Senior Curator, Department of Palaeobiology, at the Royal Ontario Museum. Sues studies late Paleozoic and Mesozoic non-mammalian vertebrates. He named the dinosaurs: Majungatholus (with P. Taquet, 1979), Ornatotholus (with P. Galton,1983), Saurornitholestes (1978), Stygimoloch (with P. Galton, 1983), and Zephyrosaurus (1980).

Sulculus An elongated latitudinal ectoaperture not situated at a pole.

Sulcus An elongated latitudinal ectoaperture situated at the distal or proximal pole of a pollen grain.

A sulcus has the same shape as a colpus, but differs in orientation. Sulci are essentially latitudinal apertures whereas colpi are essentially longitudinal apertures. Sulci may be distal (anasulcate), proximal (catasulcate) or extend right around the grain (zonasulcate).

Sumer A region in the southern part of ancient Mesopotamia. The Sumerians arrived about 5000 B.C.; the civilization was made up of independent walled city-states, and declined around 1700 B.C. with the rise of the Babylonians.

Sundance A renewal and purification ceremony fairly common among North American plains Indians. Generally, a man would pledge to give a Sundance if he or a relative was guilty of some misconduct or if the hunters had been unable to locate buffalo. The ceremony often lasted four days and nights and was marked by much singing, dancing, self torture and mutilation as well as more purely social activities such as courting, gambling and playing of games.

Supercontinent The joining together of land masses to form a larger continent, such as Gondwana or Pangea.

Supernatural Relating to phenomena that cannot be described by natural laws, cannot be tested by scientific methodology, and are therefore outside the realm of science.

Supernatural beliefs A set of beliefs found in all societies that transcend the natural, observable world.

Superposition This describes how layers are usually laid down according to their age: the oldest layer is found on the bottom, and the most recent layer is on top. So, if a layer is on top of another layer, it is probably more recent.

Supra (prefix) A prefix for above.

In palynology this prefix is mostly used for features on top of the tectum, as for example in suprareticulate, suprarugulate, suprastriate.

Supratectal Indicating the position of features, such as spines, on the top of the tectum.

Surface collection The recovery of artifacts from the ground surface; the artifacts themselves. These are generally of limited interpretive value as their original spatial relationships have been disturbed. Nonetheless it may be sufficient to determine the age and cultural affiliation of the original site. If done systematically, it can also indicate the size of the site and the location of specific activity areas.

Surface finish In the study of ceramic artifacts, the mainly decorative outer elements of a vessel.

Surface scatter Archaeological materials found distributed over the ground surface.

Surface structure Particular arrangement of words that we hear or read.

Survey 1. The investigation of an area to locate archaeological sites and to acquire a preliminary understanding of its prehistory. This latter aim is most commonly achieved by means of surface collecting and the excavation of test pits.

2. To systematically map and grid an archaeological site. Surveying instruments such as the transit and the theodolite are generally used.

Suture The irregular line of joining between two bones, especially between the bone plates of the skull in vertebrates.

Swan river chert A multi-coloured (white, gray, pink, yellow, orange) chert with a glossy, waxy or dull lustre with concoidal fracturing properties composed largely of quartz with chalcedony as a cementing agent. Swan River Chert is quite common in the western half of the province and was a popular material for the production of stone tools throughout its prehistory.

Sylvestris-type Bisaccate pollen grains in which the outline of the sacci in polar view is discontinuous with the outline of the corpus so that the grains seem to consist of three distinct, more or less oval parts. Examples: *Pinus sylvestris, Abies* (Pinaceae).

Symbiosis When two very different organisms live together and each receives what it needs to survive from the other. For example, Lichen is an algae and fungus together.

Symbol Something that can represent something distant from it in time and space.

Sympatric speciation Speciation via populations with overlapping geographic ranges.

Sympatry Living in the same geographic region. Compare with allopatry.

Symplesiomorphy (versus synapomorphy) Shared "primitive" (= plesiomorphic) similarities that are regarded as homologous are termed symplesiomorphies. These symplesiomorphies do not represent any evidence for a close phylogenetic relationship (monophyly) of the groups that share these characters. The term symplesiomorphy is only used if one is referring to the shared similarities of several compared taxa, e.g. the presence of legs in all lizards (symplesiomorphy), compared to the secondary absence of legs in all snakes (synapomorphy.). Equally the primary absence of feathers and hairs in all groups that have traditionally been united as "reptiles", has to be regarded as a symplesiomorphy of these organisms. A certain problem is the application of the term "homologous" to the absence (primary or secondary) of structures or behaviours, since contrary to other homologies these "absence-homologies" cannot be recognized a priori by similarities of certain features, but only by an interpretation of the total character pattern on a well-supported phylogenetic tree (cladogram) using the principle of parsimony (compare reversals). Nevertheless the secondary absence of legs in snakes has to be regarded as "homologous" since it is not regarded as a multiple convergent reduction, but as a synapomorphy of all snakes, that is defined homologous derived similarity.

Syn A prefix indicating the fusion or anastomosis of features.

Synapomorph 1. Apomorphic features possessed by two or more taxa in commo If the two groups share a character state that is not the primitive one, it is plausible that they are related in an evolutionary sense, and only synapomorph character states can be used as evidence that taxa are related. Phylogenic trees are built up by discovering groups united by synapomorphies.

2. in cladistics, an apomorphy that occurs in two related clades and thereby supports grouping of the two clades into a single larger clade —synapomorphy — synapomorphic.

Syncolp(or)ate Describing a pollen grain with two or more simple (or compound) colpi the ends of which anastomose at the pole. Example: *Primula farinosa* (Primulaceae).

The term was used in a wider sense by Iversen and Troels-Smith (1950) for other forms with fused apertures such as spiraperturate and parasyncolp(or)ate.

Syncretism Absorption of one God into the cult or persona of another.

Synrugoidate Describing a pollen grain with six colpi of which three are long and meeting at one pole and three are short and not meeting at either pole. Example: *Schisandra grandiflora* (Schisandraceae).

Syntax The rules by which words are combined to form grammatical sentences.

Systematic sampling The selection of sample units according to some regular interval, such as every tenth one.

T

T shaped tetrad A **uniplanar tetrad** in which two of the members are perpendicular to the other two so that tetrad has the shape of the letter "T". Example: *Typha* spp. (Typhaceae).

T type tetrad A **tetrad** in which all four members are fully developed. Example: *Pentachondra pumila* (Epacridaceae).

Taconite A fine-grained sedimentary rock which occurs naturally in the Lake Superior region and which is a suitable material for the manufacture of stone tools.

Tactile pads The tips of the fingers and toes of primates; area richly endowed by tactile nerve endings sensitive to touch.

Taenia One or more strap-like, more or less parallel strips of exine on the proximal and/ or distal sides of the corpus of certain fossil gymnospermous pollen grains. Examples: *Striatites, Vittatina.*

Taiga The environmental zone dominated by coniferous tree species which lies immediately south of the tundra in the Old and New Worlds. In North America, this equates with the Boreal Forest.

Tang A long, narrow, pointed projection at the base of a tool or weapon to facilitate hafting.

Tape measure A tape measure is used to lay out a grid over an archaeological site, to measure each unit to make sure it is square, to measure the depth of each level, and to measure the distance of each feature from the corner of the unit. This tool is essential to archaeologists.

Taphonomy The study of the means by which organic materials become part of the archaeological (or fossil) record. A study of an object with an eye to determining hoe it has been preserved or why it has decomposed in the way it has can tell us something of the conditions prevalent at the time it was deposited.

Tapir A browser related to rhinoceroses and appearing at the same time. Abundant in North America during the Pleistocene epoch and still found in South America and Asia.

Tarquinia An ancient Etruscan city north of Rome.

Tarsal Pertaining to or constituting the bones of the human ankle and heel, or one of the corresponding bones of the hind leg of other animals.

Tarsier One of three species of small nocturnal primate belonging to the genus *Tarsius*, found in Sumatra, Borneo, Celebes, and the Philippines. They have a naked tail, 130-270 mm (about 5 - 11 inches) long, that makes up about half the total length of their bodies

(220-460 mm or between 8 and 19 inches). Tarsiers have enormous eyes, large hairless ears, and gripping pads at the end of their digits. They are mainly arboreal, using both hands to seize insects and small vertebrates such as lizards.

Task groups Behavioral components composed of sets of people (not households) that carry out the bulk of a society's activities.

Taxon Any named taxonomic group, such as the family Felidae, or the genus Homo, or the species Homo sapiens. Also, a formally recognized group, as distinct from any other group (such as the group of herbivores, or the group of tree-climbers).

Taxonomy Naming, describing, and classifying organisms into different categories on the basis of their appearance and other diagnostic characteristics as well as their evolutionary relationships. The biological sciences primarily use the Linnaean classification system for this purpose.

Techno-function The role of an artifact in a society's technology.

Technological types Artifact types designed on the basis of techniques and stages of manufacture.

Technology l. the knowledge and methods necessary to produce useable or consumable artifacts or products of similar materials by similar methods (eg. Paleo-Indian bone technology).

Tectonic movement Lateral and/or vertical movement of segments of the earth's crust. The continents and the ocean floors consist of tectonic plates that are moving relative to each other. On average, such movements are about 1 inch per year.

Tectum The layer of sexine, which forms a roof over the columellae, granules or other infratectal elements.

Tectum imperforatum With a continuous tectum, without perforations.

Synonym of eutectum.

Tectum perforatum A tectum with perforations smaller than 1μm in diameter.

Tectum solidum Synonym of eutectum.

Tegillum Synonym of tectum.

Tell In the Near East, a large mound built up from trash and the remains of mud-brick architecture.

Temper Solid particles such as sand, grit, crushed shell, plant fiber or crushed pottery (grog) which is mixed with the wet clay before the vessel is fired. Temper adds strength to pottery and makes it less likely to crack when drying or while being baked.

Temple mound culture A less frequently used term for the Mississippian.

Temporal types Artifact types designated on the time of manufacture.

Tent ring The circle of stones used to hold down the edges of a tent and which remains behind in a slightly disturbed form after the tent is removed.

Tenui A prefix for thin.

Erdtman often used this prefix in combination with other terms (e.g. tenuimarginate) but thinness is subjective; it is preferable to use alternative descri-ptions.

Tenuitas A general term for a thinning, that has been applied to many different situations in palynology.

The term was used by Potonié to describe endoapertures, by Potonié and Kremp (1955) for sulci and has sometimes been used for the distal thin parts in the exine of the Circumpolles group (*Corollina*).

Tephra Volcanic ash. In the Mediterranean, for example, deep-sea coring produced evidence for the ash fall from the eruption of Theta, and its stratigraphic position provided important information in the construction of a relative chronology.

Tephrochronology Dating and stratigraphy of volcanic ash (tephra) beds.

Termite stick A tool made and used by chimpanzees for collecting termites for food.

Terms of address The terms people use when they address their kin directly.

Terms of reference The terms by which people refer to their kin when they speak about them in the third person.

Terra cotta Fired (baked) clay.

Terra cotta soldiers The more than 7,000 terra-cotta figures that were found in the tomb of the Emperor Qin Shihuangdi, China's first emperor (221-206 B.C.). The tomb is located central China's, Xi'an province.

Terrace The former banks of a river or shores of a body of water exposed due to a drop in water level. When terraces occur as a series of steps, the highest is taken to be the oldest as it would have been the first to be exposed.

Terrestrial Referring to land habitats in distinction from water (aquatic) habitats.

Tertiary The first period of the Cenozoic era (after the Mesozoic era and before the Quaternary period), spanning the time between 65 and 1.8 million years ago.

Tesserae Small cubes of stone, glass or ceramic used to create a mosaic.

Test pit A unit excavated to determine the presence or absence of an archaeological site, or the nature of the deposits.

Testing The first stage of excavation, normally undertaken to investigate the way a site formed and sometimes to estimate the population parameters of artifacts.

Tethys ocean A small ocean that existed from the Triassic to the Jurassic; as Pangea was split into Gondwana and Laurasia in the Jurassic, an arm developed westward called the Tethys Seaway or Tethys Sea.

Tetrad A general term for a group of four united pollen grains or spores, either as a dispersal unit or as a developmental stage.

Tetrads may be uniplanar, with all members lying in the same plane (for example, linear, rhomboidal, tetragonal and T-shaped tetrads) or multiplanar, with members in more than one plane (for example, decussate or tetrahedral tetrads).

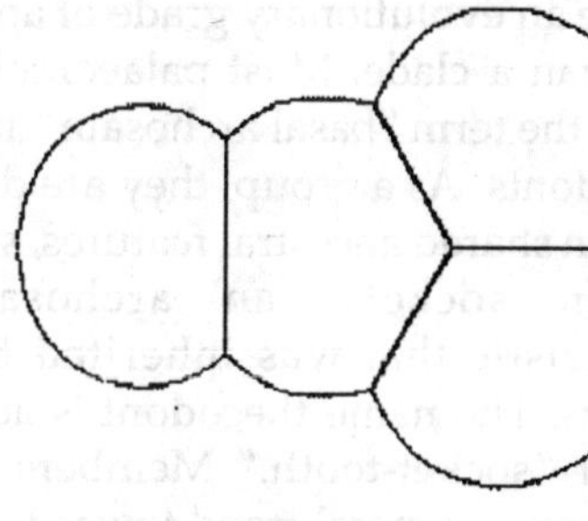

Fig. Tetrad

Tetrad mark The monolete or trilete mark on the proximal face of a spore or, more rarely, a pollen grain.

Tetrad stage The period during post-meiotic development when the four microspores or megaspores are united by the presence of a temporary special wall.

The tetrad stage ends at the start of the free spore stage, when the special cell wall is reabsorbed.

Tetragonal tetrad A uniplanar tetrad in which all four members are in contact at the centre of the tetrad so that, in the correct orientation, the adjacent walls form a cross. Example: *Uvariastrum hexaloboides* (Annonaceae).

Tetrahedral tetrad A **multiplanar tetrad** in which each member is in contact with three others, so that the centers of the grains define a tetrahedron. Example: *Erica* (Ericaceae).

Tetrapod A member of the group made up of amphibians, reptiles, birds, and mammals.

Textile A fabric produced from the weaving of fibers of animal or vegetal origin.

Thanatocoenose An assemblage of fossils that formed at the actual place of death.

Thecodont The thecodonts were a diverse group of Triassic reptiles that included large four-legged carnivores, armored herbivores, small, agile two- and four-legged forms, and crocodile-like aquatic reptiles. They gave rise to crocodiles, dinosaurs, and pterosaurs. The term Thecodontia is no longer used, as they are a paraphyletic group. The thecodonts are therefore an evolutionary grade of animals, rather than a clade. Most palaeontologists now use the term "basal archosaur" to refer to thecodonts. As a group, they are defined by certain shared ancestral features, such as teeth in sockets, an archosaurian characteristic that was inherited by the dinosaurs. The name thecodont is actually Latin for "socket-tooth." Members of the group show a general trend toward a more upright, less sprawling stance, with the hindlimbs especially being progressively positioned more directly beneath the body, until some could walk upright on two legs.

Theism Belief in one or more gods of extrahuman origin.

Theodolite A surveying instrument capable of measuring vertical and horizontal angles and consisting of a tripod-mounted telescope.

Theory A well-substantiated explanation of some aspect of the natural world that typically incorporates many confirmed obserations, laws, and successfully verified hypotheses.

Theory of acquired characteristics The concept, popularized by Lamarck, that traits gained during a lifetime can then be passed on to the next generation by genetic means; considered invalid today.

Theory of evolution Charles Darwin's theory developed in A.D. 1859, in his book Origin of Species. Darwin theorized that humans evolved from a lower order of animals, such as primates.

Therian mammal Mammal within the marsupial+placental clade.

Thermal prospection A remote sensing method used in aerial reconnaissance. It is based on weak variations in temperature which can be found above buried structures whose thermal properties are different from those of their surroundings.

Thermoluminescence A method of dating that measures the energy given off from the breakdown of radioactive elements. This energy is trapped in pottery and given off as light. Older objects give off more light.

Thermoluminescence dating (TL) A radiometric dating method based on the fact that trace amounts of radioactive atoms, such as uranium and thorium, in some kinds of rock, soil, and clay produce constant low amounts of background ionizing radiation. The atoms of crystalline solids, such as pottery and rock, can be altered by this radiation. Specifically, the electrons of quartz, feldspar, diamond, or calcite crystals can become displaced from their normal positions in atoms and trapped in imperfections in the crystal lattice of the rock or clay molecules. These energy charged electrons progressively accumulate over time. When a sample is heated to high temperatures in a laboratory, the trapped electrons are released and return to their normal positions in their atoms. This causes them to give off their stored energy in the form of light impulses (photons). This light is referred to as thermoluminescence (literally "heat light"). A similar effect can be brought about by stimulating the sample with infrared light. The intensity of thermoluminescence is directly related to the amount of accumulated changes produced by background radiation, which, in turn, varies with the age of the sample and the amount of trace radioactive elements it contains.

Thermoremanent magnetism Magnetic fields that are altered or formed in atoms of iron by heat in excess of 1100° F. (600° C.). Such fields line up with the magnetic field of the planet at the time of the exposure to a high temperature. They will remain oriented to that direction indefinitely despite the fact that the true position of magnetic north wanders over thousands of miles around the rotational north pole and even reverses with the magnetic south pole over longer periods of time.

Theropod The theropod (meaning "beast-footed") dinosaurs are a diverse group of bipedal dinosaurs. They include the largest terrestrial carnivores ever to have lived, and many quite small species. Theropods typically share a number of traits including hollow, thin-walled bones and modifications of the hands and feet (3 main fingers on the hand, and 3 main (weight-bearing) toes on the foot.) Most theropods had sharp, recurved teeth for eating flesh, and claws on the ends of all of the fingers and toes. Some of these characters were lost or modified in some groups later in theropod evolution. Theropod fossils are fairly rare and often fragmentary. Fossils of small theropods are especially rare, since small bones are harder to find and are weathered away easily.

Theropoda ("beast feet") Major grouping of dinosaurs that is comprised of carnivorous dinosaurs and birds -theropod.

Thomsen, christian jurgen Pioneering Nineteenth Century Danish archaeologist. Thomson organized his exhibitions at the National Museum in Copenhagen on the basis of three ages: the Stone Age, Bronze Age and Iron Age. His scheme is still used today.

Thrace A region and ancient country in the southeast Balkan Peninsula north of the Aegean Sea. In the seventh century B.C.E., it was colonized by Greeks and later came under the control of Rome, Byzantium, and Ottoman Turkey. It is now divided between Greece and Turkey.

Threat gesture A physical activity used by one animal to threaten another animal. Some threat gestures are staring, shaking a branch, and lunging toward another animal.

Three age system A classification system devised by C.J. Thomsen for the sequence of technological periods (stone, bronze, and iron) in Old World prehistory. It established the principle that by classifying artifacts, one could produce a chronological ordering.

Threshing Separating grain from the husks and straw by beating or rubbing.

Thule The Latin name which ancient geographers gave to the most northerly land in the inhabited world. More recently, the term has been applied to an archaeological culture which blanketed the North American arctic shortly after A.D. 1000. The artifactual inventory of these coastally adapted people included rectangular winter houses, soapstone lamps and cooking vessels, trace buckles and swivels for dog harnesses, the ulu, bone and antler arrow and harpoon heads and carved ivory bird figurines. Judging from the time slope indicated by numerous radiocarbon dates, Thule culture emerged in Alaska and spread rapidly across the arctic as far as Greenland. Their relationship, if any, with the earlier Dorset peoples is unknown but there is general agreement that the Thule is directly ancestral to the modern Inuit.

Thunderbird In mythology, large bird that produced thunder by flapping its wings. Belief in the thunderbird was formerly widespread among native North Americans and in Asia.

Thunderbird nest. The general term for a pile of boulders with a central depression. The function that these served is essentially unknown and in fact they may have been constructed for a number of different purposes. In the popular imagination they are places where young men went to receive a vision and/or a spirit helper.

Thyrsos Staff made of pinecone topped fennel stock carried by followers of Dionysos.

Till fabric Alignment of elongated pebbles within till. Thought to be an indicator of direction of ice movement when till was deposited. Both transverse and parallel fabrics commonly developed.

Till General term refering to any kind of sediment deposited directly from glacier ice with little or no influence by running water; typically unstratified and unsorted; sometimes called boulder-clay. May originate in various manners: lodgement (basal) till is deposited below active ice, ablation till is let down to the ground by melting of stagnant ice, and flow till forms as mudflows off the edge of stagnant ice.

Time slope A series of dates which become progressively older in a given direction through space, thus pointing to the place of origin of the dated complex or trait.

Time-depth The life span of a culture (sense 1), language or family of languages. Thus a culture which is recognized in the archaeological record as beginning in 500 B.C. and terminating at A.D. 500 would have a time-depth of 1000 years.

Tinkler (or tinkling cone). The (usually) metallic cones or cylinders which were used as ornaments on clothing. Brass sewing thimbles were sometimes used for this purpose. The artifact received its name from the sound produced by the many cones striking one another with each movement of the wearer or dancer.

Tipi ring 1. The circular arrangement of stones used to hold down the edges of a tipi in those areas where wood for stakes was unavailable. When camp was broken and the tipis disassembled, the stones would simply be rolled off and left behind thus maintaining the roughly circular arrangements.

2. A circular arrangement of tipis within a camp.

Tipi. The conical tent of hide over a wooden frame used by plains indians.

Toga Woolen mantle worn by Roman men as a sign of their citizenship.

Toggle. A device consisting of a rod or peg, pointed at one or both ends with a line attached near its centre. The toggling principle may be applied to buttons, or more familiar to archaeologists, the harpoon. The toggling harpoon is characterized by a detachable tip with a line fastened to a hole near its centre. When the point enters an animal, the shaft falls away and tension on the line causes the point to turn sideways thus firmly embedding itself.

Tokogenetic relationships (versus phylogenetic relationships) The parental (genealogical) relationships between the individual organisms within a species. Contrary to the phylogenetic relationships between different species, the tokogenetic relationships within biospecies of biparental organisms are not hierarchical but reticulate. In uniparental organisms (agamo-species) the parental relationships of the individual organisms are hierarchical, too, so that a distinction between tokogenetic and phylogenetic relationships is not possible. Since synapomorphies can only correctly diagnose monophylic groups if the relationships are strictly hierarchical, the terms "synapomorphic character" and "monophyletic group" cannot be applied to a single biospecies or even within such species. The only exception is a group of allopatric populations, that are classified as subspecies within a single species because of their potential (!) interbreeding capabilities, since in this case the relationships between these populations can be hierarchical, too.

Example: Grey mice a reared in a laboratory for several generations. Because of an albino-mutation, one day a white mouse *Blondy* is born. The albinism is inherited according to the laws of mendel to the descendants of *Blondy*. Finally the laboratory population includes a certain percentage of white mice that are all descendants of *Blondy*. Although the albinism originated only once in this population, and consequently has to be regarded as derived and homologous, this character cannot be called a synapomorphy, since all white mice do not form a closed community of descent (monophyletic group), that means some white mice are closer related to some grey mice than to some other white mice.

A population of interbreeding biparental organisms, that is separated from other such populations, is developing "genetical exclusiveness" after a sufficient number of generations. This means that at a given point of time every individual member of the population is closer related to any other member of this population than to any individual organism in other populations. For this phenomenon I here suggest the new term "tokophyly" and "tokophyletic". Monophyla and biospecies mostly are tokophyletic, but also populations within a biospecies can become temporally tokophyletic at least.

Toltec Native Americans from the city of Tollan.

Tool-kit As used by archaeologists, the full range of artifacts of a single complex.

Topography Description of the physical characteristics of an area that shows relative positions and elevation.

Torah The five books of Moses in the Hebrew scriptures. The term also refers to the study of the Jewish tradition.

Torus An arcuate invagination or protuberance of the exine more or less paralleling the laesura of a spore in the interradial area.

Total group A stem-based clade defined using two crown clades. For example, the total group Crocodylotarsi is defined as the crown clade Crocodylia and all taxa sharing a more recent common ancestor with the Crocodylia than the crown clade Neornithes.

Totem A plant or animal whose name is adopted by a clan and that holds a special significance for its members, usually related to their mythical ancestry.

Trace fossil Evidence of former life preserved as a mark on a rock. A good example would be dinosaur footprints.

Trace Fossil signs left behind such as footprints, nests, and burrows.

Traces All the physical characteristics of an artifact.

Tradition A style, technology or lifeway which persists for a long period of time within a given region.

Trait An attribute or character of an individual within a species for which heritable differences can be defined.

Transcription The process by which messenger RNA is read from the DNA forming a gene.

Transcriptome Set of activated genes associated with specific tissue, which will vary over time.

Transfer RNA (tRNA) A type of RNA that brings the amino acids to the ribosomes to make proteins. There are 20 kinds of transfer RNA molecules, one for each of the 20 main amino acids. A transfer RNA molecule has an amino acid attached to it, and contains the anticodon corresponding to that amino acid in another part of its structure. In protein synthesis, each codon in the messenger RNA combines with the appropriate tRNA's anticodon, and the amino acids are arranged in order to make the protein.

Transformism The evolutionary theory of Lamarck in which changes occur within a lineage of populations, but in which lineages do not split (i.e., no speciation occurs, at least not in the sense of the cladistic species concept) and do not go extinct.

Transgression/regression Advance of a feature (sea or ice) over the land. Regression is the opposite, the retreat of a feature from the land.

Transhumance Seasonal movement of livestock between upland and lowland pastures.

Transhumant. Of or pertaining to the seasonal movements of livestock from one grazing or browsing range to another.

Transit A sophisticated optical surveying instrument similar to an alidade, except that it is mounted directly on a tripod, rather than resting on a plane

Transition A mutation changing one purine into the other purine, or one pyrimidine into the other pyrimidine (i.e., changes from A to G, or vice versa, and changes from C to T, or vice versa).

Transitional fossil A fossil or group of fossils representing a series of similar species, genera, or families, that link an older group of organisms to a younger group. Often, transitional fossils combine some traits of older, ancestral species with traits of more recent species (for instance, a series of transitional fossils documents the evolution of fully aquatic whales from terrestrial ancestors).

Translation The process by which a protein is manufactured at a ribosome, using messenger RNA code and transfer RNA to supply the amino acids.

Transversal furrow Synonym of endocolpus.

Transverse Synonym of oblate.

Transverse has also been used to indicate the orientation of latitudinal features. It is also commonly used to indicate a section cut through the equatorial plane of a pollen grain.

Transversion A mutation changing a purine into a pyrimidine, or vice versa (i.e., changes from A or G to C or T and changes from C or T to A or G).

Travelers Hunter-gatherers who follow a regular yearly round, occupying a series of campsites for brief periods when a valued resource is available in the vicinity of each site (a logistical pattern).

Travois Two long, parallel poles which serve as frame to which baggage or supplies may be attached while in transit. One end of the poles is fastened on either side of a draft animal while the opposite ends drag behind. There is some evidence to suggest that the dog-drawn travois was a Middle Prehistoric Period innovation.

Trema Synonym of aperture.

Treme A suffix synonymous with aperture.

The -treme system of aperture classes suggested by Erdtman and Straka was intended to replace that of Iversen and Troels-Smith (1950). The system included atreme, monotreme, ditreme, tritreme, tetratreme, pentatreme, hexatreme, polytreme, anomotreme, pleotreme and stephanotreme.

Trench An excavation recovery unit, in the shape of an elongated rectangle, often used to expose the layering of deposits at a site

Tri A prefix for three.

Triangulation A method of surveying (sense 2) in which an area is sub-divided into triangles of which two angles and the length of one side are known. The lengths of the other two sides are calculated using trigonometric functions.

Triassic A period of time stretching from 248 to 206 million years ago. During this time the first dinosaurs began to evolve, which later dominated the earth during the Jurassic and Cretaceous periods.

Triassic period Between 248 and 206 million years ago. Period characterized by the appearance of the first dinosaurs and mammals on earth.

Tribe In popular usage, any non-literate society. For anthropologists, however, the term usually takes on a more specific meaning. Like a band, a tribe is a fairly small, non-literate society, but it makes use of institutions (clubs, age grades, secret societies) to hold it together in addition to kinship networks. Many tribal societies subsist through horticulture or the raising of livestock rather than hunting and gathering.

Trichotomocolpate Describing a pollen grain with a three-armed colpus. Example: *Trapa natans* (Trapaceae).

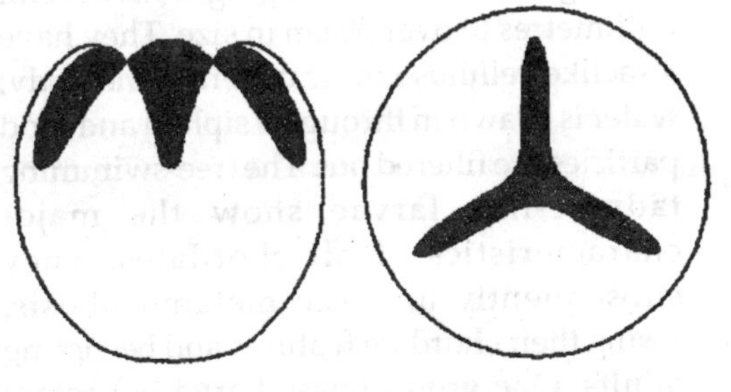

Fig. Trichotomocolpate

Triclinium Dining room of Roman house.

Tricolpate, tricolporate, triporate Describing pollen grains with three ectocolpi, three compound apertures or three pores.

The terms belong to the system of pollen classes introduced by Iversen and Troels-Smith (1950). Trifolium A three-bladed, proximal feature arising from the proximal pole of a megaspore. Example: *Capulitriletes*.

Trilete Describing a spore with three laesurae, thus showing a trilete mark. Example: *Pteridium* (Hypolepidaceae).

Trilete mark The triradiate mark of a trilete spore.

Trilobite An extinct marine arthropod common from the Cambrian to Permian eras (570-245 million years ago). Trilobite fossils are abundant in rocks of this period. Trilobites were 10-675 mm long, and their flattened oval bodies were divided into three lobes by two longitudinal furrows. They had a single head shield, which bore a pair of antennae and in many species, insect-like compound eyes. This was followed by more than 20 short body segments, each with a pair of forked appendages. Many trilobites apparently burrowed in sand or mud, preying on other animals or scavenging.

Fig. Trilobite

Trilobites A former group of marine arthropods which were characterised by a hard exoskeleton and threefold division of the body.

(Adapted from 'The Atlas of the Prehistoric World', Douglas Palmer.)

Trim bit A piece that is trimmed or pared off during manufacture.

Triprojectate Describing projectate pollen with three projecting arms on which the apertures are situated, as for example in the Triprojectacites group of fossil pollen grains. Example: *Aquilapollenites*.

Tropical Referring to climatic conditions like those found in the region on the earth today between the tropic of Cancer and the tropic of Capricorn; it includes high temperature and humidity and abundant rainfall.

Trowel A small hand tool consisting of a metal blade (sense l) attached to a handle. The mason's trowel, having a flat blade.

Troy City in western Turkey and site of the war between the Greeks and the Trojans.

Truncate To shorten or blunt by cutting off an end or through breakage.

Tryphine A material deposited on the surface of pollen grains by the breakdown of the tapetum and differing from pollenkitt in that it contains membraneous components derived from organelles.

Tuberculate Covered with knobbly projections.

Tuberculosis An infection of the lungs, accompanied by fever and a loss of appetite, caused by the bacillus *Mycobacterium tuberculosis*.

Tubulus A general term for a small channel, applied in palynology to a channel through the nexine.

Tuckahoe. A green (when wet) or black (when dry), underground fungus which is generally found as large, moist nodules while ploughing a field. These are commonly and erroneously identified as (petrified) pemmican.

Tuff Deposits of volcanic ash that have formed a crust-like layer over the underlying land.

Tula A sexinal inflation in gymnospermous pollen at the end of the axis of the distal sulcus or leptoma. Example: *Ovalipollis*.

Tumescence A gradual increase in thicknes of the wall layer(s) from a point in the equatorial interradial region to the germinal aperture. Example: *Megatriopollis*.

This term is used in descriptions of the fossil Normapolles group.

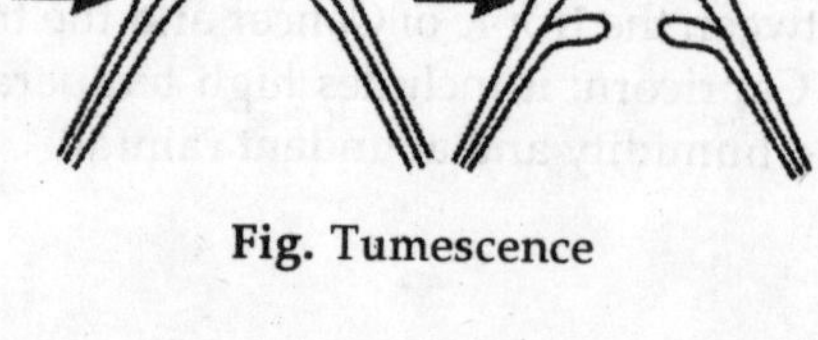

Fig. Tumescence

Tumulus A mound of earth or loose stones covering a burial.

Tunaxa. A cultural tradition of the northern plains which includes the mckean, hanna, pelican lake and avonlea phases (reeves 1983).

Tundra The vast treeless regions of North America, Europe and Asia which lies between the polar ice cap and the northern edge of the coniferous forests. The subsoil here is permanently frozen, but the surface soil may support mosses and lichens.

Tunicate A group of simple chordates, including sea squirts (class Ascidacea) that live attached to rocks, and the salps (class Thaliacea) that float in the sea. Tunicates are small marine animals, cylindrical, spherical, or irregular in shape, ranging from several millimetres to over 30 cm in size. They have a saclike cellulose tunic covering the body; water is drawn in through a siphon and food particles are filtered out. The free-swimming tadpole-like larvae show the major characteristics of all chordates. They subsequently undergo metamorphosis, losing their chordate features and becoming adults. One group (class Larvacea) retain their larval characteristics throughout life.

Turbidite Deposit of sediment from a turbidity current. Such currents are dense, muddy suspensions produced on steep slopes in lakes or seas by slumping and sliding of soft sediment. Turbidites usually indicate conditions of rapid sedimentation in unstable environments.

Turma An artificial suprageneric grouping of form-genera of fossil spores and pollen.

The following groups are recognized in the system: anteturma, turma, subturma, and infraturma.

Turonian European stage of the Upper Cretaceous, spanning the time between 91 and 90 million years ago.

Turriform Biform sculptural elements consisting of a capitate basal portion surmounted by a sharply contracted distal spine. Example: *Dibolisporites.*

Two creeks interval. The interstadial within the wisconsin glaciation which preceded the valders re-advance. This period of somewhat milder climate persisted from 12,800 to 11,500 years ago.

Type site An archaeological site which has produced the artifacts which are considered to be typical of a particular complex. Very often, the site at which a complex was first discovered or defined is considered the type site.

Typological viewpoint The idea that if two fossils look slightly different, they must be categorized as being from two different species. This approach emphasizes minor differences. People who maintain this approach are generally referred to in the biological sciences as "splitters".

Typology The classification of artifacts into groups (types) on the basis of method of manufacture, form, decoration, etc.

Tyrannosaurs The name given to a group of land dwelling predatory dinosaurs that lived during the Cretaceous period.

U

U.t.m. abbreviation for universal transverse mercator; a rectangular grid system found on all standard military maps which has been advocated for use in site designation and location by some archaeologists.

U.T.M. abbreviation for Universal Transverse Mercator; a rectangular grid system found on all standard military maps which has been advocated for use in site designation and location by some archaeologists.

Ubisch body A distinctive, orbicular granule of **sporopollenin**sporopollenin produced by the tapetum, particularly in plants with secretory tapeta.

Ulculus A rounded ectoaperture not situated at a pole. Example: Poaceae.

Ulcus A rounded ectoaperture situated at the distal or proximal pole of a pollen grain. Examples: *Sparganium* (Sparganiaceae), *Typha* (Typhaceae).

An ulcus has the same shape as a pore. Ulci may be distal (ana-ulcerate) or proximal (cata-ulcerate).

Ulu. A half-moon shaped eskimo woman's knife. Typically, it is made of ground slate, and is fitted with a wooden handle.

Umiak A large, open, flat-bottomed boat consisting of walrus hide over a spruce frame and used primarily by Eskimo women.

Unconformity The surface of a stratum that represents a break in the stratigraphic sequence.

Underwater archaeology The process of excavating archaeological material covered by fresh or seawater.

Underwater reconnaissance Geophysical methods of underwater survey include 1. A proton magnetometer towed behind a survey vessel, so as to detect iron and steel objects which distort the earth's magnetic field;

2. Sidescan sonar that transmits sound waves in a fan-shaped beam to produce a graphic image of surface features on the sea-bed;

3. A sub-bottom profiler that emits sound pulses which bounce back from features and objects buried beneath the sea floor.

Underworld books A textual and pictorial compositions that is found in New Kingdom tombs. It follows the daily passage of the sun god across the sky and through the underworld.

Unequal crossing-over A crossing-over in which the two chromosomes do not exchange equal lengths of DNA; one receives more than the other.

Uniface A lithic artifact which has been worked or retouched on one face only.

Unifacial flaking The removal of secondary flakes from only one surface of a stone nucleus.

Uniformitarianism The principle that maintains that processes seen operating today also operated in the past.

Unilineal descent group A kin group in which membership is inherited only through either the paternal or the maternal line, as the society dictates.

Unilineal evolution A pattern of cultural progress through a sequence of evolutionary stages; the basic premise of the early cultural evolutionists.

Uniplanar tetrad A **tetrad** in which the individual members lie more or less in one plane.

Unit One of the squares excavated on an archaeological site; a pit.

Unit production The manufacture of an artifact by a craftsman who performs all the operations in its production.

Universal time scale A time scale, or calendar, that can be used any place in the world since it has a finite beginning point from which any earlier or later event can be related exactly. All chronometric dates are given in terms of a universal time scale.

Unstructured interview An ethnographic data-gathering technique usually used in the early stages of one's fieldwork in which interviewees are asked to respond to broad, open-ended questions.

Uplifting Caused when pressure pushes rock strata upwards until it breaks, leaving the rock layers exposed on their end. The rock strata are now laying vertically instead of horizontally.

Uraeus A symbol of Egyptian kingship. A rearing cobra was worn on the king's forehead or crown. The cobra was associated with the "eye" of the sun. It was a protector of the king, spitting out fire.

Uranium series dating A dating method based on the radioactive decay of isotopes of uranium. It has proved particularly useful for the period before 50,000 years ago, which lies outside the time range of radiocarbon dating.

Uranium thorium dating A **radiometric dating** technique based on the rate at which uranium-238 and thorium-230 fission. This method has been used to date organic marine sediments, bone, wood, coral, stone, and soil from deep water, cave, or land fall areas. The time range that has been dated with this method so far is less than 300,000 to about 1,000,000 years ago. The half-life of U-238 is 4.468 billion years and the half-life of Th-230 is 75,380 years. This is one of several uranium series dating methods.

Urbanization The proportionate rise in the number of people living in cities in comparison to the number living in rural areas.

Urbanized society A society in which a majority of people live in cities.

Urceolate Describing a type of ornamentation consisting of urn- shaped elements situated on the footlayer. Example: *Pinanga aristata* (Palmae).

URN A pottery vessel, usually rather large, deep, and without handles. Urns were most often used for holding the ashes and bones of the dead and were sometimes buried.

Use life The average time that artifacts of a particular type remain in use.

Use wear The gradual attrition or accumulation of materials that occurs on an artifact during use.

Ushabti Literally translated it means "to answer." It is a small mummiform figure placed in tombs to do work in the afterlife on behalf of the deceased. In some tombs of the late New Kingdom whole gangs of ushabti workers were included with different tools for doing different work. A complete collection would consist of 401 ushabti: one for each day of the year, 365 plus 36 foreman.

Utilitarian Pertaining to the characteristics of an artifact determined by the physical requirements of the job it was made to perform.

V

Vacuoles Rounded to elongated spaces within an equatorial faeture. Example: *Vallatisporites ciliares.*

Synonym of dissections.

In botanical sence, vacuole is a general term for a with liquid filled cellular component In latin it means a hollow space, and in this sence this term is used.

Valanginian European stage of the Lower Cretaceous, spanning the time between 131 and 122 million years ago.

Valders a re-advance of Wisconsin ice in the Lake Superior and Michigan basins at approximately 11,800 B.P.

Valley A county in nebraska which has lent its name to a plains woodland pottery style and the complex and phase (originally focus) with which it is associated. Valley vessels are characteristically conoidal with vertical or oblique cord roughened exteriors and straight rims. Decoration is restricted to oblique and horizontal cord-wrapped stick impressions near the lip, exterior bosses and less frequently, incised lines. Sites are most comon in nebraska and kansas, but valley sherds have been found as far north as southwestern manitoba (syms 1977).

Vallum Synonym of murus.

Sometimes used to describe a single, broad, raised feature in certain fossil spores.

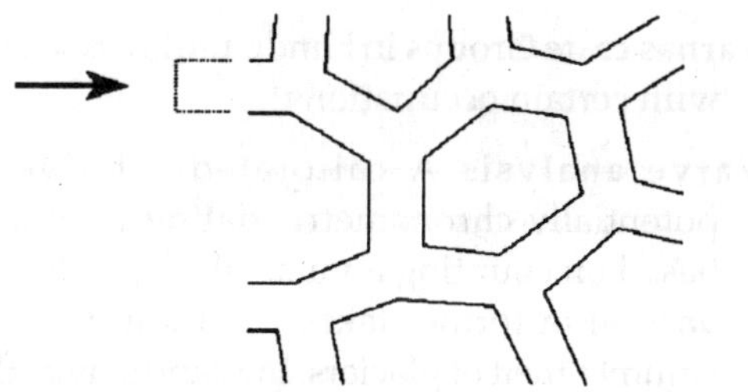

Fig. Vallum

Valva Radial thickenings in the areas beyond the ends of the laesurae of trilete spores. Example: Triquitrites tribullatus.

The term auriculate (Potonié and Kremp, 1955) refers to an extremely valvate condition.

Van valen, Leigh An evolutionary biologist who came up with the model of the Red Queen – the living chess piece that Alice encounters in *Through the Looking Glass* who must keep running as fast as she can to stay in the same place – as a metaphor to explain evolutionary patterns. His studies involve genetics and systematics, and involve a wide range of topics, including the evolution of biotas and of mammals.

Vandalism The malicious and intentional destruction or defacing of property.

Variable Any property that may be displayed in different forms.

Variance A measure of how variable a set of numbers are. Technically, it is the sum of squared deviations from the mean divided by (n-1) (the number of numbers in the sample minus one). Thus, to find the variance of the set of numbers, 4, 6, and 8, we first calculate the mean, which is 6. We then sum the squared deviations from the mean (4 - 6)2 + (6 - 6)2 + (8 - 6)2, which comes to 8, and divide by (n-1) (which is 2 in this case). The variance of the three numbers is 8/2 = 4. The more variable the set of numbers, the higher the variance. The variance of a set of identical numbers (such as 6, 6, and 6) is zero.

Varnas caste Groups in Hindu India associated with certain occupations.

Varve analysis A calibrated relative or potentially chronometric dating technique based on counting annual silt deposits (i.e., varves) in former lakes resulting from the summer melt of glaciers. In Scandinavia, this geological clock has been used to date associated archaeological evidence from the melt years of the last ice age, which began its final retreat about 15,000 years ago.

Varves Glaciolacustrine laminations deposited as seasonal couplets (summer and winter); each couplet represents one year, and varve sequences can be used for dating purposes.

Velum A feature of a monosaccate pollen grain in which the saccus is convoluted. Example: *Tsuga* (Pinaceae).

Vendian The latest period of the Proterozoic era, spanning the time between 650 and 544 million years ago. Sometimes referred to as the Ediacaran period, the Vendian is distinguished by fossils representing a characteristic collection of complex soft-bodied organisms found at several localities around the world.

Ventifact Facetted or triangular-shaped stone formed by sand-blasting in strong winds, often polished to a high degree.

Ventral. 1. The flat or concave face of an artifact.

2. That face of an artifact which was nearest to the centre of the core from which it was manufactured.

3. Of or pertaining to the front of the body; towards the belly.

Venus figurines Small Upper Paleolithic statues characterized by exaggerated breasts and buttocks and very stylized heads, hands, and feet.

Venus Roman name for Aphrodite, goddess of love and beauty.

Vermeij, Geerat J. Biologist at the Center for Population Biology of the University of California at Davis, and author of *Privileged Hands: A Scientific Life*. Vermeij, blind since age 3, combines autobiography and description of the evolutionary "rams race" between intertidal predator and prey species. Wider research interests include economic relationships between organisms and ecosystems and their implications for human organisms.

Vermiculate A general descriptive term used to describe winding features.

Vermiculate has been used by Kosanke (1950) and Harris (1955) to describe depressions (fossulae). Others use the term for raised structures (muri) in rugulate pollen and spores.

Verruca A wart-like sexine element, more than 1µm wide, that is broader than it is high and is not constricted at the base. Example: *Plantago* (Plantaginaceae).

Verrucose Synonym of verrucate.

Vertebrate A member of the subphylum Vertebrate; possesses a bony spine or vertebral column.

Vertebrates The group (specifically, a subphylum) of animals, descended from a common ancestor, that share the derived character of an internal skeleton made of bone or cartilage.

Vertical angle In mapping, the angle of sight measured on the vertical plane.

Vertical circle With major surveying instruments, the graduated vertical table around which the sighting telescope rotates; used to measure the vertical angle.

Vertical clinging and leaping A method of locomotion in which the animal clings vertically to a branch and moves between branches by leaping vertically from one to another. The animal moves on the ground by hopping or moves bipedally.

Vertical datum A base measurement point from which all elevations are determined.

Vertical distance The measurement of distance (or elevations) on a true vertical plane.

Vertical provenience The vertical position of objects within a site determined in relation to a vertical datum or datum plane, as well as to the local ground surface.

Vesiculate Synonym of saccate.

Vestibulum A separation between layers of the exine forming a cavity between the inner and outer pores. For example: *Betula* (Betulaceae). Vestigial Any structures that have been greatly reduced in size and function over evolutionary time to the extent that they now appear to have little or no current function.

Vickers-rich, Patricia Dr. Patricia Vickers-Rich is an Australian geoloogist, paleontologist and author. Vickers-Rich has unearthed many important finds in Victoria, Australia, Alaska, USA, and Patagonia, South America. She co-named (with her husband T. H. Rich) the dinosaurs Leaellynasaura (1989) and Timimus(1994) (both named for his children), Atlascopcosaurus (1989), Qantassaurus (1999), Tehuelchesaurus (1999, also with Gimenez, Candúneo, Puerta, and Vacca), the important early Australian mammal Ausktribosphenos, and others.

Victoriapithecidae Family of Early and Middle Miocene Old World monkeys from north and east Africa.

Vikings A culture originating in Scandinavia (now Norway, Denmark and Sweden) around the mid-8th century A.D. The Vikings were fierce conquerors, brave explorers, and skilled craftspeople; they invaded and settled countries throughout Western Europe. They were the first Europeans to discover America (in about A.D. 1000), almost 500 years before Columbus.

Virolocal Pertaining to the pattern in which a newlywed couple takes up residence in the home, village or territory of the husband's family. Virilocality generally results in patrilocality.

Virulence The disease-producing ability of a microorganism.

Virus A kind of intracellular parasite that can replicate only inside a living cell. In its dispersal stage between host cells, a virus consists of nucleic acid that codes for a small number of genes, surrounded by a protein coat. (Less formally, according to Medawar's definition, a virus is "a piece of bad news wrapped in a protein.")

Viscin thread A general botanical term (Jackson, 1928), applied in palynology for an acetolysis resistant, sporopollenin thread arising from the exine of a pollen grain, usually from the proximal surface. Example: *Oenothera* (Onagraceae).

Viscosity A measure of resistance to flow in a liquid (water has low viscosity while honey has a higher viscosity.).

Vision pit. A subterranean pit in which a youth would wait while on a vision quest. Some thunderbird nests may have served essentially the same function.

Vision quest Among some native North American peoples, a puberty rite in which young boys would seek visions and/or a spirit helper. Physical hardships such as lack of water, food or sleep and inadequate clothing in the winter were commonly part of the experience.

Vistulian/saale/elster glaciations Traditional sequence of glacial stages in northern Europe.

Vitamin A A member of a chemically heterogeneous class of organic compounds that are essential, in small quantities, for life.

Vitreous Pertaining to, or resembling glass.

Volant Describes an animal capable of flight.

Volcanic ASH Layers of airborne pumice resulting from violent volcanic eruptions. Provide valuable dating markers when found in sites.

Von Huene, Friedrich Friedrich von Huene (1875-1969) was a German paleontologist who named: Altispinax (1922), Avipes (1932), Antarctosaurus (1929), Betasuchus (1932), Cetiosauriscus (1927), Coeluroides (1932), Compsosucus (1932), the family Dicraeosauridae (1956), Dolichosuchus (1932), Dryptosauroides (1932), Erectopus (1922), Fulgurotherium (1932), the family Halticosauridae (1956), Halticosaurus (1908), Iliosuchus (1932), Indocuchus (1933), Jubbulpuria (1932), Laevisuchus (1932), the family Lambeosauridae (1948), Laplatasaurus (1927), Loricosaurus (1929), Magyarosaurus (1932), the family Melanorosauridae (1929), Ornithomimoides (1932), the family Podokesauridae (1914), the infraorder Prosauropoda (1920), Proceratosaurus (1926), the family Procompsognathidae (1929), Rapator (1932), Saltopus (1910), the suborder Sauropodomorpha (1932), Sellosaurus (1908), Thecocoelurus (1923), Velocipes (1932), and Walgettosuchus (1932).

Von Meyer, Hermann Hermann von Meyer was a German paleontologist who named and described Archaeopteryx; (1861), Rhamphorhynchus (1847), Plateosaurus (1837), and Stenopelix (1857).

Von Mutius, Erika A pediatrician and allergist, Dr. von Mutius's research interests include the epidemiology of childhood asthma and allergies with a focus on environmental predictors and gene-environment interactions.

Vrijenhoek, Robert A senior scientist in the areas of evolutionary biology, marine biology, and conservation, Vrijenhoek studies the ecological and evolutionary consequences of genetic diversity in animals. His research efforts have focused on the evolutionary and ecological consequences of sexual and asexual reproduction in Mexican poeciliid fish (genus *Poeciliopsis*), as well as invertebrates in deep-sea hydrothermal vents.

W

Wadi Arabic word for a rocky watercourse. Dry except in the rainy season.

Wake, David A professor in the Department of Integrative Biology at the University of California at Berkeley, Dr Wake's research emphasizes analysis of evolutionary patterns and the processes that produce them, ranging from functional morphology to evolutionary genetics and population ecology. Amphibians and reptiles are the focus of his work.

Wallace, Alfred Russel A British naturalist and contemporary of Charles Darwin. Wallace conducted research on the Amazon River and studied the zoological differences between animal species of Asia and Australia, developing a theory of evolution similar to Darwin's.

Wanaman. The cree name for red ochre.

Ward, Peter Douglas Professor of geological sciences at the University of Washington in Seattle, where he is also adjunct professor of zoology and of astronomy. Author of several books on biodiversity and the fossil record, including *Rivers in Time: The Search for Clues to Earth's Mass Extinctions* and *Rare Earth: Why Complex Life is Uncommon in the Universe* (with Donald Brownlee). He is the principal investigator for the University of Washington's portion of the NASA Astrobiology Institute.

Ware The largest and most general classification into which pottery can be grouped. Wares are usually defined on the basis of such attributes as surface finish (eg. Smooth, corded, etc.), composition of the paste (e.g. Shell-tempered, grit tempered, etc.) And vessel form (globular, conoidal etc.). A ware generally includes a number of types (often defined on the basis of decorative elements) and these in turn may include a number of varieties (based upon the manner in which the decorative elements are arranged).

Water table The level of groundwater below which all cavities are filled and permeable rock formations are saturated with water.

Wattle & daub A construction technique using mud plastered over a framework of cut branches.

Wealth The accumulation of material objects that have value within a society.

Wear pattern. The distinctive way a tool is worn or abraded through use. Examination of wear patterns can often and more reliably identify the function a tool served, than can consideration of size and shape alone.

Weathering The disintegration and breakdown of rocks at or near the earth's surface.

Mechanical weathering – occurs when rocks are broken into smaller and smaller pieces. This can occur by the activity of plant roots breaking rocks apart; by freezing and thawing that produces wedges in rocks; by physical banging of hard materials; or by rubbing against the rock by sediments found in water or the air.

Chemical weathering – occurs as water carrying other chemical elements alters the rocks. An example is carbonic acid. Water combines with carbon dioxide to produce a weak acid (carbonic acid). The carbonic acid is carried to the point of contact by the water. When the carbonic acid comes in contact with limestone rock a chemical reaction is produced which dissolves the rock.

Weathering zone In pedology, the depth to which soil processes are operational.

Wegener, Alfred A German climatologist and geophysicist whose book, *The Origins of Continents and Oceans*, was the first to propose the concept of continental drift (the forerunner to the theory of plate tectonics), as well as to suggest a supercontinent called Pangaea, which Wegener suggested had fragmented into the continents as we know them today. His ideas remained controversial until the 1960s, when they became widely accepted as new evidence led to the development of the concept of plate tectonics.

Weir A barrier constructed across a stream or on a tidal flat to trap fish.

Welded tuff A rock formed of consolidated pumice or volcanic ash. Occasionally used as a raw material for lithic artifacts.

Wentworth scale A particle size scale ranging from "boulder" (greater than 256 mm) to "clay" (less than 0.0039 mm) with intermediate ranges to allow the definition of cobbles, pebbles, sand and silt. Sieves with correct mesh sizes are available so that the proportions of these various particles can be readily calculated.

Wetmore, Alexander Alexander Wetmore (1886-1978) was an ornithologist who was an expert on the birds and bird fossils of Central and South America. He named many species of fossil birds, including Plegadornis, 1962 (now Angelinornis). He was put in charge of the National Museum (the Smithsonian), the National Gallery of Art, and the National Zoo in 1925. He collected a lot of birds (which were stuffed for the Smithsonian's collection) and fossils. In addition to his field work and administrative duties, he was famous for being extremely formal. While in the rainforests collecting specimens, he would always wear a tie, and he insisted that the tents, chairs and tables were always arranged perfectly in a particular linear fashio The Cretaceous Period fossil bird Alexornis (meaning "Alex's bird") was named by Pierce Brodkorb in 1976 in honor of Wetmore.

Wheeler box-grid An excavation technique developed by Mortimer Wheeler from the work of Pitt-Rivers, involving the retaining of intact baulks of earth between excavation grid squares, so that different layers can be correlated across the site in the vertical profiles.

Whetstone A sharpening stone.

White, Tim A paleoanthropologist with University of California, Berkeley's Laboratory for Human Evolutionary Studies, White is known for his meticulous fieldwork and analysis investigating early hominid skeletal biology, environmental context, and behavior. With an international team of colleagues, he discovered and named *Ardipithecus ramidus* and *Australopithecus garhi*.

Whiteshell focus. The most ancient (3000 to 1500 b.c.) of the cultural-historical periods in macneish's (1958) southeastern manitoba chronology. The only diagnostic projectile points of this focus, however, were those of the mckean complex. Recent re-excavation of the type site failed to produce any material to support the existence of a whiteshell focus. This term is no longer used by manitoba archaeologists.

Whiteshell side-notched. A projectile point style defined by macneish (1958) on the basis of his investigations in southeastern manitoba and attributed to the nutimik and manitoba foci (laurel and blackduck respectively). As originally defined, these specimens are rather long (42 to 58 mm) and bear narrow but deep side notches above a convex or less commonly a straight base. Maximum width occurs just above or immediately below the notches.

Wickiup A beehive-shaped hut of grass or brush most commonly found in the American Southwest.

Wiffen, Joan Joan Wiffen is a self-trained amateur paleontologist who pioneered dinosaur hunting in New Zealand. Her discoveries, beginning in 1974, greatly changed scientists' views of New Zealands' paleontological history. Wiffen discovered fragmentary fossils of late Cretaceous period dinosaurs, including an ankylosaur (probably a nodosaur), a carnosaur, and a sauropod. These are the first dinosaurs found on New Zealand; it had been previously thought that no dinosaurs had lived on this long-isolated island. Wiffen's discoveries are from Mangahouanga, North Island, New Zealand. Joan Wiffen and her husband M. A. (Pont) Wiffen have also found mosasaurs (*Mosasaurus flemingi*, *Prognathodon overtoni*, *Rikisaurus tehoensis*, etc.), a plesiosaur (a plesiosaurid), a pterosaur, a protostegid (a large marine turtle), and many other fossils on New Zealand.

Wigwam A dwelling structure consisting of bark, matting or hide over a frame of arched poles.

Wild type The genotype or phenotype, out of a set of genotypes or phenotypes of a species, that is found in nature. The expression is mainly used in lab genetics to distinguish rare mutant forms of a species from the lab stock of normal individuals.

Wilford, John Noble A *New York Times* reporter and winner of two Pulitzer Prizes for his national reporting of science topics, and for his work on the Challenger explosion and the aftermath. While at the *Times* he served as science correspondent, assistant national news editor, and director of science news.

Williston, Samuel W. Samuel Wendell Williston (1851-1918) was a vertebrate paleontologist (specializing in ancient reptiles), dipterologist (a scientist who studies flies), author, medical doctor and professor. He served as assistant to Othniel Marsh at the Yale Peabody Museum from 1876 to 1890. Williston searched for dinosaur fossils for Marsh in Colorado and Wyoming.

Wilson, E.O. A biologist and professor at Harvard University since 1955. Wilson has won two Pulitzer Prizes for his books *On Human Nature* and *The Ants*, and has received numerous honors for his research and conservation efforts.

Windbreak A rude screen or fence intended to provide shelter from the wind.

Winkley, William William Winkley (1900-1976) discovered the first Pachycephalosaurus fossil in 1938 while herding cattle on the family ranch outside of Ekalaka, Montana, USA. Winkley also found a Tröodon skull on his ranch.

Winnipeg fabric-impressed ware. The ceramic ware more commonly known to archaeologists as selkirk ware.

Winnipeg ovoid. A projectile point styel defined by r.s. macneish (1958) on the basis of his investigations in southeastern manitoba. These crudely chipped specimens are teardrop shaped and range in length from 35 to 78 mm. As maximum thickness ranges from 5 to 60 mm, it is more than likely that many of these are not projectile points at all, but rather quarry blanks. Macneish attributed the winnipeg ovoid style to the whiteshell and larter foci.

Winnowing. Separating grain from chaff by means of a wind or air current.

Winter count. Among native peoples of the north american plains, a pictorial history painted on buffalo hide.

Wisconsin (an) glaciation The latest major episode of glacial advance in the Pleistocene of North America; from about 70,000 to 10,000 B.P.

Wisconsin/illinoian/kansan/nebraskan glaciations Traditional sequence of glacial stages in central North America.

Wisconsinan. The last period of glaciation of the north american pleistocene. The wisconsinan began approximately 70,000 years ago and ended with the commencement of the holocene some 10,000 or 12,000 years ago.

Wobble The ability of the third base in some anticodons of tRNA to bond with more than one kind of base in the complementary position in the mRNA codon.

Woese, Carl A molecular biologist, Dr. Woese's identification of the Archaea as a distinctive group of organisms changed the way life is classified on Earth and transformed our view of biology.

Wonders of the world A list made by Greek writers in the second century B.C.: 1. The Pyramids at Giza,

2. The Hanging Gardens of Babylon,

3. The Statue of Zeus at Olympia,

4. The Temple of Diana at Ephesus,

5. The Mausoleum at Halicarnassus,

6. The Colossus of Rhodes, and

7. The Pharos of Alexandria lighthouse. Only the Pyramids at Giza survive.

Woodland period. The most recent of the three "stages" in the prehistory of the eastern forests of north america. In accordance with the trend which began with the earlier archaic period, the woodland witnesses increased regionalism and the proliferation of local cultures. As this period is often defined by the appearance of pottery in local assemblages, and because these cultures adopted pottery at different times, no single date marks the beginning of the woodland period. In the american southeast, pottery has been found which dates as early as 2400 b.c., while some northern peoples never adopted it. It southern manitoba, a date of 200 b.c. or thereabouts is often applied to the archaic/woodland boundary. A number of other innovations are often associated with the woodland: the use of the bow and arrow (as opposed to the atlatl), the construction of effigy and burial mounds, possibly the birch bark canoe (as opposed to dugouts) and in some areas, horticulture. The woodland period has been subdivided in a number of different ways. In the technical literature, the terms early, middle and late woodland are frequently used. In canada, the initial woodland equates with the early and middle periods, while "terminal" is used instead of "late". By definition, a terminal woodland culture (eg. Blackduck, selkirk, mistikwas, mississippian) is one that can be traced to the historic period and identified with a known culture. Presumably this is not possible with the earlier initial woodland (eg. Laurel, hopewell) cultures.

Workday The culturally established number of hours that a person ideally spends at work each day.

World system A term coined by the historian Wallerstein to designate an economic unit, articulated by trade networks extending far beyond the boundaries of individual political units (nation states), and linking them together in a larger functioning unit.

Worsaae's LAW The law that artifacts deposited together in a grave were in use at the time.

Wrangham, Richard A primatologist, Dr. Wrangham's central interest is in the significance of chimpanzee behavior, ecology, and life-history for understanding the common ancestor between chimps and humans and subsequent human evolution.

Wristguard. Bracer; wristlet, a rectangular plate of bone or stone which is bound to the inside of an archer's wrist to protect him from the recoil of the bowstring.

Würm/riss/mindel/günz glaciations Traditional sequence of glacial stages in Alps of central Europe.

X

Xeric Pertaining to, or growing in dry conditions.

Xerophyte A dry-loving plant such as cactus.

Xerothermy Aridity; dry heat.

Xibalba Maya underworld.

Xiphiidae Family of cetaceans known as Beaked Whales.

X-ray diffraction analysis A technique used in identifying minerals present in artifact raw materials; it can also be used in geomorphological contexts to identify particular clay minerals in sediments, and thus the specific source from which the sediment was derived.

X-ray fluorescence The bombardment of a sample for chemical analysis with X-rays. The light which the sample emits indicates which chemicals are present and the relative proportions in which they occur.

Y

Y mark Synonym of trilete mark.

Young chung chien Young Chung Chien is a Chinese paleontologist who named the following dinosaurs: Chingkankousaurus (1958), Dianchungosaurus (1982), Lufengocephalus (1974), Lufengosaurus (1941), Lukousaurus (1948), the family Mamenchisauridae (with Zhao Xijin, 1972), Mamenchisaurus (1954), Omeisaurus (1939), Sanpasaurus (1946), Sinocoelurus (1942), Sinosaurus (1948), Tawasaurus (1982), Tienshanosaurus (1937), Tsintaosaurus (1958), the family Yunnanosaurid (1942), and Yunnanosaurus (1942).

Young earth creationism The belief that the universe came into being only a few thousand years ago. Most young Earth creationists interpret the Bible literally, including not just the special, separate creation of human beings and all other species, but the historicity of Noah's flood.

Yucca Several plants that are members of the lily family. The plants grow throughout the Southwest at elevations of 1500 to 6000 feet. Plants are characterized by a clump of thin, pointed leaves at the base and a single flowering stalk. White flowers appear on the stalks each spring. The plant is important not only for its fiber but for soap obtained from the root and for the use of its seeds. fruits. and flowers as food.

Z

Z layer Synonym of exintine.

Zhao xijin Zhao Xijin is a Chinese paleontologist who named the following dinosaurs: Chaoyoungosaurus (1983), Chinshakiangosaurus (1986), Dachon-gosaurus (1986), Damalasaurus (1986), Klamelisaurus (1993), Kunmingosaurus (1986), Lancangjiangosaurus (1986), the family Mamenchisauridae (with Young Chung Chien, 1972), Megacervixosaurus (1983), Microdontosaurus (1983), Monkonosaurus (1990), Monolophosaurus (with P. Currie, 1994), Ngexisaurus (1983), Sangonghesaurus (1983), Sinraptor (with P. Currie, 1993), Oshanosaurus (1986), Xuanhuasaurus (1986).

Zhong (Chinese) Bell

Ziggurat A type of step-pyramid temple built by the Sumerians 4000 years ago in southern Mesopotamia, from sun-dried mud bricks.

Zodiac The Babylonian and Greek signs of the zodiac were introduced into Egypt in the Greco-Roman Period. They were adapted into Egyptian imagery and used to decorate ceilings of tombs and temples, and coffin lids. Other cultures have also divided the heavens into zodiacs for related uses.

Zona A thin outer structure of a spore that projects at the equator, but does not extend over the distal face or proximal face. Example: *Cirratriradites saturni.*

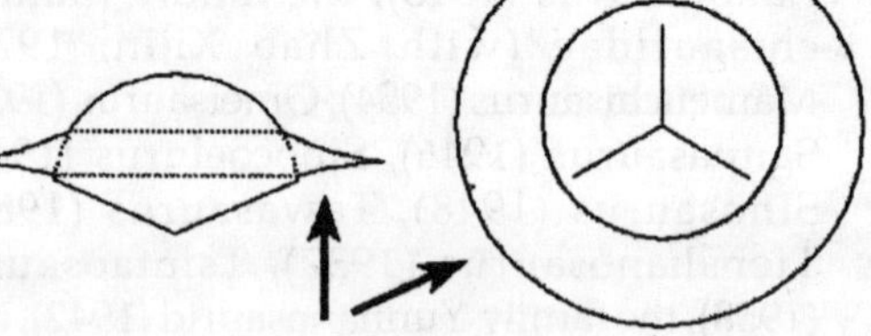

Fig. Zona

The term is especially applied for many representatives of fossil spores. Example: *Kraeuselisporites.*

Zona aperturate Describing a pollen grain with a ring-like aperture.

The range of ring-like apertures includes for example: anazonasulculus 1. A ring-like sulculus in the distal hemisphere; catazonasulculus, a ring-like sulculus in the proximal hemisphere; zonasulculus, a ring-like sulculus around the equator; and zonasulcus

2. A meridional ring-like sulcus perpendicular to the equator.

Zonasulculus A ring-like sulculus around the equator. Example: *Nymphaea violacea* (Nymphaeaceae).

Zonasulcus A meridional ring-like sulcus perpendicular to the equator. Example: Laurelia novaezelandiae (Monimiaceae).

Zone 1. A geographic area characterized by some distinctive feature such as the flora, fauna, climate etc.

2. A particular cultural, geological or pedological layer or level.

3. An area on the surface of a ceramic vessel which is modified differently (eg. punctates, rocker stamping, painted, etc.) than the adjacent areas.

Zoni Prefix indicating a latitudinal orientation.

Zono A prefix indicating features located equatorially.

Zonoaperturate Describing a pollen grain with apertures situated only at the equator.

Pollen classes with zonoaperturate pollen include, zonocolpate, zonocolporate, zonoporate. Originally Erdtman and Vishnu-Mittre (1956) introduced the prefix zoni-, but later Erdtman and Straka (1961) changed the ending of the prefix to zono-.

Zonorate Describing a pollen grain with a continuous endoaperture (os) around the equator.

Synonym of endocingulum.

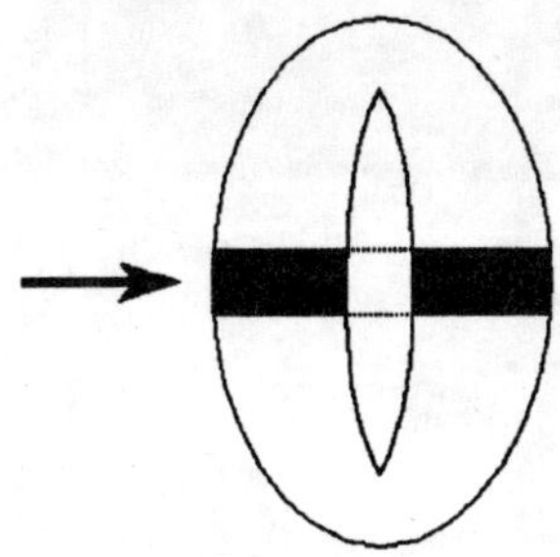

Fig. Zonorate

Zooarchaeology The study of faunal remains found in archaeological sites and their cultural significance.

Zoomorphic "animal-like". refers to art-work or decorated objects with an animal motif or appearance.

Zoroastrian One who accepts Zoroastrianism, a Persian religion founded in the sixth century B.C.E. by the prophet Zoroaster. It is characterized by the worship of a supreme god, Ahura Mazda, who requires good deeds to help in his struggle against the evil spirit Ahriman.

Zun (Chinese) Wine vessel.

Zwischenkörper A lens-shaped body below the aperture of certain unacetolysed pollen grains.

The feature resembles an oncus but is treated as distinct because some pollen grains have both features.

Zygote The cell formed by the fertilization of male and female gametes.

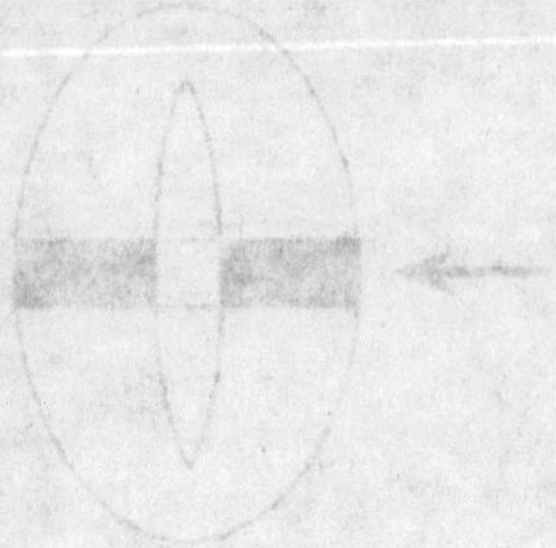

Zonasulculus A ring-like sulculus around the equator. Example: *Nymphaea violacea* (Nymphaeaceae).

Zonasulcus A meridional ring-like sulcus perpendicular to the equator. Example: Laurelia novaezelandiae (Monimiaceae).

Zone 1. A geographic area characterized by some distinctive feature such as the flora, fauna, climate etc.

2. A particular cultural, geological or pedological layer or level.

3. An area on the surface of a ceramic vessel which is modified differently (eg. punctates, rocker stamping, painted, etc.) than the adjacent areas.

Zoni Prefix indicating a latitudinal orientation.

Zono A prefix indicating features located equatorially.

Zonoaperturate Describing a pollen grain with apertures situated only at the equator.

Pollen classes with zonoaperturate pollen include, zonocolpate, zonocolporate, zonoporate. Originally Erdtman and Vishnu-Mittre (1956) introduced the prefix zoni-, but later Erdtman and Straka (1961) changed the ending of the prefix to zono-.

Zonorate Describing a pollen grain with a continuous endoaperture (os) around the equator.

Synonym of endocingulum.

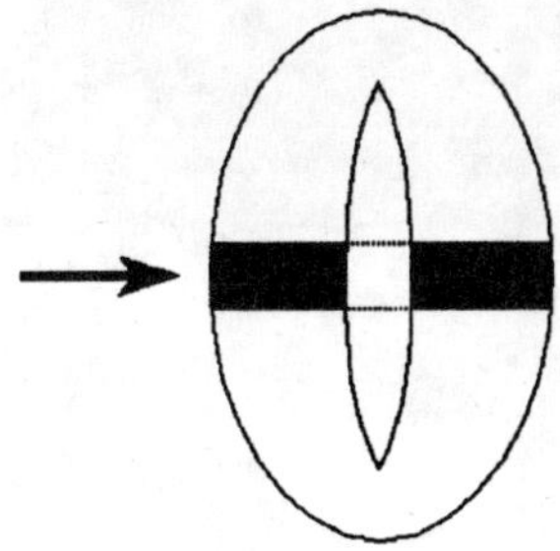

Fig. Zonorate

Zooarchaeology The study of faunal remains found in archaeological sites and their cultural significance.

Zoomorphic "animal-like". refers to art-work or decorated objects with an animal motif or appearance.

Zoroastrian One who accepts Zoroastrianism, a Persian religion founded in the sixth century B.C.E. by the prophet Zoroaster. It is characterized by the worship of a supreme god, Ahura Mazda, who requires good deeds to help in his struggle against the evil spirit Ahriman.

Zun (Chinese) Wine vessel.

Zwischenkörper A lens-shaped body below the aperture of certain unacetolysed pollen grains.

The feature resembles an oncus but is treated as distinct because some pollen grains have both features.

Zygote The cell formed by the fertilization of male and female gametes.

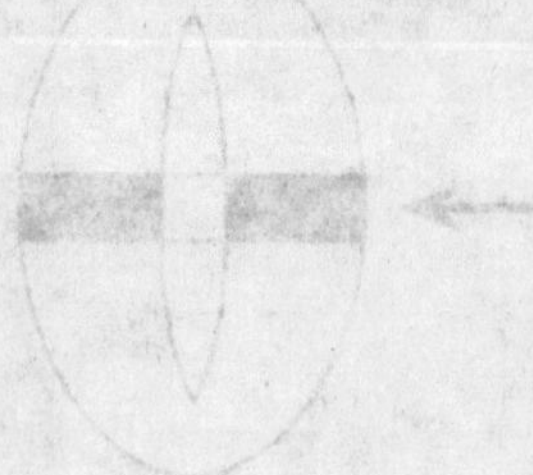

